T0256450

THE ANIMAL GAME

THE ANIMAL GAME

Searching for Wildness at the American Zoo

DANIEL E. BENDER

Harvard University Press

Cambridge, Massachusetts
London, England
2016

To Piya and Jo

Second printing

Library of Congress Cataloging-in-Publication Data
Names: Bender, Daniel E., author.
Title: The animal game : searching for wildness at the American zoo / Daniel E. Bender.
Description: Cambridge, Massachusetts : Harvard University Press, 2016. | Includes bibliographical
references and index.
Identifiers: LCCN 2016017880 | ISBN 9780674737341
Subjects: LCSH: Zoos—United States—History. | Zoos—Employees. | Wild animal trade—History. |
Endangered species—History.
Classification: LCC QL76.5.U6 B46 2016 | DDC 590.73—dc23
LC record available at https://lccn.loc.gov/2016017880

CONTENTS

INTRODUCTION

The Zoo Parade

W<small>E FOUND</small> the prints in the wet Indian jungle loam. We are in search of the rare Asian rhino to capture for an American zoo. In the mud we see the heavy imprint of a three-cloven hoof. It must be our prize. We are finally on the game trail, but our two biggest challenges still lie ahead: how should we capture our prize, and how should we bring it back alive?

My six-year-old daughter and I talk it over. We decide that a metal trap could never hold an angry, panicked rhino. Finally, we decide to dig a mud pit—success! Braving torrential rains, we have hauled in our first catch, but how should we bring it home? Rail is just plain silly, and we chuckle at the idea of a rhino in an airplane. We decide to send him by ship. The rhino arrives alive, and this is lucky; many animals perished on the way from jungle to zoo. We are off in search of a tiger next.

It was all just a game—the animal game—to those Americans who for a half-century made their living collecting the world's animals to sell to zoos and, in the process, entranced ordinary adults and their kids with the excitement of jungles and veldts. For my daughter and me, it's an old board game I purchased while tracking zoos and their animals as they evolved over the past century. She and I trap animals for American zoos while playing the 1954 Marlin Perkins' Zoo Parade game. Her grand-parents tell her how they remember waiting for Perkins each week on television.

By the 1950s Perkins was a household name, a "zoo man" and a celebrity. Perkins was born in rural Carthage, Missouri, in 1905, and by the time young Marlin was my daughter's age, he had begun his own private menagerie, collecting insects and reptiles. Like many early zoo officials, his love of collecting led him to university. For Perkins, that was the University of Missouri, where he intended to study zoology. Also like many zoo workers, he never finished and instead found a job at the lowest rungs of the St. Louis Zoo. While he swept the pathways of the recently constructed zoo, his zeal for snake collecting caught the attention of George Vierheller, the zoo's director. With a flair for the spectacular, Perkins—the first to turn the forced feeding of pythons, the only way they would eat, into a public event attended by hundreds—was launched on a career. He went on to direct the Buffalo and Lincoln Park Zoos before returning to St. Louis as Vierheller's obvious successor. Along the way, he turned the drudgery of zoo work into televised animal adventures, first with *Zoo Parade* (1945–1955) and then *Wild Kingdom* (1963–1984). He lent his name, his job, and his thirst for adventure to the board game we're playing.

You can play the game as well. (I put it online: https://digitalscholarship .utsc.utoronto.ca/ZooParade/) The game is just like the zoo. It's supposed to be educational, but it appeals because it helps us imagine encounters with wild animals. To collect your animals, you must first answer three questions about their habits and habitat. Does the zebra make a "loud snort"? (Yes.) Is the rhino's horn made out of hardened hair? (Also, yes.) Then, trained, you are in the field, facing the tests and hardships of the animal game. Is "freshly killed game" a sign of a lurking tiger, or do claw marks scratched on a tree suggest an elephant? (Yes and no.) As well, the game is supposed to replicate the thrills and dangers of the tropics. The animal is frightened or has escaped a poorly chosen trap. (Go back two spaces.) A storm crashes through the jungle. (Go back and choose a new trail.)

Experience, as did so many Americans who played the game in the 1950s, the dubious thrills of "bringing 'em back alive," and ask the same question my little one did: where do all those animals in the zoo come from today? In fact, they asked the same question in 1905. "Are you frightened, Winnie?" asked Aunt Jane when they entered the zoo's tiger house (so the children's storybook *Trapping Wild Animals* went). "No, auntie," he said, "I was wondering how they caught those big wild tigers." Tigers stumbled into traps, she said, and hippos, like rhinos, blundered into pits. Of course, there was more little Winnie needed to know: "What's the jungle, auntie?" Simple enough, it was a large forest "where the men who catch animals" lay their traps.[1]

We'd give different answers today. Perhaps we might talk about species survival or explain the idea of "endangered species." Maybe we'd head to the zoo to see newborn tiger cubs. It's still front-page news when cubs are born. Just sixty years ago, the arrival of a newly captured rhino sent reporters scrambling to meet its boat. The men and a few women who captured wild animals became celebrities, America's best-known adventurers, even if today they are mostly forgotten. The world has changed and the animal trapper, American, African, or Asian, is a poacher, not a filmmaking hero. Winnie needed reminding that tigers live in India. My little one knows that, but she also knows that really there aren't many left.

In 2013, my daughter joined 181 million visitors who go to (accredited) American zoos. I take her because, like so many other children, she wants to go. Almost a third of all zoo visitors are children.[2] She wants to see the same animals she hugs in her sleep: an elephant she calls Ditto and a tiger she calls Little Tiger. She also enjoys the tender moments when animals seem cute and cuddly. Once in the Bronx Zoo we paused by the Père David's deer and I explained that they were very rare, alive only because they were among the first species American zoos deliberately bred. She seemed more interested in a mother deer playing with her foal. More often, though, she rushes past the slow, somnolent animals. She and I joke about what it would be like to own a pet hippo, but she never wants to visit the hippo exhibit. She paused only briefly at the new display at the St. Louis Zoo, my childhood hometown, where you can watch the hippos walking under water. She always likes the giraffes, but I have a hard time enticing her into a reptile house. The big apes are slightly unnerving. When she was very young, she dashed into my arms, frightened when a chimpanzee suddenly ran toward the glass. I still believe that she recognized herself in its anthropoid form; she's never been frightened by the crocodile's open mouth.

As we hunt for that tiger, I realize we truly want to catch it. We are together experiencing part of what made zoos enjoyable a half-century ago: the thrills of the animal trade, the anger of cornered beasts, the lure of the tropics. We're having fun and we're a little nervous. We pause over each question. Are those really a tiger's claw marks on a tree? Is that the tiger's spoor on the way to a water hole? Is that a new kill alongside the trail? As we pack up the game pieces, I explain why kids don't play Zoo Parade anymore. We talk about endangered species and the disappearance of animals, like her elephant and tiger. I tell her about safaris and the leisure sport of big game hunting, and she's a little stunned. Unlike Winnie, she needs no explaining to identify a vast number of species. The game collects rhinos, giraffes, tigers, elephants, jaguars, zebras ("it looks like a horse, but it's got stripes"), and more—and she knows them all.

Here's the paradox of wild, tropical animals in American history: the more rare animals became in the wild, the more common they became in American cultural life. On their way to becoming the nation's most popular attractions, zoos were buffeted by global politics, economic depression, the rise and fall of European empires, and war. Zoos competed with other forms of popular entertainment in the business of animal exhibition, and workers, visitors, and animals all challenged the educational mission of zoo founders. Animals in the zoo (truth be told) are always a little disappointing. At the zoo, they sit; they sleep. Occasionally, they mate. But they don't prowl. Once I saw a zoo leopard that had killed a wild rabbit that had wandered into the leopard's enclosure. He tossed it about and pounced again. The crowds gathered and they were enthralled. Here, at last, was the leopard of our fantasies: predatory, deadly, dangerous.

Africa Is Closer than You Think

This is a book about how we have learned to look at faraway places, environments, and peoples through the lens of animals on display at zoos and for sale in the animal business. It is about global history as it appeared at the zoo through the daily life of the animals, the keepers who mucked out their cages and reared their young, the traders who captured animals and the imagination of the American public, and the zoo officials like Perkins who helped make the idea of animal endangerment a key indictment of contemporary societies.

Once my daughter and I went to the zoo and she looked at a chimpanzee leaning against the class, scratching its eyes. ("Daddy," she looked closely, "it has fingernails.") Meanwhile, I watched both a father asking his young son to pose for the camera next to his primate cousin and a keeper nearby whose T-shirt promised that Africa was "closer" than you think. It took some cajoling to get my daughter to enter the ape house. Despite the hanging vines meant to convince visitors they had penetrated the "Jungle of the Apes," the overpowering sensation is a sour pong. In a skirmish of the senses, sight won out. Overcoming strange odors, my little one gazed at a primate. Despite trepidation, she ran her fingers over the glass when the ape approached. She was startled to see the apes pluck fruit treats with their fingers.

For a century, American zoos have implored us to look at animals, but zoos are multi-sensual places. In feeding the animals (or watching them eat), gazing at real animals and fake scenery, recoiling at strange aromas of bodies and dung, touching rough fur and cold scales, visitors imagine faraway lands, peoples, and nature. Through sight, taste, sound, smell, and touch, zoos and the animal trade created relationships between Americans and tropical animals. Animals had arrived in America for visitors to ride, tease, feed peanuts, breed, pet, train, and admire. Those relationships helped ordinary people dream of life in the tropics. They brought Africa closer but in the process made its peoples and politics seem exotic—so near but so different. Zoos and the animal trade helped ordinary people sense life in tropical lands. In the process, they fractured humankind into distinct peoples and races, more different than alike.

Social and economic processes that unrolled thousands of miles away shaped zoos. Zoos and the animal trade helped make imperialism and then decolonization tangible, even sensible. Empire privatized and enclosed lands where both animals and peoples lived. Jungles and veldts became ranches, mines, and plantations. Animals had long been hunted for food. With the rise of European and American empires, they became commodities desirable by customers far away. In the animal trading business, some animals became valuable (if endangered) commodities. Private property abroad fostered fenced displays at home, but zoos never have reconstructed a rubber plantation. In the jungle at the zoo, visitors penetrated the wild tropics: strange, exciting, dangerous, and profitable. Since the end of empires, visitors also learned about the threats faced by animals. Maybe the tropical wild lives best in American zoos.

Zoos and the animal trade created a vision of the tropics as exotic to enjoy from afar, then up close as tourists. But this took labor and lots of it. Putting animals on display required workers to capture, domesticate, care for, and feed the animals. Thousands of laborers, assistants, shopkeepers, and sailors made possible the global trade in wild animals. Once in zoos, animals depended on keepers to bring their food, clean their enclosures, and shovel their manure. Even as they recognized that theirs was caring work, the first zookeepers fiercely defended their jobs as men's work, more military than mothering. Trainers taught some animals, chimpanzees especially, to perform. During the Great Depression, relief workers paid by New Deal programs reconstructed crumbling cages into open enclosures and monkey mountains. In the postwar years zoos refashioned themselves as familial homes for endangered species. Then they required another form of caring work, albeit unpaid. Keepers' wives often reared animal babies in their own homes.

Perkins had risen through the ranks of the zoo workforce. Years later, a television celebrity, he must have recalled the hard labor needed to keep animals alive and on display. He also knew that the animal game—the capture, buying, and selling of animals—was a global commodities business employing thousands of workers from Asia to the Americas. Much is missing in his board game, including the tensions between trappers and laborers over pay and meat rations. A zoo director as well, Perkins also knew that keepers and zoo managers clashed. Strikes, mutinies, and more subtle forms of human and animal resistance disrupted the easy passage of animals from jungle to zoo.

Anyone who pays admission on a crowded summer weekend gets a glimpse of how chaotic zoos can be. Trash piles up, kids whine, and animals hide in cool shadows. Yet the men who dreamed of the first American zoos treasured order, first in the focus on taxonomy (natural order) and then later in enclosures, lush, fenced spaces and a pretend tropical nature in which animals were wild, natives were exotic, and governments colonial. This was the nature twentieth-century zoos wanted to show and save. As zoos expanded beginning in the late nineteenth century, ordinary Americans developed closer knowledge of tropical animals, especially those from regions colonized by European powers. Displaying the taxonomy of exotic nature, such zoos never achieved the biological and social order their founders so cherished. Workers, animals, and visitors did not behave in ways that matched the visions of zoo officials or founders. Visitors fed the animals, littered, even poached. In fact, they typically behaved in ways that would shock our kids today. Yet some things haven't changed. Visitors sought tales of animal adventure more than science lessons.

Tigers in the Bed

Visitors have long wondered whether animals are happy in the zoo. Does an animal's pacing mean that it is stalking imaginary prey, or is it bored? Or, worse, is it psychologically damaged? Wildness is hard to find at the zoo, and we rely on fantasies of natural adventure to make enclosures exciting.[3] We tell stories to our children, encouraging them to imagine motherly love and friendly affection. We spin tales of what it would be like if the lazy lion stalked instead of sleeping or pacing. Once when I was a child, a tree branch fell into the cheetah enclosure and the frightened cats sprinted in circles. The excitement has never left me: that's what it must be like when cheetahs chase an impala in Africa, I thought.

   With my daughter, I embarked on a "journey along a mythical waterway through four continents." We walked past wallowing rhinos and an elephant splashing in a waterfall. It was the St. Louis Zoo's new River's Edge, a recent innovation in how to display animals in their "naturalistic environment." Soon, we'll return to visit the Sun Bear Forest.[4] Decades ago it was easy—and often still is—to find fault in the way animals were kept. No matter how much zoo directors promised a glimpse of the wild jungle, animal houses smelled of dirt, damp, and dung. Zoos have been in a state of constant change since they first appeared in America at the end of the nineteenth century. Our ancestors wanted to see, ride, pet, feed, tease, and imagine animals in ways disapproved of by zoo directors.

These first zoo directors designed bland, small cages in which animals had little room to move around. Visitors were meant to examine the specimen closely, its form, fur, bones, and evolutionary similarity to other animals. Such exhibits left wildness to museums, which could restore animals to wild life through naturalistic taxidermy.

Like animals, zoos age. By the Great Depression, buildings that seemed palatial to earlier generations looked and smelled decrepit. Animal odors seemed ingrained in their very foundations. Even worse, from the perspective of zoo directors, there were more exciting places to view animals. On screen, snarling tigers stalked native humans and gigantic pythons fought in mortal combat with black leopards. If zoos had become places of somnolence and smells, films introduced jungles of "fang and claw." Supported by federal relief dollars, zoos began imagining new, modern exhibitions and animal entertainments. Cages became enclosures, promising illusions of nature. Over by the towering monkey islands, cramped crowds of monkeys wrestled and crunched on tossed peanuts. On the new zoo stage, chimpanzees rode bicycles and performed other human tricks to the delight of packed audiences. In some places, zoo shows outdrew the local baseball team.

As naturalism became a goal, many zoos experimented with grouping some animals together in vistas wild enough that visitors could pretend to be fearless adventurers gazing across the open savanna. Visit the zoo and see not just hippos or elephants but wild Africa. Zoos also built children's zoos where kids could pat the lion cub or cuddle the chimpanzee baby. Wildlife could become part of family life. During the Cold War as the nation retreated to the security of hearth and home, zoos promoted animal families. See the affectionate monkey cuddle her infant. But sometimes in the stress of captivity, animals didn't want to mother. So animal babies went home with keepers to be raised by their wives.

By the 1960s and 1970s, the concrete walls crumbled again. The animals seemed as decrepit and damaged as their enclosures. Racially tinged condemnation of inner cities as ghettos helped zoo directors and critics explain the crisis of urban zoos. The animals in dilapidated cages were not wild, just neurotic slum dwellers. Some zoos fled from city to suburbs to build safari parks. Here was tropical Africa and Asia, born again as a wild sanctuary.

Through a century of changing zoos, animals that were rare and endangered became familiar and entwined in our daily lives, Over the course of my daughter's lifetime, the animals that inhabit her toy chest will come to live mostly, if not entirely, in zoos. Animals prowled in zoos, circuses, pulp fiction stories, explorers' memoirs, and films but steadily disappeared from their native habitat. The tigers on her bed are a prime example: the World Wildlife Fund recently reported that tigers likely will become extinct in the wild in the next dozen years.[5] We exist at a remarkable, if tragic, moment in our human and animal histories. Tigers populated American popular culture during a century in which their wild numbers decreased 97 percent.

If animals have an uncertain future, they have a conflicted history. Animal and human lives dramatically collided during the twentieth century. Local populations in Asia and colonial officials alike feared tigers and condemned them as hindrances to the expansion of jungle plantation industries, such as rubber. Tigers were also in demand. Americans devoured stories of tiger attacks, hunts, and captures, and the big cats fetched strong prices for American animal traders (long before they became high-priced medical ingredients). Animals from the tropics had value. They were commodities with a spirit and a personality. One of many products ripped from the tropics in the long history of imperial economies, animals lived, breathed, and fought back.[6] Animals brought to America from tropical lands became omnipresent in our culture and urban lives. They shaped fantasies about the wild exotic and are becoming extinct.

That's why we don't play Zoo Parade anymore. Some animals, like gorillas or orangutans, didn't even make it into the game. By 1954, it just didn't seem right even to pretend to capture such rare apes. No one was collecting bison anymore, either; they were already bred in zoos. Other animals, like pheasants, might have been rare, but few visitors paused at their cages. Peter Ryhiner wrote his memoirs about his struggling career in the animal business around the same time Perkins released his board game. Everything was changing. The golden age of the animal business had passed. "Animal collecting is becoming too difficult," Ryhiner mourned. Still, "it has been a great period and I am happy to have seen it."[7] From "bring 'em back alive," the zoo had become a place to "breed 'em alive." And that change needed a whole new set of stories about love, affection, and human women tenderly raising animal babies in their own homes. All this made for good media reports but lousy board games or movies. Instead, we rely on older histories of safari and wild adventure to overcome the uncomfortable reality of tigers whose real habitat is the zoo.

The Wild Zoo

I grew up within walking distance of the St. Louis Zoo. Even to a growing boy, the zoo stood out because it was free and changing. I remember a dank, barred, and odoriferous ape house giving way to a new spacious primate display. Its concrete trees evoked for me—though not necessarily for the orangutans—a jungle. The St. Louis Zoo's mammal curator Martha Fisher recently reflected on Raja, the zoo's male Asian elephant, and his twentieth birthday. "We hear that all the time," she said, "that people remember when he was born and how excited they were." She's right. I do remember it, just as I recall Cardinals' World Series wins. The local newspaper called both "touchstone" events.[8] At the zoo, I patted the domestic animals, rode the zoo railroad, marveled at chimpanzees' human likeness, sought out the poisonous snakes, and came home with stuffed animals (on occasion).

The zoo taught me, as it teaches my daughter, about foreign places, peoples, land, and culture. Yet when we eventually went to India, it looked nothing like the tiger enclosure in the zoo's Big Cat Country. My daughter and I are not like most Americans—our contemporaries or our predecessors throughout the twentieth century—because we have been able to travel widely in the tropical world. I have visited India five times, partly for academic work and partly to visit my wife's family. When I come home, friends, relations, colleagues, and neighbors ask an identical question: did you see a tiger? In a country of a billion people and fewer than 1,411 tigers (the 2008 census count), Americans still associate India with jungles and big cats. I have, in fact, seen tigers in India—in a zoo in Hyderabad. Whether we admit it or not, over the past century of American zoos, the tiger has become as native to the United States as to India or Malaysia or Sumatra.

The insatiable desire of Americans to see the most rare and dangerous of animals has profoundly altered real human and animal lives. Workers in Asia and Africa perished trying to capture terrified elephants or tigers. Zookeepers worked in dangerous conditions for little pay. Native hunters and trappers sold animals to American traders who transported them long distances into conditions that zoo officials could only guess would keep them happy and healthy. Some animals became national celebrities, like Bushman, the sullen but long-lived gorilla at the Lincoln Park Zoo in Chicago, or Jo Mendi, a cognac-swilling, performing chimpanzee in Detroit. Others lived more anonymously but left behind traces of their difficult lives, such as a lion in Washington that faced a vengeful public demanding his death after he killed a young girl.

In recent years, historians and other scholars have realized that the past engages more than just *Homo sapiens* and have explored the encounters of humans with other animals. The reality is that humans have always lived in close intimacy with other animals.[9] We depend on them as pets for affection and as domesticated beasts for labor and food. We shoot them for pleasure and to mark our own social status.[10] We exterminate them as pests, and images of animals are threaded throughout our sacred lives and symbolic language. Despite all that familiarity and for centuries, Americans and Europeans were enthralled with animals from the tropics. From Egypt to China, emperors, kings, or nobles as far back as antiquity kept private menageries of animals. By the age of European expansion in the fifteenth century, ships returned from Africa or Asia with exotic animals. Unfortunately, they rarely lived long in Europe's less than salubrious climates. However short-lived, those animals were rarely seen by the public. By 1662, King Louis XIV had installed a collection of animals in his palace at Versailles, one of the grandest of menageries owned by European nobles and rulers.

Royal menageries in Europe slowly evolved into public zoos by the nineteenth century. The London Zoo was initially designed to house a private collection of scientific specimens but was finally opened to the public in 1847. Europeans constructed the first zoological gardens— zoos—before the Americans. In fact, the first American zoos found their inspiration in zoos in Britain and Germany.[11] Still, by the nineteenth century, sideshows as well as circuses and their menageries introduced Americans to an array of animals from elephants to pythons to big cats.

Flurries of construction in the first decades of the twentieth century and then again during the Great Depression birthed zoos in virtually every American medium and large city. Philadelphia claims to be the first, opening its gates in 1874, followed in the coming decades by most large and many medium-sized cities, including Cincinnati, New York, St. Louis, Chicago, and San Diego. The interwar period saw the construction of new zoos in cities like Chicago and Detroit and the modernization and expansion of many more, everywhere from Toledo to Tulsa.[12]

These were zoological gardens, insisted their founders, an austere title evoking sober education and science. Visitors, instead, called them zoos. As our board game reminds us, the lines between zoos and other forms of wild animal display from circuses to vaudeville to films to television have always been blurred.[13] Animals that began their captive lives in circuses or acts might end up zoos. In fact, the first elephants in American zoos were often hand-me-downs from circuses. Aged, often irascible, they gave rides in the zoo rather than performances under the big top. Performing chimp acts introduced on vaudeville stages—the variety shows popular in the early decades of the twentieth century that often relied on racist humor—were later perfected in zoos. Wildly popular adventure films introduced Asian jungles and African plains to ordinary Americans but starred animal traders whose job it was to sell animals to zoos. And Perkins became one of the biggest stars of early television.

Over the course of the past century, we've seen animals in more exciting ways than in zoos. Yet we keep coming back to zoos even as they competed against first circuses, then jungle movies, then television. Visitors could poke the animals in the circus menagerie, and the elephants did tricks.[14] On film, jungle animals stalked and battled their enemies in villainous rage. On television, great herds of animals stampeded across vast natural vistas while predators prowled. At the zoo, a handful of specimens grazed placidly.

By the millions, we still visit zoos. Why are we so fascinated by tropical animals? What was it like to work capturing, breeding, and caring for zoo specimens? Why have animals become so crucial to our shared popular culture? Zoos would much rather narrate their own story. It is a rare zoo that opens its archives to researchers or the public. Instead, history retold by the zoo becomes a formula of early mistakes combined with a longstanding commitment to animals and their survival. Zoos explain the animal game as mistakes of a less enlightened age. The zoo, writes one typical guidebook, is "inextricably linked to its conservation vision—an outgrowth of the need to educate society about the severity of the environmental crisis before time runs out."[15] Another uses its long history as a plea for conservation: "All of us, working together, can arrest and even turn back the tide of extinction. Therein lies the special mission of zoos."[16]

That history is more complex, weaving together the decline of animal populations, the rise and fall of empires, and shifting relationships of animals and people. Over their history, American zoos have fashioned themselves into institutions for animal conservation and breeding. Along the way, they helped define what conservation entailed and which nations and peoples were best equipped to save the world's animals.[17] For Americans over the past century, words like *wildness, savagery, primitive, jungle,* and *veldt* evoked something strange, distant, and exciting. They still do. The popularity of zoos and, indeed, the very nature of animal conservation are forged on the backs of the sometimes violent, sometimes intimate stories told about animals, their native habitats, and the people who caught, fed, and trained them. Animal traders who grappled with angry leopards and zoo officials who handled cobras on live television fascinated Americans with little chance to travel. Expeditions to bring back live rhinos from remote India, pythons from Malaya, or gorillas from the Congo helped create an understanding of the tropical, non-Western world as exotic—fundamentally different, savage, economically backward, and in need of help to preserve its wildlife. Our fantasies of the exotic sowed relations between peoples and between humans and other animals that sometimes produced affection across species lines and sometimes tragedy. Despite campaigns to "phase out the zoo" that accelerated in the 1960s, Americans still want zoos—as sites of leisure and centers of conservation.

Tellingly, zoo officials often compare zoo and sports attendance. In 1964 about sixty million Americans went to zoos, boasted one zoo official. He was particularly proud of the comparisons with another "spectator sport": 21,280,346 Americans went to baseball games, while the two football leagues drew a paltry 5.9 million. "Zoo attendance staggers the imagination," he said. That year, one in three Americans visited a zoo, even as others questioned their ethics.[18] Partly, the comparison makes sense because the numbers match. In 2009, in St. Louis, the zoo recorded its highest ever attendance of more than three million, about the same as the Cardinals baseball team. Partly, that comparison reveals the entwining of strategies for species survival with spectatorship and spectacle. Discarding dour instructions to examine the animals in an orderly fashion that matched the hierarchy of nature, zoos promoted animal shows, dramatic exhibits, riding elephants, zoo railways and monorails, and photo safaris. We visit jungles, veldts, savannas, deserts, and rain forests at zoos and gaze at the cuddly, the strange, and the ferocious in ways that ape the practices of an earlier era.

1

THE ELEPHANT'S SKIN

Animals and Their Visitors

A STORY of two elephants and their skin presents an atypical history of the formative era of zoos in the last years of the nineteenth century and the first decades of the twentieth. Their life and death help trace the distance that separated the visions of zoo founders, visitors' desires, and the experiences of animals on display. Jennie died a peaceful death in Philadelphia after years of patient display and service to a struggling zoo; she became a wallet. In New York, Gunda died a rogue's death, shot by the nation's leading big game hunter after years of rage and imprisonment. Reduced finally to a corpse, Gunda was the casualty of an agonized debate about the failures and successes of the zoological garden. Jennie was a pioneering star in the Philadelphia Zoo. Her exhibition helped the zoo distinguish itself as a true, scientific zoological garden and not simply a menagerie, a haphazard assortment of animals meant more to amuse than educate. Gunda was first a prize and then a rogue for the Bronx Zoological Park—the Bronx Zoo, visitors called it.

There would not have been zoos in America without elephants like Jennie and Gunda. Visitors enjoyed lions or tigers and thrilled to the poisonous snakes, but they flocked to the elephants. Baltimore, for example, knew that the "movement for a real zoo is growing" when it was finally able to purchase an elephant.[1] As observers in Milwaukee noted in 1906, when the elephant arrived, the zoo had truly opened its gates. "The elephant was the moment that persuaded Milwaukee to become zoo-minded," and some of its first funds came from "The Elephant Show" sponsored by a local fraternal lodge.[2]

Exotic animals—elephants especially—brought polish to growing American cities. Elites in the nation's biggest cities competed to build urban zoos that emulated those in European imperial capitals. The Philadelphia Zoo still boasts of being the nation's first zoo. Its zoological society was organized in 1859, but the zoo displayed its first permanent collection only in 1874, a fifteen-year gap that showed how hard it actually was to turn zoological dreams into houses occupied by animals. New York exhibited animals in the Central Park menagerie (1864) and Chicago in the Lincoln Park Zoo (1868). Cincinnati opened its gates in 1875 with a small collection including grizzly bears, monkeys, a tiger, a pair of elk, and, most important, an elephant donated by a circus. The National Zoo opened in Washington, DC (1889), and the New York Zoological Society built the Bronx Zoological Park (1899). The Smithsonian Institution displayed birds in a free-flight cage at the 1904 St. Louis World's Fair, and the city purchased the exhibit at the fair's conclusion as the foundation for what would eventually become a zoo. San Diego exhibited its first animals in 1916.[3]

The life and death of elephants in America revealed the fragility of such zoos. Zoo founders and funders dreamed of majestic zoological parks that would match the wonders of nature and mark the cosmopolitan status of their cities. They sought refined zoological parks but could only assemble incomplete collections that, despite renderings of palatial animal houses, often looked and smelled like plebian menageries. Elites donated animals—Andrew Carnegie gave a Barbary lion to the Bronx Zoo, Jacob Shiff presented a cheetah—but too often, young zoos depended on circuses, especially for their elephants.[4] Zoo patrons dreamed of palaces of science but were often the charity cases of circuses, that most pedestrian of entertainments.

Still rare and exciting to American audiences around the turn of the twentieth century, many elephants in America were born wild and as infants passed through the hands of animal traders to circuses, from wild life into performing life. The most tractable animals learned tricks for big-top acts while those that ignored commands went to the menageries accompanying itinerant shows. Circus promoters had little interest in preserving wild character, even for those animals relegated to menageries. Elephants were trained to perform tricks, give rides, and—above all—subordinate their awesome strength to human masters. The life span of many elephant performers was limited, however. Too dangerous or obstinate, their worth plummeted. No longer able to follow the wandering life of the circus, some made their way into a new life as a zoo star. Circuses realized that not only could they rid themselves of a dangerous animal, they could win public acclaim when they donated an elephant "gone rogue."[5]

Those urban elites who dreamed of zoological parks fantasized that displays of biological order would beget social order. From the start, though, visitors demanded closer interactions with tropical animals. They wanted zoo stars like Jennie and Gunda that could allow them to imagine faraway lands as exciting, adventurous places. Typically, zoo founders were urban elites, but visitors didn't always recognize their authority. Visitors just didn't behave as they were supposed to. When Americans rejected haughty zoo directors and founders and the austere vision of science they promoted, they climbed inside the skin of elephants like Gunda, imagining the experience of chains and the tedium of the cage. In better times, visitors sought out elephants, like Jennie, to ride, feed, and scrutinize. Visitors to the first American zoos forged closer, more intimate connections with elephants than with other, even more exotic beasts. Perhaps this was because elephants became the focus of zoo entertainment despite officials' plans for zoos to be parks for science lessons. Perhaps it was because elephants' size and strength exaggerated responses that ranged from what seemed to be affection to what was clearly fury. Perhaps it was because visitors, after tossing handfuls of peanuts, could run their hands over the elephant's trunk, reaching across the species divide to feel their rough skin.

The elephant's death was sometimes natural. On other occasions, the rogue elephant—too angry, too vengeful for the life of the cage—died of poison or bullets. In the moment of death, as a huge life ebbed, plans for the elephant's skin began. Life did not end with death for the captive elephant. Skin, bones, and ivory all lived on as valuable materials and as mounted taxidermy specimens. More than a century after her death, Jennie's skin is still there for touching.

Jennie: Philadelphia

Jennie, an Asian elephant, lived at the Philadelphia Zoo since its opening in 1874. When the fifty-year-old elephant died in 1898, she was skinned, tanned, and turned into souvenir wallets. The last of Jennie lies in a plastic folder in an archival box in the zoo library. The skin is rough to the touch, leathery but dimpled with a coarseness even a casual zoo visitor would recognize as an elephant's skin.[6]

Jennie's life, death, and afterlife embodied the vast distance that separated the expansive dreams of the nation's first zoo founders and the reality of animal life in the zoos they actually managed. The Philadelphia Zoo still claims to be the nation's first, a boast that reaches into the heart of the question of what separated zoos from the other places where ordinary Americans could examine wild and tropical beasts. These included natural history museums, menageries, and circuses. Natural history museums and zoos often shared the same patrons: wealthy urban elites, often dedicated hunters, impassioned by biology and eager to instruct the nation's masses.[7] In cities like New York or Chicago, natural history museums had mastered the art of taxidermy. Sometimes they displayed animals on their own as aseptic representatives of their species. Other times, they mounted just their heads in galleries that evoked the private grandeur of an English hunting estate. Increasingly, museums also mounted animals in naturalistic dioramas.[8] For many Americans, the rare mammals and birds that would come to evoke life in distant jungles and savannas were long-dead hunting trophies.

Others first encountered wild tropical animals as menagerie prizes or circus attractions. In New York, visitors could examine animals cramped into dirty cages in the middle of Central Park. Critics dismissed that collection as a menagerie, a mere assortment of wild animals, not a dignified scientific zoo. At the end of the subway lines to Coney Island, New Yorkers could also visit Frank Bostock's Animal Show. For decades, traveling hucksters had brought wild animals from town to town to exhibit to curious masses. Long before William T. Hornaday became director at the

Bronx Zoo, he longed for the call of the menagerie promoters. "The thrill of civilized children," he recalled, "is broadcast in one shrill and piercing cry: 'The Elephants Are Coming!'"[9]

For a fee, patrons of these menageries could touch the snakes or mangy lions. They might feed or tease the animals and listen to the outlandish claims of the menagerie owners, more folktales than accepted science. They could watch the elephants dance or ride on their backs. In Philadelphia, the zoo's founders dismissed such menageries as pap entertainment. Zoo founders were, instead, animated by grand fantasies of displaying the biological hierarchies of the animal kingdom all in the service of social order. As is so often the case in the long history of zoos, plans and dreams didn't match up with reality. Even backed by elite philanthropy, zoos were mere beggars compared to other animal businesses. Circuses were a regular attraction in small towns and growing cities, and the biggest circuses had budgets for animal purchases that dwarfed those of fledgling zoos. Circuses employed animals to perform tricks. They also owned menageries as side attractions to complement the big top. Circuses, not zoos, were the largest American consumers of the wild animals traders shipped via Europe, especially Germany at the turn of the twentieth century. Asian elephants like Jennie came to the United States as circus treasures, not biological specimens. Only when elephants became too ornery to carry paying customers on their backs or perform tricks did they become useless to circus owners.[10]

Surplus, Jennie and many others were donated to zoos. Like many animals, she had migrated from the wild into a traveling circus menagerie. Too old for safe display, she retired to the zoo. After death, the beloved elephant was too valuable simply to bury, and she became a souvenir wallet. Often elephants and even other zoo animals passed from captive life to taxidermy display. It was a difficult life for the captive elephant in America. Paraded from city to city in circuses, elephants like Jennie could amble only from boxcar to the big tent. Zoos offered little more freedom or space. In their more tractable youth, elephants marched in circles with children perched in howdahs. Perhaps these children imagined themselves stampeding through Indian jungles, but for elephants, life in the zoo evoked little memory of the jungle. Elephant houses were woefully cramped. In the long, wet winters of cities like Philadelphia, there were few chances for a walk outside. Many elephants, especially those that had become destructive or dangerous, spent their lives in foot chains. Those very chains became the source of the crisis that engulfed the Bronx Zoo and led to Gunda's early death.

Jennie was hardly the wild and stupendously strong "*elephas indicas*" promised in the zoo guide. In the tight space of an odoriferous elephant house, she stood very still. Maybe she swayed and zoo directors assumed movement helped keep elephants comfortable.[11] When Jennie was outside, she carried children on her back. Inside the dour enclosure, though, the captive elephant stood too still to seem wholly alive. Yet even so, because visitors insisted animals were more than simply representatives of their species, the Asian elephant specimen became Jennie, an urban star and a zoo favorite.

In Jennie's passage from circus life into zoo death, the tragic irony of the first American zoos was revealed. Jennie's biography, then, becomes enmeshed in the history of the infant zoo where she was a star attraction. Zoos were born of elite dreams of natural order; their life was one of disorder because of what the animal trade could offer, what visitors desired, and how animals behaved. Early zoos promoted an orderly world and systematic science in competition with other forms of animal display like circuses and menageries. Cage bars alone, though, couldn't make visitors into scientists or animals into specimens for contemplation. When animals rebelled—attacked, fled, or simply died—and visitors misbehaved— fed, touched, taunted, or killed the animals—the tidy scientific park became untidy urban entertainment.

Elephant Day: St. Louis

For urban boosters, the strange calls of animals echoing from zoos through surrounding neighborhoods were music to their ears. Animal noises were evidence that the proud metropolis had come into its own as the equal of any European capital. In 1910, the elephant Jim's trumpet in St. Louis was triumph, not annoyance. St. Louis children had collected their pennies to buy the elephant, and his calls announced the fruition of the city's long-held hopes for a proper zoo. The mayor even joined schoolkids in celebrating "Elephant Day" by riding around Forest Park, home of the new zoo's grounds, in a howdah.[12] Three years later, though, Anna Sneed Cairns, president of St. Louis's Forest Park University, worried about how the 100 female students in her dormitories would sleep. It wasn't rattling streetcars or factory whistles that kept these women up nights in the heart of the city; lions roared and ostriches grunted. The "calls of the wild" were disturbing the neighborhood. Even the local South Side Forest Park Association went to the zoo to complain. (The newspapers reported that of the 150 protestors, only two were women, proving men "were as much frightened and worried as women.") "The blood-curdling defiance of the king of beasts" had made American urban life unbearable.[13]

Cairns' complaints reveal an odd reality about American life at the turn of the twentieth century: exotic animals, like the lion, ostrich, and elephant in St. Louis, had become part of the fabric of the growing American metropolis. Elephants, though, like many other tropical prizes, were still more common in circuses and menageries than in fledgling zoos. Most Americans saw their first elephants in traveling circuses or menageries that offered a haphazard collection of specimens. In smaller Little Rock, Arkansas, the "dense crowd" visited a circus's "wonderful collection" with its array of tropical prizes: a two-horned rhino, a hippo, and, of course, "a huge elephant." Circuses toured from town to city and brought with them menageries and animal acts. In Dayton, Ohio, locals enjoyed the drama of a circus railroad car that broke in two because the elephants were simply too heavy.[14] To be a first-class show, circuses needed elephants, however difficult they were to control and transport. The Sells Floto Circus, for example, owned a small herd, including the famous Alice. Alice gave birth four times, but likely because of the "annoyance of circus travel," she refused to rear any of her children.[15]

While crowds in Little Rock, Dayton, and beyond encountered exotic animals like Alice in menageries and circus shows, a select few, the very wealthiest, went to Asia and Africa and saw tropical animals in person. Other wealthy elites also traveled to European zoos and came home both jealous and inspired. Their European journeys helped launch America's eagerness for zoos. Dr. William Camac was one of a handful of American travelers who returned from their tours of "the wonderful zoo gardens abroad" only to "marvel at the paucity of like institutions" in their native cities. Camac was a rich man with enough leisure time to resurrect an almost moribund Philadelphia zoological society. With the approach of the American centennial celebrations in 1876, Camac found new inspiration and new life for a dream of a Philadelphia zoo modeled after the private London Zoological Society. Supported by the city's financial and scientific elite, the proposed zoo would boast the "air and general appearance of famous long-established like institutions in Europe."[16]

Camac may have been a pioneer, but he had his followers. By the beginning of the twentieth century, urban elites formed zoological societies clamoring to build a zoo that could educate, entertain, and compete with zoos in other cities. The cream of New York's social register, from Andrew Carnegie to J. Pierpont Morgan, added their names and wallets to the campaign for a zoo that could challenge London's. The zoo was, in the mind of such donors, part of the natural evolution of the American metropolis from a chaotic admixture of peoples, buildings, and industry to a civilized metropolis. First came dwellings, then schools and colleges, and finally zoos and other museums. Even before laying the foundations of the first buildings, the New York Zoological Society lauded the zoo as "the high-water mark of civilization."[17]

The alchemy of wealth and zeal for biological taxonomy helped urban boosters forge a zoological park out of a menagerie. In mid-sized Buffalo, New York, the city's campaign to turn a mere menagerie of North American animals into a proper zoo encapsulated its cosmopolitan ambitions. The year 1892 was a "feeble start" for a zoo that displayed just two bison and a handful of elk. Still, the fledgling zoo was a popular attraction. By the beginning of the new century, on warm summer Sundays, 25,000 people crowded in front of a mere 100 specimens. Then came an Indian elephant, and the menagerie became an urban zoo.[18]

In 1910, the "Million Population Club" in St. Louis gazed jealously at Cincinnati, Buffalo, Detroit, and Denver, which already had zoos. In a typical claim audible from coast to coast, boosters argued that truly great cities needed a zoo of "instructive or entertaining character." In Buffalo, the crowds had turned out to see the elephant. St. Louis and many other cities wanted zoos with elephants as well. Soon the director of Chicago's Lincoln Park Zoo traveled to St. Louis to add his pro-zoo voice. "Give St. Louis a zoo," Cy De Vry said, "and you make her the envy of all the large cities of the country."[19] In 1912, Dallas and Boston also sought to grow their menageries into zoos befitting large cities. In Boston, a wealthy booster endowed the zoo with a million-dollar gift.[20]

Tropical animal prizes boosted the cosmopolitan polish of any metropolis. Around the time St. Louis declared its "Elephant Day," even smaller cities had begun to measure their status by their zoo ambitions, even if they were more likely to stock their zoos with local or North American animals, such as bears, elk, or wolves. By 1910, many large and small American cities had some kind of collection of animals for display. Los Angeles had three menageries, each with a haphazard collection of animals from parrots to seals. Smaller cities, like Fall River, Massachusetts, and Rochester, New York, also exhibited animals. In Rochester, the city's zoo offered a range of "common" animals like goats and sheep and a small selection of tropical animals, including four baboons. Parrots were displayed in cages alongside robins and bluebirds. Fall River's residents could visit just two young bears in a tiny menagerie.[21] In some cities, like Youngstown, Ohio, a streetcar company might assemble a handful of animals into an attraction anchoring the end of the line. Often, as in Milwaukee and Memphis, animal collections were an extension of the development of city parks. The Park Commission in Memphis by 1910 had begun construction of carnivore and elephant houses to augment a collection of 65 different animal species.[22]

In Richmond, Virginia, a local enthusiast mourned, as late as 1923, that despite being a city of 200,000 and the former Confederate capital, "we lack a zoological garden." He imagined a zoo dotted with aviaries, dens, and houses as the highlight of a 100-acre park.[23] In Los Angeles, the local Adventurers' Club sought a proper zoo, but a "certain element," likely neighbors, worried about animal calls in the dead of the urban night, foul odors, and animal escapes.[24]

Education for the Masses: Philadelphia, New York

That "certain element" had reason to worry. Zoo founders imagined their parks as orderly, but visitors and animals together ensured they were anything but. When zoological societies doubled as the social register, zoo founders imagined that biological order could beget social order. The

vision of nature they presented was one of natural hierarchy with animals
in their cages representing genus and species. Zoological gardens, one
booster declared, "form the most agreeable places of popular resort, while
their importance as a means of education for the masses is an interesting
and useful branch of natural history is fully recognized . . . by the most
cultured men of the age."[25] Perhaps the hierarchy of nature, in all its order,
might offer a model to the urban masses?

As if to match dreams of social order forged through biological education, the Philadelphia Zoo offered imaginative sketches of model animal
houses. The carnivore building, for example, appeared as a European-
style palace, its angular pediments softened by lush but well-tended
foliage. In this verdant resort paradise, couples and families promenaded
in Sunday best. In the "Interior of the Monkey House," the well-dressed
gazed through the bars at climbing monkeys. They stood removed from
the cage, respectful of rules. One child pointed; the other listened to her
older sister's explanations. Others waited patiently for their turn in front
of the cage. It all seemed so dramatically different than the circus where
animals performed tricks or offered rides to squabbling children alongside tumbling acrobats or clowns and even more removed from menageries where children and adults jostled to poke strange animals while a
hawker spun outlandish tales.[26]

Zoo planners hoped tidy taxonomy would produce social order. The
problem was it was impossible. Even elites did not have bottomless
pockets for pet projects like a city zoo. Instead, zoos still depended on
visitors to buy tickets, spend at restaurants, or pay to ride an elephant.
From New York to Cincinnati, zoo administrators yearned for streetcars
or railways that could bring visitors directly to the zoological garden. "A
crowd at the Garden is an attraction in itself," noted the Cincinnati's zoological society president as he celebrated the construction of a "steam
railroad."[27]

The wealthy cherished the order of biological taxonomy and hoped it
would inspire quiet contemplation. But they couldn't shoulder alone the
cost of the zoo. Dependent on admissions and concessions, zoos were at
the mercy of economic downturns and even bad weather. In the 1870s and
1880s, the Philadelphia Zoo had profited from economic good times. In
February 1893, a financial panic led to a run on banks and a collapse in

stock prices. From the prosperity of 1892 to the depression of 1893, attendance in Philadelphia fell about 20,000, almost 10 percent. Even worse, declining gate receipts only exacerbated the "serious condition"; the zoo was badly in debt with few resources to procure new animals. Animals aged and died, and even worse, visitors were unwilling to spend their meager wages to view empty cages or familiar attractions. With the "financial troubles of the year" affecting even the richest, the zoo's campaign to recruit zoological society members from the ranks of the city's elite failed miserably. "At this hopeless juncture," the zoo's leaders abandoned "hope of saving the Garden by private means." They appealed to the city—the first of numerous times private zoos turned in bad times to the public for help. In exchange for free tickets to public schoolchildren, the city filled the budget gap.[28]

The Philadelphia Zoo cautiously backed away from the precipice of bankruptcy, but in New York, Hornaday, the Bronx Zoo's inaugural director, still dreamed of a clean, ordered zoo.[29] As the Philadelphia Zoo struggled to rebuild its diminished collection, Hornaday and his wealthy patrons offered grand plans, unlike the much-maligned Central Park Zoo whose cages spread their stench in the lower corner of the park. With its haphazard collection of birds and mammals, the Central Park Zoo was simply a "boarding place" for menagerie promoters. The zoological society set out to tame a "wild and unkempt" Bronx Park, gifted by the city. The park was still a "jungle" of fallen trees entangled with dogwood and poison ivy. Mosquitos swarmed. Minks preyed on the songbirds. According to Hornaday, the park teemed with human riffraff, vegetal weeds, and animal vermin. For years, Italian migrants who lived at the park's southern boundary dragged firewood out of its forests. Even worse, weed-tangled undergrowth had become a haven for passing tramps. It was the jungle "primeval," and Hornaday, who had hunted in Indian jungles and western plains, never entered the zoo's future grounds without his pistol.[30]

By 1898 as the economy recovered, workmen had hacked away weedy underbrush to polish a forest of majestic trees. Under their shade the first animal houses took shape. Hornaday, meanwhile, was in Europe, visiting British and continental zoos, and in his notebook, a vision matured of a zoo that would be the pride of the urban elite. The proposed elephant house copied its design from the Antwerp Zoo's Palais des Hippopotames.[31] The proposed bear pits matched those from Germany. The urban boosters who imagined grand American zoos looked with envy on European capitals. In London, the zoo, even if restricted by its limited Regent Park real estate, was the living embodiment of a grand empire. Travelers, soldiers, merchants, and diplomats all returned with live trophies from the edges of empire.

In New York, boosters coveted the easy access to wild tropical trophies that came with the rise of empire. Wild nature, captured, tamed, commodified, and displayed, became a mark of a cosmopolitan metropolis that valued social order. New York, Hornaday reasoned, is "overcrowded" and the "poorest people suffer most." The zoo, a scientific feat of philanthropy and education, would construct a "beautiful natural world in miniature" where the city's huddled masses would realize their debt to European zoologists who gave "the world the modern zoological gardens."[32]

From forest into zoological park, Hornaday imagined more than a construction project of weeded forests, walkways, cages, and buildings. Rather, it was the "evolution of the order, and system, and general polish which . . . should characterize everything in . . . our Zoological Park."[33] The zoo was hewn from the sordid wildness Hornaday believed reflected the primitive character of the immigrants who lived at the forest's edge. The display of animal taxonomy was not simply putting animals into different spaces separated from each other and from their native environments by walls and cages but the ordering of the city that now surrounded and gaped at them. Despite rain, cold, and incomplete buildings, 90,000 people visited the new Bronx Zoo in its opening two months in 1899.[34] Yet those numbers alone can't tell the full story. The zoo was well liked, but the reasons for its popularity proved a constant frustration for Hornaday. The zoo never achieved quite what it promised.

Walking the Wild Kingdom: Philadelphia, New York

Perfect zoo visitors were like those who, on a summer day in Philadelphia, turned a father's simple error into a polite debate about the classification of nature. At least according to a passing journalist, a father pointed out a "splendid bird" to his son, but the bird was not, in fact, the American eagle but a "carrion buzzard." A "gentleman and one elderly lady" as well as the embarrassed "instructor of youth" debated whether the two birds, one a national symbol and the other a scavenger, were, in fact, related. They turned to the zoo's guidebook, a textbook in miniature of animal taxonomy.[35]

This was exactly how zoo directors and their patrons wanted visitors to behave. In directors' and donors' visions, visitors came well dressed. Affluent or not, they admired the animals in their order of natural hierarchy from the lowly reptile to the magnificent tiger to the rare and strikingly humanoid ape. In search of edification and civilization, they walked the orderly garden pathways from the elephant or lion house to the reptile house. In New York they lunched at the Rocking Stone restaurant or the boathouse restaurant. They visited to look at the animals—just to look— and placed their trash in bins, purchased a photo souvenir of the animals, and left again by subway or streetcar. They enjoyed a day out while learning biology and, somehow, left more civilized than before. In reality, zoo visitors wanted to touch the animals, hear them roar, and feed them peanuts (or worse).

Around the turn of the twentieth century, growing American metropolises offered new choices of cultural institutions from public libraries to museums. New York boasted the most of all. Its art and natural history museums were venerable institutions by the time the zoo built its first buildings. In the Bronx, at the other end of the park from the zoo, plants at the botanical garden were reaching tropical maturity. The zoo was one of many choices and, increasingly, the most popular one. By 1903, even before the Bronx Zoo had finished its initial houses, more than a million people visited. When the elevated railways and streetcars reached the zoo gates, attendance quickly doubled.[36]

When elites and zoo directors imagined the place of their zoo in the growing city, they believed science lessons taught civilization. In an ideal visitor's careful observation at the zoo "some question is certain to arise, discussion to ensue, followed by a consultation of the guidebook," hoped the Philadelphia Zoo. In New York, the zoo quickly sold out its first two editions each of 3,000 copies. By its third edition, it printed 5,000 copies.[37] Guides were deliberately dry, "an admirable synopsis of the whole subject of zoological classification," the Philadelphia Zoo boasted.[38] Guides were, in the guise of maps to the zoo, biological textbooks about evolution and taxonomy and, at their heart, a plea for order and hierarchy. The Philadelphia guide commanded visitors "to enter at the Girard avenue gate." Continue through the zoo from the Carnivora house down through the taxonomy of nature eventually to the birds and reptiles and, finally, out through the zoo gate "at the close of the book."[39]

Arthur Brown, the zoo's director and the guide's author, forged his biological reputation through the classification of the "animal kingdom."[40] He began his guide by reminding visitors that his zoological garden was for "instruction" as much as amusement. The visitor should learn about Linnaeus, the great biologist who arranged the "mere mass of living forms" into families and, ultimately, species. The guide was not simply a reference but directions on how to look—and nothing more—at caged animals as specimens, interesting for their similarities and divergences, not for their individual character or antics.[41]

The deliberately dry tone of zoo guides matched the orderly rows of the architecture. Nature reduced to its taxonomy manifested in the long rows of cages in the small mammal house at the Bronx Zoo. Visitors faced a wall of cages stacked upon each other. Leisurely walking the hall, visitors were invited, even impelled, to compare the animals and to notice patterns of form. But not of function—small spaces left little room for anything that might resemble natural behavior. The animal's body could be examined closely through barred cages while the guide described the specimen's nature. The vision of natural and social order represented in the zoo guides shaped the way zoos collected and displayed animals. The architecture of houses, ornamented on the outside, was deliberately organized on the inside to present animals as specimens, not individuals.

Visitors, however, didn't politely pore through guides, and they didn't come across town by streetcar or subway merely to glimpse a specimen. They demanded emotional and physical connections to certain star animal attractions but ignored others. Animals, too, didn't behave. Animals lived contradictory lives in zoos. On the one hand, they were meant to be specimens, mere representatives of their class, genus, and species. On the other hand, the zoo's prestige and its distinction from mere menagerie depended less on its scientific pretenses than on its display of a few star animals, above all elephants. And this is why Jennie was more crucial to the Philadelphia Zoo than any of its tended paths or ornamented buildings.

The Rubbish War: Philadelphia, Washington, New York

In 1879, reporters joined visitors in admiring the newly opened Philadelphia Zoo. Caught up in the enthusiasm of the moment, they described the still unfinished and understocked zoo as the equal of the venerable London Zoo. The lush foliage surrounding the houses suggested a well-kept garden enjoyed by well-dressed promenaders observing endless rows of cages. Behind this bucolic image of urban natural and social order, reporters accidently described something that didn't fit the image of a zoo defined by its taxonomic guide. Bears climbed their poles to beg pathetically for tossed food while a rhinoceros reached through the bars for a child's treat.

It was the rare visitors who, in fact, turned their zoo visit into an edifying debate about classification. They may have purchased a guide, but its authoritative tone couldn't force visitors to follow certain etiquette or children to wait patiently while their parents debated taxonomy. Perhaps more typically, Truman Jones, a young boy, went to the Philadelphia Zoo and simply wanted his five cents' worth. For him, the admission price gave him license to smack a zebu (the Indian Brahmin bull) on its nose.[42]

The National Zoo declared that "amusement and instruction of the people are eminently desirable"—but, to the dismay of zoo directors, visitors sought amusement more than instruction.[43] Circuses and traveling menageries had distinct advantages over urban zoos, which struggled to maintain expansive grounds through donations from elites, admission fees, and the occasional city support. Zoos' competitors were simply more "sensational." Promised more than the experience of simply looking at animals, visitors flocked to temporary traveling shows eager to enjoy "striking and varied novelties" and the close proximity to animals that circuses encouraged. Even advertising seemed fruitless for zoos. Why produce the kind of garish posters used by circuses if the zoo was a permanent collection? The zoo may have been one of the marks of a great city at the top of the pantheon of the "educational and diverting influences of

"Philadelphia Zoological Garden, the Rhinoceros." Visitors were meant to act as scientists, understanding the taxonomy of the animal kingdom. Instead, like these two young girls feeding the rhino, they wanted entertainment and interaction. "Story of the Philadelphia Zoo," *Harper's New Monthly Magazine* (April 18, 1879).

human society," but it struggled to compete with the base and crass. The Philadelphia Zoo was particularly disadvantaged by its commitment to the "simple exhibition of wild . . . animals." Its by-laws, in fact, even prohibited the sale of wine and beer.[44]

Visitors sought proximity to animals, sometimes with quiet tolerance and occasionally with deadly results. An animal romping or eating was infinitely more exciting than the animal lying still in its cage like a taxidermy display. Even the elephant's bath could be a "popular means of diversion" for visitors, admitted the Philadelphia Zoo director. The elephant's skin is rough to the touch, like a rock covered in lichen, but according to the zoo's guide, "exceedingly porous." Every afternoon in the hot summer months, the elephants were bathed in their pond to the delight of crowds.[45]

Like Truman Jones, visitors sought the close interactions with animals that circuses encouraged, and they rejected the distance that the idea of the specimen promoted. Some reached across the bars and tossed peanuts. They fed the animals so much so that the Philadelphia Zoo revoked its contract with its peanut vendor. The nuts ended up, too often, in the animals' mouths and the shells on the ground. Jennie, for example, ate "daily tokens of regard in the shape of peanuts, pretzels, and the like."[46] Zoo directors were furious, but watchmen tried fruitlessly to separate animals from visitors who demanded close encounters.

At the National Zoo a guard reproached J. B. Quinlan as he tossed an orange peel at a monkey. At home that night, Quinlan wrote directly to Frank Baker demanding the director reprove the guard. Guards, Baker lamented, come from a "rank of life where courtesy is not invariably practiced." Still, in the guard's defense, Baker noted that he was "irritated and pressed" by the crowds and it wasn't the first time Quinlan had been caught throwing objects at the monkeys.[47] Enough visitors were like Quinlan and veered so widely from the proscribed path of the zoo guide to suggest that the idyllic images of the zoological garden were merely the fantasies of elite founders. Instead of the respectful distance portrayed in Philadelphia's engravings, visitors poked, fed, even killed. "The viciousness which seems innate in some humans is displayed often at Zoological Gardens," mourned an observer in St. Louis. There, a camel died from poison from "some crank." Perhaps it was the same criminal who threw poison pills into the orangutan cages. The sea lions contended with broken bottles, and one particularly "cruel-minded brute" tossed roofing nails into the bear enclosures.[48]

When Quinlan threw orange peels or Jones smacked a zebu, the shaky façade of social order crumbled into chaos. The zoo was hardly a space of quiet contemplation. In Washington, the zoo struggled to contain "rowdy" youths like two young men in the bird house who guards reprimanded for leaping onto the canary cages and declaring "what lovely birds" whenever young women passed. The institutions supposed to bring order, polish, and civilization to the city were struggling against an element Hornaday dismissed as incorrigibly degenerate. "Warfare against . . . disorder" was the zoo's mission. "Order," he reminded keepers, "is heaven's first law; and must be ours, also."[49]

Hornaday must have stamped angrily around his zoo, glaring, maybe screaming, at the everyday visitors as they dodged off of the marked paths, tossed their picnic remains onto the lawns, teased his favorite specimens, and joked about the strange beasts. Hornaday's fury with how ordinary New Yorkers, especially immigrants, visited his zoos became a war, a "Rubbish War," he called it. "The worst offenders," Hornaday thundered, "are the lower class aliens."[50] Hornaday was especially disgusted with the Italians who settled just outside the zoo's gates. By 1916, Hornaday declared war, and the war would last, in his estimate, for six years. His ire grew. He budgeted more for watchmen and park police. Yet the peanut shells and scrap paper that dotted the pathways remained. Hornaday had hoped the zoo would bring order to the city, but he fumed that its immigrant workers had instead brought anarchy to his zoo. On a single day in May, his guards arrested 126 visitors. On Hornaday's orders they would "show grown men and large boys no mercy. Women who provided incorrigible in rubbish-throwing were also to be arrested."[51]

"Rowdy" visitors frustrated watchmen and directors alike, but ordinary visitors who kept to the walkways also quietly rejected fantasies of social and biological order. Sometimes it was simply through their choices and pleasures. In place of visual contemplation, they turned the zoo into a riotous sensorium: bestial smells became a source of childish jokes, animals something to pat or slap, and food something to be shared with animals, especially zoo stars. Even the tamest visitors selected their favorite animals, turning them from specimens back into individuals, substituting first names for scientific nomenclature. Zoos, however reluctantly, promoted their zoo stars. By 1905, the Bronx Zoo sold a postcard souvenir of Sultan, the rare Barbary Lion.[52] This first generation of zoo directors never quite became comfortable with the idea of animals with human names, stars to be touched, fed, teased, and ridden. Despite their efforts in architecture and with zoo guides to show animals as specimens of their species, they just simply had no choice. Visitors wanted to touch, ride, feed, and see animals as stars, not specimens.

In the early years of American zoos, the biggest zoo stars were elephants. Even the staid prose of the zoo guide turned to exaggeration when it described the elephants. "The elephant is very long-lived," said Philadelphia's first zoo guide. Some lived more than 200 years, it claimed, and Alexander the Great was rumored to own an elephant that lived 350 years.[53] Elephants offered the closest intimacy to visitors—even when, in their adult years, they could become murderous. Elephants were at once wild and domesticated, the ideal acclimatized animal. Elephants were civilized yet exotic and perfect for zoos that relented to demands for entertainment and proximity. Alone among the zoo's charismatic animals, they could be touched and ridden and, best of all, seemed to respond with affection. So when Jennie finally died after a long battle with rheumatism, kidney disease, and stomach abscesses, she was mourned as "the friend of the children."[54]

Maybe it was her long history in show business, but Jennie had won "the hearts of the little ones." The Philadelphia Zoo agonized over whether its restrictive by-laws permitted offering rides, but in the end the commercial potential of Jennie's back was simply too much to ignore. "To ride Jennie's generous back was a treat anxiously to be looked forward to, long remembered," especially by one little girl who burst into tears in fright when the lumbering ride began. Gently (or so the story spread) Jennie reached back with her trunk and lifted the girl onto the soft grass. Years later, that girl, now grown, brought her own child to ride on Jennie's aging back—the elephant's last passenger. "Truly," the zoo mourned, "Jennie has a left a place that will be hard, if not impossible, to fill."[55] When Jennie died, Kaiserin replaced her. A mere baby, she was still in training to carry children.[56]

When Jennie passed quietly away, her longtime elephant companion Bolivar, who had already killed three people, stamped in a rage.[57] John Manley, the zoo's head keeper, began disposing of the dead elephant. It wasn't an easy task and only partly because of her sheer bulk. Human and elephant, Manley and Jennie had known each other for years and together they had charmed visitors with their private tricks. Sometimes Jennie even dug into Manley's pockets in search of treats. Now Manley and the other elephant keepers rigged up ropes, pulleys, and a team of horses to drag Jennie from the zoo and into the hands of David McCadden, the local natural history museum's taxidermist. Dissected, Jennie was displayed again. The museum got her brain and mounted her skeleton. Her skin went back to the zoo and became souvenir wallets.[58] She represented both the zoo's success—its endurance in the face of uncertain finances— and its drift from orderly display of species to entertainment featuring zoo stars. Jennie was meant to be observed, but, instead, she was fed, touched, and ridden. Even after death, a century later, her rough, dimpled skin evokes something alive and exotic.

Gunda's death represented a more overtly tragic drift. If Jennie's life and death represented visitors' unwillingness to accede to zoo discipline, Gunda's life, revolt, and death embodied animals' resistance to zoo life. His passage from working at the zoo to enchained in his cage to dissected skin reveals how animals' actions, in particular an elephant's outsized anger, could provoke human debates about animal care, character, and display. Despite the best efforts of zoo directors, such debates ultimately would transform the zoo.

Unlike many elephants, Gunda came to the zoo directly from the wild rather than from a circus. Direct from Assam (in northeast India) in 1904, he became a star attraction and a novel source of income. Immediately upon his arrival, he was trained to accept a howdah on his back. According to the zoo, he "unquestionably took quite a commendable degree of interest in his work." At least the zoo hoped so. Even in his first season, shortened by his training, profits from elephant rides approached those of the zoo restaurant. By 1906, Gunda had more than earned back his original $2,500 cost.[59] But soon, Gunda turned rogue.

"Jennie" (ca. 1898). The souvenir wallet made from the skin of Jennie, the first elephant at the Philadelphia Zoo.

Heads and Horns: New York, India, Borneo, Montana

An elephant's skin is thick, but just behind its ears a single bullet fired by a strong rifle can fell even the largest tusker. For Hornaday, in his earlier career as a specimen hunter in 1878, his perfect shot had protected him from the rage of a wounded elephant and preserved its valuable hide for taxidermy. The tusker teetered and shivered. Branches cracked under the weight of the falling animal, and the hunter's work was done. Now Hornaday began the tedious but repugnant task of skinning and preserving the skin. Enveloped in the aroma of the death, the young Hornaday fought back malaria, dosing himself with Bass beer and port wine. His exhausted men, meanwhile, went on strike. An elephant's skin rots quickly in India's heat, and Hornaday worried when his assistants refused to work on the smelling carcass. He withheld their food and alcohol until they had completed a full day's skinning. "They had forty-eight hours between meals, and never were strikers more effectively cured," he boasted.[60]

This elephant's death marked the beginning of Hornaday's journey from jungle hunting to taxidermy to zoo directing. Still a young man, he found work with Henry Ward's Natural Science Establishment, a scientific goods business in Rochester, New York, which supplied both live and prepared specimens to university laboratories and urban museums. In 1878, Ward's sent Hornaday with just enough money and equipment to make his way through the Suez Canal into India and finally to Malaya and Borneo to amass an extraordinary collection of tropical animal specimens. He was a working hunter, far removed from the leisurely hunts of British aristocrats and American elites. After they scraped the elephant skin bare of fat and muscle, Hornaday and his workers spread arsenic salts on the hide to desiccate the last drops of moisture. Twenty men carried the elephant's head, bones, and skin from the hills down to the plains and the railroad. When he returned to North America, Hornaday reconstructed and displayed his prize—his hunting trophy claimed for science, restored to life in the Harvard Museum of Comparative Zoology.[61]

Life and death were jumbled in the era of spare small cages in zoos but naturalistic dioramas at natural history museums. Naturalism returned only after death. In zoos where mortality rates were always high, the line between life and death was thin. In Washington when the lion Lobongula finally died, it was carefully dissected and its skin returned to the lion's former owner while the skeleton went to the National Museum.[62] Similarly, loss to the Bronx Zoo represented a gain for nearby museums. Rare, valuable animals passed from life at the zoo into taxidermy death at the American Museum of Natural History or the Brooklyn Institute for Arts and Sciences. In Washington, live specimens often became taxidermy specimens at the National Museum.[63]

Here was the irony: cramped, mostly alone, into cages, zoo specimens seemed much less alive than in museum displays where animals returned to nature snarling, proud, alert. Two orangutans battled for supremacy in the jungle canopy high above a Borneo river—in Hornaday's pioneering "A Fight in the Tree-Tops." He completed this taxidermy diorama while still working for Ward's in 1879. Years later when he visited the exhibit at the National Museum, it still gave him "a thrill of satisfaction." Unlike these stuffed orangutans, the lives of zoo animals, pushed into cages where their wild traits had become merely characteristics enumerated in the zoo guide, passed into an afterlife of taxidermy where wildness again became noble. The taxidermy display, in all its detail, provided rich, naturalistic detail; the dour zoo cage turned the animal into a specimen. "It is impossible to appreciate too highly the thought, the labor, and the expense that have been lavished upon these efforts to bring wild mammals and birds to the doors of the millions of city dwellers," marveled Hornaday. Finally, ordinary people could "see all manner of wild life in its haunts." But that wild life was long dead.[64]

Zoos sought notable and rare examples of their genus and species, not breeding pairs to propagate vanishing animals. In fact, these first zoos disdained zoo-bred animals. The goal of building a collection was not the preservation of the species. Rather, it was breadth of "species of special educational value." In fact, zoos avoided collecting extra animals. Instead, like hunters, they sought trophies, like a single Tasmanian "Zebra Wolf," rarely ever on display, even if they were unlikely to survive for long. Often purchased as solitary prizes, such solitary animals could not, of course, reproduce. Before World War I when the Bronx Zoo couldn't display the finest specimens, Hornaday collected heads, horns, and hides.[65]

There was one exception: the American bison. By the first decades of the twentieth century, the largest zoos, notably the Bronx Zoo, had assembled sizable bison herds and rejoiced in new births. Biologically and historically, elephants and bison shared similar traits and fates, but at the zoo these animals had opposite roles. Both grazing animals had been commercially hunted to the edge of extinction for the commodities they offered the industrialized world. The bison's hide made the finest machine belts, and the elephant's ivory became piano keys and billiard balls. Both animals also had reputations for truculence. The similarities ended at the zoo.[66]

Hornaday owed much of his career to bison and elephants. The former shaped him as the nation's leading conservative conservationist. The latter marked him as a hunter and taxidermist. Hornaday had little formal training in biology or taxonomy. Instead, he ascended to the post of director of the nation's largest zoo because of his close personal and political ties with its elite patrons. In New York the zoo was the brainchild of sportsmen, especially Madison Grant, a founding member of the Boone and Crockett Club.[67] Grant was already a familiar figure in New York's high society. Hunting, for Grant, was a pleasure that should be reserved for members of his own class. His elite hunting club celebrated big game hunting as a rite of manly bravery and honor. Yet the club and its wealthy founders linked hunting to the "fair chase," a set of complex traditions and rules that governed the way an animal should be killed and, above all, who could kill them. Grant believed hunting was a sport, reserved for the wealthy, that targeted only the most perfect specimens.

The club was effusive in its admiration for British colonial game laws that severely restricted native gun ownership and subsistence hunting in favor of a system of expensive game permits that reserved big game for those, like Grant or Theodore Roosevelt, eager and wealthy enough to test their manhood in fair fight. Roosevelt famously left for a grand African hunting safari when he completed his presidential term in 1909.[68] Grant's dream of a New York zoo was born equally from his passion for hunting and his suspicion of the American working class, especially its immigrants. Though he would not write his virulently anti-immigrant book *Passing of the Great Race* until 1916, he had already linked his concern for animal conservation with his disgust for those races and peoples he considered inferior. The British had passed laws to keep guns out of native hands, but Grant bemoaned their easy access in America.

"The story," Grant wrote praising British laws, "is everywhere the same." Give the lesser sorts, whether Africans or the immigrants in America, access to "modern weapons" and game disappeared into meat and hides. Conservation for Grant, and equally for Hornaday, was a matter of race solidarity and support for empire. "We Americans must come to the financial assistance of those members of our race who are settling new countries," Grant declared. In particular, game refuges abroad and zoos at home should be expanded and founded. The marketing of game "must everywhere be prohibited."[69] Hornaday shared with Grant an enthusiasm for a kind of big game hunting open only to the cultured and wealthy and governed by strict regulations and moral codes of a fair chase. He also shared Grant's antipathy to immigrants at home and native peoples abroad.

There was a difference, though, between the wealthy Grant and the workaday Hornaday. Grant hunted for pleasure; Hornaday for a living. Before he became a zoo director, he was a specimen hunter, collecting big game not for a personal heads and horns collection but for scientific study. His specimen hunting took him from the jungles of India, Malaya, and Borneo to the plains of the American West. In Asia, as an employee of Ward's, he collected animals, including the adroitly killed tusker. In the American West, he tracked the last American bison. In Asia, he enjoyed the company of colonial elites, luxuriating in the privileges of the imperial hunt. In America, he arrived after the hunting hordes had reduced huge herds to bleached bones. For Hornaday, the bison slaughter was the degeneration of civilization itself.

In 1887, after his two years in the jungle, Hornaday went west, armed with a gun and pen to preserve in hide, bone, and words the last of the bison herds. "The buffalo-hide hunters," he realized, "had practically finished their work," and Hornaday, now working as a taxidermist for the United States National Museum, begged to take gun in hand again. "Hide-hunters, reckless cowboys and poachers," the riffraff of hunters, had virtually exterminated the great American herds by the time Hornaday headed west to seek bones and hides to preserve the species, at least in taxidermy. As he crossed high plains to the river bottoms that had once sheltered the bison, he discovered only bones. "[Y]ou could see where the millions had gone." The bleached skeletons lay exactly where they had fallen—except when scavenged by bone gatherers, erstwhile hunters with nothing more to kill. Here were seventeen skeletons where a hunter had killed a small herd. Nearby lay another fourteen where the skinners had left the hide on perfectly preserved shaggy heads. Hornaday was hunting in a graveyard, and the idea of adding to the toll was "exceedingly unpleasant."

Yet the thrill of the chase was too intoxicating, and despite himself, he enjoyed this "last buffalo hunt." The herds had dissolved into straggling individuals and families hiding in the wildest spots. There, Hornaday waited. "By Jove, there's another!" he cried and "banged away." He was "determined to kill all that came in sight." As fall became Montana winter and the bison grew their wooly coats, Hornaday saw the bull that could become the centerpiece of his taxidermy exhibit. The old bull was "built on a grand scale." He was "without the slightest flaw or blemish," and Hornaday was determined to kill him. For a moment, he observed the grand beast as he pawed the snow. Hornaday took a mental snapshot of his body and character for the taxidermy to follow. Then he shot. As with the elephant, it was a clean, fatal shot.[70]

Hornaday associated his bison hunt with the darker side of the advance of civilization in America. It was the sad result of what happened when the lower sort, those ignorant of the sporting hunt, had easy access to guns, six million of them, in fact. The elephant and other tropical prizes were a different case. They were entertaining, exotic, and rare. It is worth pausing on the different meanings of *rare* for Hornaday and so many other zoo officials and visitors. *Rare* could mean vanishing, like the bison, or uncommon in collections, like the elephant. Even Hornaday, vigilant in his defense of the bison, viewed the elephant as part of the still-abundant jungle. Their peril lay only in the future with the advance of civilization.

Zoos' policy toward bison contrasted sharply with how they housed and collected tropical animals, even distant biological kin. The contrast between the temperate bison and the tropical elephant reflected the class pretensions of the American zoo and the way ordinary Americans imagined vast differences in the nature, animals, and peoples of the temperate and tropical zones. The bison had become a totem animal, a tragic symbol of the dark side of democracy. Wanton slaughter of bison herds embodied the overrun of temperate nature by the masses ungoverned by morals or

"The Group That Opened the Road" (1886–1887). William T. Hornaday's taxidermy display at the National Museum of bison he killed in the American West. In death, they appeared more natural than animals on display in the first zoos.

rules. The elephant was the totem of rules and regulations, the civilized hunt in the tropics.

Years later, as the newly installed director of the Bronx Zoo, Hornaday insisted on assembling a sizable bison herd culled from their last stand in Yellowstone National Park. Other zoos followed with their own substantial collections. He assembled the Bronx herd to be restored to the Great Plains. In 1905, Hornaday and the zoo implored the federal government to fence the Wichita Forest Range in Oklahoma, and in return, the zoo promised to donate eighteen of its own animals.[71] Further gifts of animals followed, and the zoo's herd traveled west. By contrast, elephants were single animals, riding favorites to be bought and sold. If bison breeding marked the zoo as civilization's corrective, the elephant, frustrated and maddened, was the zoo's undoing. As Hornaday would soon learn, keeping solitary males could paralyze a zoo.

The Bronx Zoo was born from the hunt. Hornaday would use his position as director to rail against the democratization of hunting and, closer to the park itself, to halt the killing of songbirds even within the zoo's fences. "So bold had these bird killers become," Hornaday fumed, "that they actually invaded the lands of the Zoological Park and shot birds."[72] In an age where guns were affordable, "apparently no species is too large, too small, too worthless or too remote to be sought out and destroyed," he said, and the very worst hunters were those who commercially provided taxidermy heads and horns so even common folk could turn their parlors into sad simulations of a manor hall.[73]

Yet his zoo also maintained a collection of heads and horns. The similarities between the elite practice of collecting trophies and plebian capturing of songbirds in the Bronx Park never seemed to bother Hornaday or Grant. "A collection limited to personal trophies won by the owner is quite another matter," the zoo explained. Where the zoo's live collection fell short and where specimens bore the scars of their capture, the zoo sought to collect for study and display perfect mounted specimens felled by elite hunters—even as the zoo railed against indiscriminate and plebian hunting. The nation's elite hunters brought their trophies to the zoo. "Only particularly fine specimens will be admitted," such as Emerson McMillin's famous collection of Alaskan heads and horns or Grant's white mountain sheep. George L. Harrison Jr. presented fourteen African antelopes. The heads and horns collection made the zoo look like a manor hall, and sportsmen patrons of the zoo hunted as a matter of privilege so that no one else need to. The pride of the Bronx Zoo's collection was the world-record elephant tusks. They were eleven feet five-and-a-half inches long.[74]

There was a close relationship between taxidermy and the emergence of American zoos, not least through Hornaday and others who hunted to preserve and stuffed to study. Taxidermy was the fine art of taxonomy; zoos were its architecture. Taxidermy, Hornaday insisted, was a new art form, born from the scientific urge to classify and document and the keen observation of nature. His turn to conservation grew both from his journeys to the American West and from his suspicion of the American masses. An armed elite was governed by the sportsmen's ethos. The masses, he believed, were driven by a thirst for blood. The decline of American animals was the grizzly by-product of civilization, but the collection of taxidermy specimens was the mark of the elites and the scientific knowledge they championed. As much as the zoo and its officials hated hunters and those pitiable dealers who mounted animal heads for crass consumers, they blurred the boundaries between live animal specimens and disembodied heads. The actual life of zoo elephants, however, complicated the easy distinction between bison and elephants and, in the tragedy of their death, gave the unruly public an opportunity to talk back to zoo directors who sought to educate and order.

Only the Finest Specimens: Philadelphia

Long before Jennie became wallets, she worked in a circus. Elephants, the animals zoos desired above all, frequently came as gifts from circuses. It was an uncomfortable detail for zoo founders who disdained circuses and traveling menageries. She arrived in Philadelphia as a hand-me-down from the Robertson circus.[75]

Her arrival had marked the zoo as a major American scientific institution, but her past was troubling. After all, the zoo's founders frequently stressed their differences with the traveling menageries and circuses that occasionally visited the city. It is easy to imagine the zoo's founders looking with disdain on sideshow hawkers proclaiming the uniqueness of their beasts and inviting visitors to poke and tease. Animal entertainment was even banned in its initial by-laws. For the zoo, the strange exoticism of the animal was, in theory, an anathema. The zoo promised rarity, but only of fine examples of what existed in wild nature, and zoos cherished wild-caught animals above all others. New York, unlike smaller city menageries, was wealthy enough to seek primarily wild-born animals. The "finest animals" were wild while "imperfect specimens" were often captive bred, leading to their "degeneracy," noted the Bronx Zoo.[76] Elephants, though, took circuitous routes from jungle to zoo.

Zoo founders and directors wanted animals to instruct, not entertain or tantalize. They sought the finest specimens—even if, as when they accepted the castoffs of circuses, they were forced to compromise. Yet they refused freak animals—the two-headed, hermaphroditic, and otherwise abnormal animals so often offered to zoos. In its early years the National Zoo was desperate for animals to display. Its collection was haphazard and dependent on the generosity of circus owners and diplomats posted abroad and the occasional gift. The zoo refused some gifts like a two-headed snake. The zoo displayed ideal, wild specimens (when it could) but would never stoop to exhibiting "monstrosities" or "freaks."[77] The zoo was not a circus or a sideshow. When Buffalo graduated from menagerie to zoo, it stressed its "perfect specimens" by killing or disposing of a "number of cripples and freaks." It also sold off its duplicate animals.[78]

Yet as the National Zoo's trials demonstrated, the ties between circuses and zoos were closer than zoo founders might have wanted. Zoo directors disdained circuses, even as they were dependent on them for animals and for the training they provided so many keepers. Despite the wealth of their founders, zoos needed circus castoffs. The National Zoo suffered, especially in its infancy, with tight budgets. Despite its dreams of becoming a zoo worthy of a capital city, the zoo depended on haphazard gifts and the magnanimity of the Adam Forepaugh circus that camped its animals in the zoo in the off-season.[79] After a long, freezing, terrible winter, Hornaday also learned a bitter lesson: even in New York, the zoo didn't have the buildings to house its biggest tropical prizes. The zoo couldn't even protect a Sumatran rhinoceros, one of the rarest prizes of the jungle. Hornaday quietly sold the rhino to the Ringling Brothers circus for its menagerie.[80]

When circuses offered elephants as gifts or reduced sale, zoos, big or small, succumbed to the temptation to trumpet beasts like Jennie. The two elephants in Washington that made the National Zoo a reality, its "nest egg," had once traveled the country performing. Then, like so many older elephants, they became intractable. The pair that arrived in Washington were damaged beasts, too resentful even for traveling cages. Philadelphia also received its elephants, starting with Jennie, as circus castoffs. The zoo learned, however, that in accepting unemployed elephants, it sparked a conflict that would eventually cost animal and human lives and reveal for public criticism contradictions in the zoo's search for order.

Bolivar, who lived alongside Jennie at the Philadelphia Zoo, arrived as a gift from the Forepaugh circus. He was an impressive tusker, valuable for his size alone until the day he killed a spectator who singed him with a lit cigar—even the elephant's thick skin is sensitive. After the incident, Bolivar's temper increased. Soon only Forepaugh and a single trainer could handle him securely. The circus owner realized the zoo was an ideal place to leave behind a dangerous and surplus animal. Everyone—with the likely exception of Bolivar—seemed to gain.

It wasn't the first time Forepaugh had donated a rogue. He had also given Tip to the Central Park Zoo. William Snyder, the zoo's head keeper for twenty-seven years, understood better than anyone why "bad" elephants came from circus to zoo. Tip hated Snyder from the start and tried twice to kill him. "The last time," Snyder recalled, "he stoved in my chest and broke my collar bone. Then we decided to kill him." A small crowd gathered in Central Park for the novelty of an elephant death. Snyder tried to tempt Tip with a cyanide-laced apple, but the elephant, wise to the end, "dropped it like a hot coal." He also refused a poisoned carrot before succumbing to laced bran mash. Tip lunged and rent his heavy chain before collapsing. "I had lost my love for Tip," Snyder claimed, "but it was hard to see such a fine, big fellow die." Still, even after death, Tip was on display. Skin separated from bone, he moved across Central Park to the American Museum of Natural History.[81]

Snyder, meanwhile, had discovered the "favorite stunt with the circus men." It was a lesson the Philadelphia Zoo was soon to learn. "I'd rather sell stock to circuses than take presents from them," Snyder recalled. "I have my fingers crossed," Snyder declared, "when any circus man tries to make me a present. You can believe it's loaded." When he gave elephants, first to the Central Park Zoo and then to the Philadelphia Zoo, Forepaugh had rid himself of deadly animals and, at the same time, won valuable publicity for his generosity.[82] In Philadelphia, though, the zoo was desperate for an animal to live alongside the aging Jennie. It would soon come to realize, though, that even removed from the circus, the angry elephant remained dangerous.

With fang, claw, tusk, and trunk, zoo animals often refused to live as docile specimens, and when elephants rejected the discipline of the enclosure, the results could be deadly. Despite zoos' efforts to display only perfect specimens, their collections highlighted mostly juveniles. Older animals were difficult to capture and even harder to display. They "either kill themselves by struggling or die of melancholy." With death rates that often approached 25 percent per year, many animals on display did not survive to adulthood.[83] Zoos quickly learned to substitute "acclimated" for "wild" in describing their specimens. Animals had to be trained to eat, drink, and exercise in captivity, and the scars of their "forcible subjugation" tainted their perfection.[84] Pythons were particularly hard to acclimatize. They simply refused to eat. They could be kept alive only when several keepers forced dead rabbits and pigeons down their throats.[85]

Acclimatization prepared animals for life in the cage. But it had a metaphorical meaning as well: it was the civilizing of animals, accustoming them to urban life. Some animals simply couldn't adapt to life in cages in the heart of cities. Opening day at the Philadelphia Zoo, for example, was marred by an injury to a kangaroo. The kangaroo, startled by the sound of a passing train, broke its legs on the bars of its cage.[86] Sometimes animals escaped, turning quiet observation into urban hunts that caricatured the elite safari. The early years of the Philadelphia Zoo witnessed a number of animal escapes. As in the case of two fleeing monkeys, news reports were often comical. They chuckled over a monkey who ended up in a coal chute, terrifying a woman who lived a half mile from the zoo gates. On other occasions, escapes ended in tragedy. In 1886, the zoo's watchman on his morning rounds discovered a leopard had escaped. Quickly, the head keeper and the superintendent closed the zoo and padlocked its gates. As they spied the leopard heading under a bush, they hoped briefly to catch the specimen alive. Soon, though, they abandoned nets for revolvers, rifles, and shotguns. As the leopard pounced, eight keepers shot simultaneously, and the force of the bullets splashed the corpse into a lake. The escaped cat became the "handsomest leopard-skin rug in Philadelphia" in a zoo clerk's house. Another leopard, this time in Cincinnati, was loose for two days, evading and mauling its captors. "Posses" roamed the city's green spaces looking for the cat; it too was shot. Wild dogs poached chickens during their weeklong tramp. A sun bear fled its traveling crate, and a lioness burst the bars of its cage to pounce on a passing donkey.[87]

These animals had refused their civilizing, but according to zoo officials, they were no longer wild. Instead, they had become "rogues of the zoo." A jaguar in New York remained a fearsome killer. For days, zoo workers attempted to introduce a cage mate. They placed the cage of a new cat next to the bars of the zoo's older cat. The two cats smelled each other and paced. Yet when the bars were slid open, the older jaguar pounced, and its jaws shattered the spine of the younger, female cat. The zoo was convinced the tragedy was not caused by poor acclimatization techniques that offended a wild cat's sense of territory. Rather, it was the fault of a rogue.

Call Me a Rogue: New York

The "king of the rogues" was an angry elephant. Gunda was once the pride of the Bronx Zoo and, whether Hornaday liked it or not, both its economic lifeline and a tragic example of the limits of a grand American zoological park. "Time was and only a few years ago," the zoo recalled, "children used to ride on Gunda's back." At the time, it seemed merely playful when he kicked over benches or a soda fountain. In retrospect, the rogue's foreshadowing seemed clear. Gunda had arrived in the zoo with torn ears: a sure sign someone had reprimanded the ornery calf.

The zoo had been warned. Hassan Bey, the native Indian mahout brought to New York originally to care for Gunda, seemed convinced Gunda simply was a "genuinely bad elephant," just like the Central Park Zoo's Tip. The zoo hoped Gunda would be less lonesome if a human from his native land trained him. Yet from the moment Gunda was released from his crate, Bey watched helplessly as the elephant destroyed his stall. Already, he earned his Hindi name. *Gunda* meant "villain," noted Bey, wondering if the zoo had purchased a "murderer" elephant. Hornaday and Frank Gleason, one of the elephant keepers, were not willing to admit failure. Not yet—instead, they were convinced the "whole trouble was due to the native Indian keeper." They accused Bey of laziness and homesickness and shipped him home. For a while, Gunda seemed to take to his new American trainer. He begged for peanuts and accepted his howdah. The "elephant which daily amuses hundreds of youngsters" was not simply a source of needed funds; he appeared for the moment as a triumph for the zoo and its scientific methods of acclimatization.[88]

"Riding Animal. Indian Elephant 'Gunda'" (ca. 1905). This souvenir postcard
depicted Gunda carrying children in a howdah on his back before he turned rogue.

Gunda grew. Within the year he gained almost 700 pounds. By the time
he attacked his keeper the first time, he had grown to 9,000 pounds. His
tusks were some of the longest of any captive elephant, but they shattered
during the attack. By 1909, Gunda no longer seemed safe to carry children. He disliked one keeper after another. By 1914, only Walter Thurman
could safely enter and clean his cage—until one afternoon. Thurman was
spreading straw in Gunda's small outdoor corral when he heard a snort.
He whirled around to confront the angry elephant. Gunda charged with
his tusks lowered and his trunk raised. Thurman dived into a small depression in the ground carved by Gunda's heavy feet just as his tusks crashed a
few inches from the cowering keeper. One of the tusks shattered, and splin-
ters of ivory flew as Thurman darted beneath Gunda's legs. For the last
time, Thurman was able to calm the elephant and lead him inside—toward
his next life. In his cage, keepers wrapped his legs in chains. The riding
elephant and zoo prize had become a zoo crisis.[89]

The howdah permanently retired, Gunda joined a sad fraternity of
enchained male zoo elephants. Like Bolivar in Philadelphia, heavy chains
bound Gunda in the confines of the elephant house. Bolivar, notably, had
spent twenty years in his cage, swaying gently back and forth, and some

observers estimated the still-young Gunda would also spend decades alone in his cage. His anger seemed unabated, and even filling his water bowl had become a dangerous challenge. Keepers had to dart close to the bars to scrape away the collected dung and refresh his water before Gunda could swing his trunk.[90] Thousands of visitors flocked to the zoo to see the angry elephant with the mutilated tusk. The zoo was frightened Gunda might actually break the chains "at any minute." Soon, they added a second chain, yet still he tried to smash keepers with his trunk.[91]

When Gunda's biological nature collided with the limits of his enclosure, his anger transformed a fine specimen into a mad monster. Madness, in turn, led many visitors, to Hornaday's exasperation, to question the zoo's orderly vision like so many tawdry words of a circus hawker. Gunda enchained launched an urban debate about the place and role of zoos that would end with a bullet.

That debate began with a poem. One morning a few verses appeared attached to the bars of Gunda's cage. No one was quite sure who attached the poem. Perhaps it was a visitor. Maybe it was a disgruntled keeper. The keepers joked to newspaper reporters it was Gunda himself, bemoaning his condition. Either way, the poem remained for visitors to read: "Lord, save me from my friends!" it read. Gunda added his own drama to the mystery poem when he managed to loosen one of the chains from the cage's stone walls, ripped the poem from the bars, and trampled it into shreds. The reporters hurried to the zoo to see a raging animal venting his anger at a poem. "Call me a 'rogue' elephant if you like," he seemed to say.[92]

The poem's lines were haunting. They turned Gunda from a criminal rogue into an indictment of the zoo, its architecture, mission, and director. Hornaday offered his own explanation for Gunda's fury. The elephant was in must, his years of sexual maturity, Hornaday explained in the dry language of the zoo's report. Gunda had gone rogue. Scientific explanation alone, however, could not capture the violence of his anger and the array of human responses it provoked. Visitors were less convinced that Gunda was simply a sexually mature elephant with an evil nature. Instead, they turned their ire on the zoo itself.

When Gunda attacked again, letters flooded into the newspapers. The *New York Times* even took the unusual step of printing a full page of letters. Ironically, the imprisonment of an elephant had pierced the zoo walls, enabling public questioning of the zoo and its pretensions. "Seldom has a story of animal suffering aroused such widespread interest and sympathy," marveled the editors. Children, architects, engineers, poets, zoologists, animal traders, and women suffragists all debated "Gunda's punishment." Hornaday insisted Gunda's "weaving" in his cage was comforting. A little nine-year-old girl disagreed. "I think Gunda ought to be helped," she wrote. She understood why "he is ready to kill any man." Others began to think about the fate of the wild animal as a captive at the zoo. The "human animal" might like a "sedentary life," but Gunda was a "wild creature, utterly unaccustomed to restraint." Some even questioned Hornaday's "fitness."

At stake was the future of the zoo and the vaunted position of its director. When Gunda was shackled, visitors began to question the role of the zoo. "Why have a Zoo at all?" asked letter writers. Why not replace the live animals in chains with "a collection of stuffed specimens?" (which, of course, was actually a natural history museum). Surely taxidermy could offer as much "instruction" as live animals. What truly was "the purpose and intent of maintaining a zoological park"? Certainly it "can afford pleasure" and the chance to study animals. But, as the Society for the Prevention of Cruelty to Animals (SPCA) came to inspect the zoo, one frequent visitor insisted that unless animals can be afforded "some degree of freedom and contentment," then the zoo is "better left unattempted."[93]

Like a maddened elephant, Hornaday raged at visitors, reporters, even the SPCA. Of course, Gunda was "deprived of his liberty," but so too were "murderously inclined human beings." Hornaday may have unintentionally admitted the elephant house was a jail, but he refused to condemn the zoo. "We are frequently hounded by persons who attack our management," he complained.[94] Gunda's imprisonment had given such critics new license to imagine the zoo in ways Hornaday wouldn't countenance. To him, all their suggestions were simply "an attack on the Zoological Park." Hornaday published his own letter to zoo critics. He was dismissive, angry, even stubbornly resentful as he tried to return science and order to the relationship between the zoo and its public. He refused to kill an animal he denigrated merely as a rogue. Nor would he rebuild the zoo. "The day has come," he raged, "when fervid Oh-Lord poets and letter writers feel cocksure that they know more about managing dangerous animals than we do."[95]

Though he taunted his "enemies" as "mental incompetent[s]," Hornaday was also right. Poets and letter writers did believe they knew more than zoo officials about what made both observers and animals happy. Gunda's attack had sparked a new "movement" that questioned the authority of the zoo director and his vision of the ideal zoo. Some simply wanted a larger elephant house where Gunda could roam into a protected yard. One particularly fanciful critic suggested returning Gunda to Africa (though he had, of course, come from India). Maybe his brush with civilization could help him in "uplifting" his wild cousins.[96] Others offered new visions of zoo design. Why not build Gunda his own "jungle" in the park? In his large "enclosure," Gunda could shiver and vent his fury, separated from keepers and visitors by a moat. Build this "moated jungle" and give him a mate, then humans could all go to "see the genuine home life of elephants in freedom." This was a vision of zoo design well ahead of its time. And it was precisely the model for zoos across the country rebuilt a decade and a half after Gunda's death by another perfectly placed shot.[97]

So Gunda died a rogue, an animal that refused to be either specimen or star. Early in the morning, Carl Akeley, the nation's best taxidermist and an accomplished hunter, arrived at the zoo with the elephant gun that had served him well in Africa. For a moment Gunda turned his head to look at his executioner. Hornaday and Thurman were convinced Gunda glared at them. They fled the elephant house, not wanting to witness his death. Then Akeley shot Gunda in the lethal spot behind his ears. The zoo closed the elephant house while the osteologists and taxidermists went to work. They already had plans for Gunda's life after death: the lions would eat elephant meat, and taxidermists would mount bones, ivory, and skin for the American Museum of Natural History. After it was all finished, Akeley estimated he had saved 230 square feet of elephant skin worth about $9 a square foot.

There was another side to Gunda's biography that began in colonial India, long before he arrived in the Bronx. Some animals like Jennie came to zoos from circuses. Others, like Gunda, were captured in their wild habitats. Born wild, Gunda made a long forced migration from jungle to zoo. Gunda had a hefty price tag—and he worked off his cost carrying children on his back. How did a wild elephant in one of India's more remote areas become so valuable? And who went in search of animals like Gunda to bring them back alive from jungle to zoo?

2

THE VOYAGE OF THE *SILVERASH*

The Big Business of Tropical Animals

A FEW DAYS out of Singapore, Lucile Mann watched the crowded steamer pitch in the waves and leak in monsoon rain. The *Silverash*'s cargo—1,500 animals, reporters back home in Washington, DC, estimated—was falling ill. In 1937, Lucile was enjoying an adventure many Americans desired but only a handful experienced firsthand. She was living the life of an animal trader as part of her husband William Mann's expedition to bring back specimens for the National Zoo where he was the director.[1]

A crate of fourteen pythons died and had to be thrown overboard. In the tropical sun, the cargo holds where the cages were stacked sweltered like a veritable Hades. A valuable mountain blue sheep, one of the expedition's prizes, sickened, and an orangutan began sneezing and died. Nearby, a young tiger developed sores from rubbing his nose on his cage. Four other pythons slipped their crates and killed seven birds, including a rare Sumatran owl. Amid the chaos, Lucile had become particularly fond of a small black gibbon. She brought the seasick animal into her cabin where it seemed curious, even tame, and chose a perch above the mirror. Yet Lucile was anxious; her new pet wouldn't eat. Once he swallowed a few drops of milk; another day she rubbed egg on his chin and he licked it off. After two weeks of refusing food, the poor gibbon "quietly closed his eyes and died."

There was little time to mourn a lost pet. The animals needed to be fed, and Lucile eagerly lent a hand to Layang Gaddi, a Dyak tribal they had hired to care for the animals during the ocean voyage. "He spent the whole day running his legs off up and down the deck." Some animals needed sweet potatoes, others fresh meat or bananas. Despite her own seasickness, she chopped vegetables and peeled bananas.[2] She cared for precious cargo, fragile in life but valuable alive, coerced into migration. The *Silverash* sailed from Singapore to Bombay and then through the Suez Canal into the Mediterranean and Atlantic, an odyssey that stressed the animals and exhausted the humans tasked with their care. From Singapore to Calcutta to Port Sudan, the Manns tracked animals, not in their native haunts but in small town and big city animal markets. Along the way, they had mastered the art of negotiating the natural resource economies and bureaucracies of vast empires.[3]

One journey in a new American business, the *Silverash* was one of many ships that left Asia and Africa bound for the American market loaded with animals, some alive and many dead. American traders had come to play a larger role in the global trade in animals in the years after World War I when older merchants, especially German firms, suffered the loss of their monopoly. Between the two world wars, the global animal trade was a rough-and-tumble business, only barely regulated by colonial regimes that governed the lands where animals lived, roamed, and fell captive.[4]

The animal trade reveals another side of the global economy created by European and American empires. Animals were one of many tropical products that passed through ports like Singapore.[5] Yet animals were not bales of rubber. They fought back against their commercialization with fang and claw. Sometimes they just died, making their purchase and resale a risky business for Americans who wanted to profit from the jungle. Animal traders extracted natural resources in fleeting but locally labor-intensive ways. They captured animals, transported them, and then moved on to new environments and cities. They admired and associated with stressed colonial bureaucracies and turned them into mechanisms for American commercial expansion. Through their efforts, ordinary Americans became more familiar with the animals and nature of Malaya, India, and Africa than with those of American colonies and protectorates.

Back home, it was big news when a ship like the *Silverash* arrived. Reporters and onlookers jostled dockworkers and zoo officials trying to unload tired, stressed, sick animals, prizes from the tropics that arrived on the arks to America.

Satan and Other Animals: Yellowstone, Philippines, Panama

William became the National Zoo's director in 1925, and he hoped to display one of the nation's most complete collections of animals from the East Indies—other countries' empires. Resigned to the fact that the National Zoo was never going to display animals from a growing U.S. empire that stretched, through formal annexation or military occupation, from Central America and the Caribbean to the Philippines, he went all the way to Asia and Africa. By contrast, zoos in other imperial capitals, notably the London Zoo, thrived because of animals collected in their imperial possessions. British colonial officials collected animals, and local nobility frequently made gifts of tigers, elephants, rhinos, and more.

The U.S. empire had grown with the National Zoo. In the first years of the zoo, the United States was building a settler empire in the western half of the continent. Farmers, ranchers, miners, and soldiers violently displaced native communities, and cattle replaced bison as the big grazers of the Great Plains. By the turn of the new century, the United States was building an overseas empire. The empire spread first after a war with the Spanish Empire in 1898 and later against an indigenous insurrection in the Philippines. Soon the United States claimed colonial possessions or protectorates from the Caribbean to the Philippines. The reach of American business (sometimes backed up by American military power) was even greater.

At first, the National Zoo aimed to display the empire's animals, before rules and fears made that impossible. The American West, the first habitat of the spreading empire, provided the bulk of the zoo's first wild-caught animals, including its substantial bison herd. In 1894, Major G. C. Anderson's cavalry, stationed in Yellowstone, were hard at work packing crates, cages, and iron boxes filled with trapped animals to send to the fledgling zoo. Satan, as the soldiers called the grizzly bear, nearly escaped his cage by charging his bars.[6]

As the empire expanded, the National Zoo pursued animals from the nation's "new possessions." In 1899, with the United States absorbing vast new territories, the zoo even printed its animal wish list and distributed it to army and navy officers and to 242 consular offices around the globe. Consular officials and military attachés faithfully collected and shipped animals home. Ocelots, boas, toucans, and a spider monkey arrived from Maracaibo, Venezuela. The USS *Wilmington* returned from Para, Brazil, with Amazonian animals, including a tapir and jaguar. Sailors and soldiers often purchased animals for sale in foreign ports, sometimes out of fascination, occasionally recognizing that exotic animals were worth good money.[7]

The Philippines, the biggest new American colony, offered wild buffalo, hogs, two species of monkeys, civets, fruit bats, flying lemurs, and a "remarkable and interesting" tarsier, a small primate. An army officer sent a peculiar black deer native to Mindanao that barked like a dog. Even as violent resistance to the American occupation persisted, officers and colonial officials purchased, captured, and transported "important and interesting forms of animal life." The National Zoo was eager to add Filipino animals to its collection, and the navy was equally happy to help with free transportation on its troop ships to San Francisco.[8]

At the other reaches of the growing empire, A. H. Pinney labored on the Panama Canal, tearing through hard rock and dense vegetation. Like many other workers, Pinney was curious about the animals the canal displaced. But he did not know how to care for his pet sloth and giant anteater, and in desperation, he wrote to the National Zoo offering a donation. The zoo was delighted and wondered if he could perhaps buy or capture any other mammals or snakes. E. B. Wilson was more entrepreneurial, with two South American tapirs he hoped to sell to zoos at home to augment his Canal Zone earnings. A surgeon in the Colon navy hospital simply wondered about an "unusual" animal that wandered into the midst of a Christmas party. "None of the natives in this part of the world . . . have been able to identify the animal," he noted. It was, in fact, a kinkajou, a relative of the raccoon, and the National Zoo wanted it.[9]

For a brief moment, the National Zoo introduced ordinary people at home to the exotic nature and animals now under American dominion. Animals, much like the lands where they once lived, now served as objects of study for American scientists and sources of trade for American dealers, like A. C. Robison. A San Francisco pet shop owner selling cats, dogs, goldfish, the occasional imported tropical bird, and monkeys for pets and research, Robison recognized the business potential of new colonies. He became a dealer in Philippine wildlife, handling all Pacific shipments for the National Zoo.[10]

It was a bad career move. The U.S. empire, whether in the Philippines or Panama, lacked animals that could match the tiger in ferocity or the elephant in impressive bulk. There were other reasons why American zoos suddenly stopped displaying animals from the nation's empire. Once rare and exotic prizes, such animals now seemed merely carriers of disease, especially rinderpest, a fatal virus infecting cattle and other ungulates. The trade in animals from the new empire proved brisk but brief, also a victim of American anxiety about the empire's growth. Americans had begun to worry about where the nation ended and the empire began. As they wondered whether the constitution would follow the flag to new colonies, they worried about whether the incorporation of new territory also meant including new peoples. Could new migrants—now colonial subjects—enter the country just as easily as animals?[11] Strange animals might have come to represent the potential of contamination rather than dominion over new lands. Perhaps the easy arrival of animals also embodied a border-free empire in which not only animals but also human migrants would be able to arrive freely in domestic cities.

Within a few short years, the animal life of the U.S. empire had become all but extinct in American zoos. This new reality was brought home especially at the 1904 St. Louis World's Fair. At the fair, crowds flocked to see reconstructions of Philippine villages, living exhibits of different Philippine ethnic communities—tribes, the fair organizers called them. Ironically, native peoples could be put on display, but not their exotic wild animal neighbors. The National Zoo even refused to lend its valuable Filipino buffalo—rare only because of import restrictions—for display. Despite the protests of zoo directors eager to depict new tropical possessions, lawmakers rushed to construct new borders within the empire, and the animal trade fell victim.[12]

Whatever the legal restrictions, the commerce of empire meant tropical animals still arrived, sometimes by mistake. At least once a week, reptile keepers at the St. Louis Zoo received a panicked phone call from a local grocer who found snakes coiled among bunches of bananas. In Central American ports, snakes crawled into the piled bananas in search of mice. Loaded into refrigerated ships, they sailed, virtually comatose, to the United States. "[W]hen they warm up the trouble begins," and the zoo gained new specimens. In a few weeks, the St. Louis Zoo added two baby boa constructors and a blunt-headed snake.[13] The snake was a welcome addition, but zoos still needed a reliable source of animals. By the time William became director of the National Zoo, American traders had set up businesses in other empires, in particular British and Dutch colonies in Africa and Asia. They muscled into a trade once dominated by German firms and now in transition.

To America in a U-Boat: Washington, Stellingen

In spring 1937, the journey of the *Silverash* was beginning. William looked to the East Indies to fill the zoo's cages, and happily, the National Geographic Society was eager to underwrite his animal collecting in the heart of the British and Dutch tropics. In their Washington home, the Manns began their preparations, packing pith helmets, khaki suits, and expensive traps. The outfits were ideal for photographs, but the traps would never be used.[14] The Manns followed the steps of other Americans who had entered the animal trade after World War I, and like them, the Manns would soon realize the life of the trader depended on access to local markets much more than on knowledge of animal habits and habitats.

Years before, in the midst of World War I, another ocean voyage, this time of two turkeys, was a harbinger of change in the global animal trade. At first, their trip seemed simple enough to Frank Baker, then the National Zoo director. Britain was at war with Germany, fighting on the western front and at sea in the North Atlantic, and the animal population of the London Zoo declined alarmingly. Baker offered to send the two American birds to help fill empty cages. The turkeys spent months sitting and waiting until, finally, Baker found a ship, the *Minnehaha*. But the turkeys never joined a convoy across the Atlantic; the *Minnehaha* was damaged, torpedoed perhaps, and aborted its crossing.[15]

As those turkeys remained stranded, American newspapers, with noticeable glee, recorded the hard times of German animal traders. Among the German firms, the Hagenbecks had dominated the prewar trade. Claus Gottfried Hagenbeck began the business to augment his earnings as a fishmonger. His son Karl Hagenbeck abandoned the fish business entirely, realizing the profit in selling animals to circuses and zoos. By 1905, an American reporter noted admiringly: "Give them your order and cash in hand, and they will get you any two-footed or four-footed or footless thing you want."[16] German colonies, especially in Africa, opened new lands for animal collecting, and friendly colonial officials willingly offered permits to hunters, agents, and trappers. The Hagenbecks' private zoo in Stellingen, the crown jewel of their animal empire, displayed their animals for sale in open, naturalistic enclosures.

Karl and his heir Lorenz arrived in America hawking animals and his vision of how to build zoos. In 1898, they offered an Indian elephant for $1,300, a young hippo for $3,000, and a Bengal tiger for $1,000. The Bronx Zoo bought two lions and a tiger. Years later, Karl claimed they ran to the bars and purred as he stroked their fur and called them by their old German names. Like Stellingen in Europe, the Cincinnati Zoo, built on a German model of open enclosures, served as the American holding pen for Hagenbeck animals for sale. Such close relations were understandable. German immigrants had started the Cincinnati Zoo and purchased the zoo's first animals from Karl. In 1902, Sol Stephan, already the zoo's director and, for Lorenz, "my American father," became the Hagenbecks' American representative.[17]

Even as the shooting began, Lorenz crossed the Atlantic to Germany with "a brief-case full of orders." He was still a popular figure in St. Louis (also a city with a large German population) where he was a frequent visitor to the zoo, and in 1916 with the United States still neutral, Lorenz donated two young lions. The zoo, eager for animals, wondered if the lions might travel to America inside a U-boat. By then, however, war had come to Stellingen. Most the firm's agents and animal keepers were already in the army. Unlucky agents operating in British colonies when the war broke out were interned. At home, food and coal became scarce, and animals began to die. "We had lost our seals to the fishmonger," recalled Lorenz. Then the lions and tigers died from eating spoiled meat. "One does not associate lions and tigers . . . giraffes and monkeys with wartime casualty lists," remarked an observer, "but the fact remains that while soldiers were dying by the thousands on the battlefields . . . hundreds of Hagenbeck's animals, some them unique among beasts in captivity and worth thousands of dollars apiece, were perishing." By one count, during the war Stellingen lost 74 lions, 2 rhinoceroses, 19 tigers, 200 monkeys, and 10 zebras, among 2,155 animal casualties.[18]

After the war, the Hagenbeck firm struggled to reclaim the animal trade. The Stellingen Zoo slowly rebuilt while the firm appointed new agents and distributed its price lists once more. Yet wartime hostility remained. Once a good customer, William T. Hornaday, in particular, sought alternatives to German traders. The Bronx, Philadelphia, and National Zoos even launched a joint expedition to Africa in 1916 to replace animals that once would have been provided by the Hagenbecks. The costs were substantial but the returns meager. Alden Loring, a Bronx Zoo keeper and a member of Theodore Roosevelt's prewar African safari, left with ambitious goals of creating a permanent compound for buying and shipping African animals to American zoos. Yet he returned mostly with surplus animals purchased from South African zoos and dealers.[19]

Loring was a good zookeeper and a capable hunter, but once in the field, he realized the animal trade was a hard business. Still, the war, the blockade of Germany, and the devolution of German colonies into British hands all opened up opportunities for American collectors. As Frank Buck, who would become the nation's best-known dealer, recalled: "Every German consul in the smallest port of Asia or Africa had been a Hagenbeck agent on the side. They had bought up animals, birds, and reptiles form the natives, and the Hagenbecks had distributed them to American and European zoological gardens. With the war all this suddenly stopped." Western front aside, for Buck, "it was a world colored with rosy glasses and I was just beginning to cash in on it."[20] Buck, like many others, realized that entrepreneurs in search of animals went to Singapore.

Collect the Collectors: Singapore, Johore, Makassar

The steamship to Singapore was tiring—though not as arduous as the trip home again on the *Silverash*. The Manns were grateful for Singapore's many comforts. On their first morning, after a leisurely and thoroughly

Western breakfast, they searched for animals not in jungles hides but in markets. No jungle had a higher concentration of wild animals than urban Singapore. The city's crowded markets, small shops, and busy port supplied, by one estimate, about half the animals that filled the world's zoos.[21] Ships arrived in Singapore with wild cargo, including a steamer from Siam and Penang that unloaded 250 crocodile skins and two live tigers that sold for 200 Straits dollars (around $100).[22] Often owned by ethnic Chinese traders, the animal shops concentrated along Bridge and Rochore Roads. They were plentiful with Australian, Indonesian, and Malay birds and animals, and in one shop, Lucile held the paw of a Malay sun bear cub. It was "whining in the most pathetic way."

In the afternoon, the Manns lunched at a British resident's spacious house. As they savored "a delicious curry," they gazed over the tennis court and tropical garden, cooled by the gentle breeze of a fan pulled back and forth by a "punka-wallah." The next morning, they returned to collecting, visiting Besappa's Zoo, a store that boasted an excellent collection of animals, including tapirs, cockatoos, pheasants, gibbons, and a pair of tame orangutans. At dinner, they admired the ocean and the palms and Lucile "thought all sorts of kind things about the tropics."[23]

By the time the Manns arrived, Singapore was the principal entrepôt for Britain's Asian imperial trade. At the strategic crossroads of the Indian and Pacific Oceans where the Indonesian archipelago came closest to the Southeast Asian mainland, Singapore was a flourishing port where merchants sold an array of tropical products. Rubber was one of the colony's largest exports. Animals, too, were big business. Ethnic Chinese traders, part of a large merchant community in the Straits Settlement, played key roles as middlemen purchasing animals from natives throughout Southeast Asia for resale to visiting American traders or zoo officials like William. Such merchants also owned the first of many holding pens scattered around the city. The number of animals, including rare birds, imported into Singapore for shipment across the world was staggering. In 1918, Singapore's traders imported 69,211 animals from Indonesia alone. By 1925, the number had risen to 346,516 Indonesian animals. Many more came from India, Malaya, and Siam.[24]

By the 1920s and 1930s American traders, including Buck and Charles Mayer, constructed compounds of their own. As they moved from jungle

A typical Singapore animal shop, owned by ethnic Chinese traders. Frank H.Buck, "A Jungle Business," *Asia: The American Magazine on the Orient* 22 (August 1922).

to zoo, animals gathered in Singapore waiting export to America. In the hot summer of 1924, more than 100 Indian animals, including tigers, elephants, and monkeys, rested before their journey to a new collection in Long Beach, California. Nearby, Buck luxuriated in the Raffles Hotel, the landmark hotel that symbolized the comforts of colonial life, while he gathered a large consignment for American zoos. Local craftsmen constructed strong cases to hold 40 large animals, hundreds of monkeys, and 5,000 birds on their journey across the Pacific. Two young elephants were en route from Siam. Five orangutans were coming from Borneo. In his compound, Buck had already collected two Malayan tigers, two black leopards, Sumatran tapirs, and Nilgai antelopes from South India.[25]

And he never once stepped into the jungle. As his wife admitted to a Singapore newspaper, Buck "rarely goes into jungle for the sufficient reason that it is a waste of time for him to do so." Instead, following the model of Chinese traders, he visited Makassar, Banjarmasin, Palembang, Pontianak, and other small ports to visit their animal markets. Typically, natives captured the animals and sold them to Chinese traders who, in turn, hawked their animals in Singapore's wild animal and bird shops. In Singapore, an orangutan sold for between 50 and 300 Straits dollars and tigers for $150 to $200, but by the time such animals crossed the Pacific for sale to American zoos, their price rose more than tenfold. If the journey home avoided major disasters (and those were all too frequent), traders like Buck could earn a substantial profit. On one of his earliest collecting expeditions, Buck embarked from Singapore for Borneo, Sumatra, and Malaya and from there to India. In Calcutta, he traded with Parsi animal dealers. His collection of monkeys, tigers, leopards, and birds earned a profit of $6,000.[26]

Buck, Mann, and other Americans arrived in Singapore at a time when the colony's British residents and Chinese traders struggled over the future of a growing animal business. American traders would benefit from that conflict. Singapore may have been the international center for the trade in animals, but that success worried British colonial residents. By the 1930s local British conservationists and avid big game hunters, led by Theodore Hubback, urged the formation of a Wild Life Commission of Malaya to investigate the trade. In one shop the Commission discovered three Malayan honey bears crammed into rattan cages. Even the shop owner could not recall how long they had been on display. Rare Argus pheasants were in cages too small even for them to turn around. The Wild Life Commission mourned the "distress and wretchedness amongst the caged occupants in these unsuitable premises."[27]

Hubback made his name as an elephant hunter for pleasure, not profit. Now as chair of the commission, he called for the creation of a central bird and animal market to replace Chinese-owned "shophouses." They were a "disgrace to any British Colony," noted a British critic. A new market could better regulate the wild animal trade and prevent the sale of poached animals. Hubback hoped the market would wrench a profitable business out of the hands of Chinese merchants. Local dealers understood that the campaign mixed British concern about declining game stocks with hostility toward Chinese merchants. A desperate petition from Chinese traders helped forestall the creation of a central market. Yet even on the eve of World War II, critics urged banning the wild animal trade as a "Disgrace to Singapore."[28]

Southeast Asia's animals had acquired enough economic value that natives were willing to risk their lives to capture them, and British officials, American dealers, and Chinese traders jostled to control the animal trade. The geography of Southeast Asia with its jungles close to small ports in close proximity to Singapore made its animals particularly valuable. In small ports and tiny islands, traders could merely pass along word that they would buy healthy animals. A local economic network was then set in motion. In a remarkable twisting of the older European belief in the racial and sexual links between Malayan and Indonesian tribes and orangutans, Buck claimed to have seen "jungle-bred" mothers suckle orangutans on one breast with their own baby on the other: "For these mothers knew that the baby orang had a commercial value her own baby did not have."[29]

The Manns followed this migration of animal traders. Leaving the city, they crossed the Straits into Johore, the sultanate at the tip of Malaya governed in effect by a British resident. Before cocktails, the Manns visited a local dealer who offered a Sumatran gibbon, a pair of recently caught tigers, and eight orangutans—the largest one called King Kong. Then they departed for Dutch Indonesia. When they arrived in smaller towns, word spread rapidly that the Manns had arrived, eager to purchase animals. As they breakfasted in one village, a local Chinese trader brought a collection, according to Lucile, of "dirty and inadequate cages" filled with birds and small mammals. The next day another trader brought a tiger cub. In Makassar, on the Dutch Indonesian island of Sulawesi, they turned down a side street to find "a Chinamen who kept some birds for sale," and they purchased lories, parakeets, and cassowaries. Chinese traders, Lucile believed, appeared to be the "head of animal collectors here. They all haunt the shipyards and pick up what they can from sailors coming in from Ternate and New Guinea and other distant islands." In other villages, too small for Chinese traders, the Manns purchased animals—often pets—directly from native villagers. In one "nice little native village . . . small boys bring in one animal after another." It was "a most heartening collection." William simply walked around the village and "every time he found that a native had a pet he would have him bring it to us."[30]

Collecting in these villages and ports south of Singapore developed its own schedule. After a comfortable breakfast, the Manns fed their growing

"Buying Pets for the Zoo." This photo from William and Lucile Mann's 1937 animal-collecting trip to Southeast Asia shows the zoo director bargaining with local children for their pets.

menagerie and then met "a line of natives" offering birds and small mammals. For Lucile, their infrequent tramps through the jungle were a welcome relief from "sitting around the hotel, waiting for animal dealers to show up." Trapping, though, was a hopeless task, and even the jungle never realized Lucile's expectations of tropical abundance. They hoped to glimpse a tiger or an elephant but saw only a solitary monkey and a hornbill in flight. It all seemed more like a neglected rubber estate than a lush Eden.[31]

As the Manns moved west from Singapore into India, they turned to colonial officials for help in trapping the large prizes they sought most. In particular, they wanted a rhino, an ornery animal, rare and almost impossible to transport, that few American zoos owned. As the Manns left for India, they were excited with the news of a rhino captured in faraway

Nepal. Yet by the time Manns arrived in India, the rhino had died some-where between the hills and the plains. American collectors like William depended on the good graces of colonial officials, especially when they sought the biggest animals. Colonial governments regulated the global trade in tropical animals. For smaller mammals, birds, and reptiles, this regulation was more a theory than a practice. For larger animals, like the rhino William coveted, traders needed a trapping and export permit from colonial officials. The permit process rarely considered the actual wild stocks of animals. Permits were available to American traders or zoo offi-cials but never to a local Indian trader or to a Singaporean Chinese animal merchant. Instead, if Indians or Chinese wanted to enter the animal busi-ness, it was as shop owners, not exporters. Colonial officials provided the permits to American traders and even the labor needed to collect large, rare animals.

Imperial projects also turned remote habitats into prime lands for busi-ness. The roads, rivers, and rails that connected South Asia to its colonial port in Calcutta also brought animals to markets. In Calcutta, a passing journalist described a three-story white cement building with bars on the windows that recalled a jail. But this "plain-looking building" was one of the city's "hotels for animal transients"—with few vacancies. It was crammed with four elephants, dozens of leopards, a pair of Bengal tigers, and hundreds of monkeys and birds. One cage held rare pheasants from Tibet, another Burmese peacocks.[32]

Meanwhile, the Manns' traps, dragged from Singapore to Calcutta, rusted. Trapping was a "colorful" experience, but they realized it was impossible without the "cooperation of the natives."[33] Once they spent four days in the jungle—and returned with just five specimens. William concluded that their expedition was simply "to collect the collectors and then take care of what animals they brought us."[34] The jungle tempted American and European entrepreneurs with its potential riches. From the rubber trees that could be nurtured in its humid climate to the tin untapped below ground to the animals that prowled wild, the tropics promised wealth but tenaciously held onto their resources.[35] Lucile and William had gone to the East Indies and assured the National Geographic Society they would confront Asia's rare beasts in lush colonial jungles. Instead, they bartered with children for household pets, browsed among

stacks of rattan cages in animal shops, and hobnobbed with imperial officials. They also enjoyed a variety of cocktails. The real danger began
when the *Silverash* left behind Asia and headed for Africa and the Suez
Canal.

Hungry Mouths: Port Sudan, Tanganyika

As the *Silverash* docked in Port Sudan in September, Mann, though
already weary and falling prey to the sickness that would land him in a
New York hospital, ventured ashore in search of Africa's animal trophies.
He was back in Africa, and as he stepped from the *Silverash*, William
must have recalled—with certain bitterness—his earlier campaign to capture big game in the former German colony of Tanganyika.

In 1926, William had promised visitors, donors, and the Chrysler
Company, which funded the expedition, that he would trap and transport
rhinos and a pair of giraffes. In Africa, William learned the hard lesson of
the animal trade: no matter his expertise in caring for animals behind bars
and despite his graduate training, he was helpless in both jungle and
savanna.[36] Long before he sailed on the *Silverash*, William had led an East
African expedition to the brink of disaster. In these prime game regions,
the colonial bureaucracy had monopolized game for trappers and elite
hunters. William was admiring: he praised the "well-organized" conservation department and the guards who enforced the laws that kept game
safe from natives but readily available to trappers like himself. Because of
that colonial monopoly, animal trappers could find the workers they
needed to bring animals back alive—except when, as William realized, the
bargains between natives, colonial officials, and traders frayed. Then animals died, workers revolted, and imperial officials worried.

Whether in Africa or Asia, catching animals depended on natives and
local traders, but it was an adventure ordinary Americans longed to join.
When he announced his expeditions, William was flooded with American
applicants. Imagine the perfect shave, one hopeful promised, in the heart
of the jungle. When William set off for Africa in 1926, John Tate pleaded:
"I am forty years of age . . . and in perfect health. I can leave on a minutes
notice [*sic*]." Animal trading united John Burns's fascination with zoology
with his thirst for tropical adventure; he asked only for a collect telegram

telling him where and when to be ready. Raymond Briggs worked in a New York office but dreamed of African adventure: "I'm afraid that you'll have to take me along, Doc." In its allure, animal capture offered such ordinary Americans the chance to outdo the rich big game hunters they read about in popular magazines or books. For Charles Day, a "colored man," an African study adventure was his "life's ambition." He was a cook and promised "excellent" food.[37]

William was sympathetic to the young man bored in an office building or to the butcher's son working at his father's counter. "In every properly raised boy," he said, "there is a germ for such expeditions as ours. All of us in our earlier imagination travel in Africa, slaughter elephants, dodge spears and suffer delightful hardships."[38] Even so, the National Zoo prepared a form letter quashing hopes of would-be adventurers; William was going to hire natives to capture animals. The African animal trade was structured differently from the bustling Asian business centered in Singapore. The animals from the Dutch and English Southeast Asian colonies flowed into Singapore's animal markets along long-established trade routes. In 1936 in Borneo, for example, 200 natives worked for three days to capture a single orangutan; seven workers were severely injured. After they drove the targeted ape into a tree, natives denuded the surrounding trees and vegetation to prevent its escape. They baited a net trap with fruit beneath the isolated tree, but even so, the ape escaped the net four times and bit natives trying to subdue him. Finally, exhausted and hungry, the ape was crated and shipped by Chinese traders to Singapore and, from there, by steamer to the United States.[39] As native workers recuperated from bites in a Borneo hospital, the ape began his journey to the Bronx Zoo. Africa, though, didn't have the same bustling game highways, and it was a long way from trapping grounds to ports.

In Africa, William could not enjoy leisurely evenings sipping cocktails and days browsing animal shops. "Safari life is pleasant," he recalled, "if you like walking and boiled guinea fowl."[40] Trappers negotiated to hire the natives who would actually bring animals to bay. When William arrived, his first task was hiring a local "white hunter," Charles Goss, who would pay locals, beg colonial officials for trapping permits, and oversee the tracking. White hunters were professional European and American hunters who lay in wait outside luxury hotels at the gateways to Africa's game lands. Once they had profited from the sale of elephant ivory; now they hoped to lead wealthy tourists on safari hunts. As safaris grew in popularity in the decades after Roosevelt's much-publicized hunt, tourists often lacked the rifle skill or tracking technique needed to bag their trophies. A good "white hunter" might sight a rifle or even fire the fatal shot—and then tactfully step back to allow the fee-paying client to claim the prize. The white hunters also hired the dozens of workers needed for any safari.[41]

William knew he would rely on Goss even more than the typical safari tourist. After all, this safari was going to capture animals rather than shoot them from a safe distance. Capturing animals required porters and skilled trappers. The hard work of animal trapping, the knowledge of local game habitats, and the management of local peoples were all far beyond the skills of an American zoo director. William realized that the hard work of capturing animals would fall to the native workers Goss managed. Once when William ventured into the bush looking for animals, "we had a hundred natives with us." But then he sent for "more to carry the animals."[42]

The expedition was in search of prizes, but only young ones. The older animals were too strong, too resistant ever to accept the long journeys from bush to port and then to American zoo enclosures. Too often trappers shot the older animals to capture the new orphans. The Tanganyika savanna was "alive with game," but capturing the young animals they sought was frustratingly slow. William and his native workers needed a full day of panicking the herds of antelope and wildebeests to tire out and then snag the young animals. Finally, they slung the "weary" young antelopes into burlap bags for the 150-mile march to base camp.[43] With his white hunter, "we will go into the 'swarming' country and catch a rhino or bust," Mann had promised zoo patrons in Washington.[44]

They would bust. "Our boys," as William called them, could swarm the baboon packs, panic the antelope herds, and net the warthogs, but rhinos and giraffes proved the expedition's undoing. Once one of William's white assistants shot a mother rhino. "SOS rhino," he telegraphed William, "am trying to capture its rather sizable calf." By the time native workers arrived with ropes, the young rhino orphan had fled. For a month, William and his expedition stalked the rhino trails. Most days they saw adults but "not a single young one." The pursuing humans agitated the rhinos. Twice, rhinos rushed the camp at night, and the boys fled into the tents occupied by the white men. William, well attuned to racial, colonial laws restricting native hunting, understood why; the white hunters alone could hold the guns. During one attack, a worker was badly injured and was rushed away for an operation.[45]

As the summer safari stretched into the harvest season, tensions were rising in the expedition's camp. William, in frustration, abandoned the rhino hunt; he suspected that he would have more success elsewhere tracking giraffes. William was confident that with ninety workers he could capture a pair of good specimens. Ultimately, he needed 500—and that proved part of the crisis. That many workers handling nets could panic giraffe herds, but capturing individual animals was almost impossible. The giraffes tended to charge the nets in a single file behind a large bull. Thus they protected their young and massed their strength against a single spot in the netting. "Day after day," hundreds of workers surrounded the giraffes only to see them escape. Finally, workers managed to corral a younger male. The fourteen-foot-tall giraffe struggled wildly against the ropes around his neck until he just "lay down and died." This was not a story of safari failure William wanted circulated back at home. "Don't like to use this for [the] public," he wrote. Just as local workers rebelled against the harsh conditions in the camp, William's (but not the giraffe's) luck turned, and his workers managed to separate a youngster from its mother.[46]

The news of the capture spread quickly back to Washington where a local newspaper ran a contest to name the giraffe—ideas ranged from Noah to Kipling to Bo Big Mann. William was appalled. The giraffe was nervous, only picking at its feed, and William knew it wouldn't survive. The names, William worried, would only mark the giraffe's "tombstone."[47] He also hadn't captured rhinos, but the failure to bring back alive captured giraffes was a worse debacle. The unnamed giraffe died; then the expedition collapsed. William had, it seemed, forbidden his workers to leave, and they were agitating to return home for the harvest. Then he cut their meat rations until they caught giraffes. Maybe feeding that many workers was too expensive with Mann's limited budget. Likely, William was growing anxious and suspicious of his employees. Whatever the exact reason, William in his harsh regime had broken the kinds of agreements about meat and money that kept workers in the field and away from their farms. By September 1926, instead of 500 workers, William employed just two animal handlers to care for animals during the trip home.[48]

William hired so many local workers to hunt rhinos and giraffes because there were fewer indigenous traders than in Asia. Even worse for William, the African animal trade frayed after the collapse of the German monopoly. In Africa, the Hagenbecks had long depended on trappers, often European adventuring men, who in turn mobilized the local workers required to capture terrified African animals. Karl built his animal empire with agents, armed with plenty of silver coins to pay native hunters. "Unlike the hunter, who is attracted only by the love of sport, the animal trader goes to work," Karl noted. "Armed with weapons both accurate and deadly," the white hunter faced "little danger." The trader, by contrast, needed higher bravery to manage "uncivilized peoples," no less wild than the beasts.[49]

Long before animals could be captured, people needed to be "secured" and Karl became an amateur anthropologist, recording the different African tribes' ability to track and trap specific animals. The Takuris, Muslim "negroes" from Darfur, were particularly adept at trapping hyenas, panthers, and baboons. Tribesmen from Taka in the Sudan were skilled with swords and specialized in capturing rhinos and giraffes. No matter the abundance of local game, Karl realized good trapping depended on the traits of local tribes.[50] More than a decade later, William made similar kinds of judgments. The Wamboro and Wambugo tribes, he explained, were fast runners who could chase down animals. They were especially helpful in tracking a swift water mongoose, one of the expedition's first prizes.[51]

Still, whatever the native workers' background, the capture of large animals was perilous for these hunters and deadly for animal mothers. On one trapping expedition in 1893, hunters encircled a mother rhino with a calf and a single "black fellow" jumped from a horse onto the "old beast" to kill it. The calf was roped, bound, calmed, and then delivered to the collector. The natives divided the meat and hide of the dead mother. This was a lesson about the distribution of the spoils of the hunt that William was, to his peril, slow to learn.[52] In Asia, William could comfortably negotiate with local traders, sometimes large-scale businessmen, other times local boys hawking pets. In Africa, by contrast, William confronted rebellious workers and dying animals.

The first Americans entering the African animal business realized collecting was a labor-intensive, if highly mobile, industry that demanded the work of collectors, agents, local assistants, carpenters, and sailors. In 1901, the animal trade had tempted John Camp, a successful river steamer captain in "the heart of the dark continent." He regularly traveled and hunted with up to fifty native workers, "a few from each important tribe." Though "flooded" with applications for work, he worried about the prospects of animal trading in Central Africa. He knew that his native workers possessed the skill and knowledge to capture even the largest elephants by spears, pitfalls, or spring traps. Yet transporting vulnerable animals from Africa to America presented almost insurmountable challenges. Perhaps he could carry young animals by steamer to Leopoldville and by lengthy rail trip to the coast. They would have to be loaded onto ships to Rotterdam and from there to New York. "This is a lovely country yet awaiting development" and rich in animal life, but Camp realized that to start in the business he needed a heady initial investment of at least $10,000 from American zoos. Maybe life on the Zambezi River was more predictable.[53]

When an expedition ran smoothly, a symbiotic relationship linked colonial bureaucracies and animal collectors. Ostensibly, colonial officials worked to regulate, even restrict, animal trading. Permits for collecting animals were hard enough for individual traders to acquire. Yet orders from American zoos could turn colonial conservation into commercial zeal. In fact, traders regularly asked zoo directors to write to colonial officials to help obtain permits, even if the zoo directors were never, in reality, going to purchase the captured animals. It was sleight of hand that helped American traders gain access to colonial game reserves.[54] William's permit allowed him to collect whichever animals he wanted; it even sanctioned him to kill the mothers in order to collect more tractable babies. Officials also helped corral the hundreds of workers needed trap and calm the large prizes William desired most, especially rhinos and giraffes.

Most important, the same laws that "gave us permission to capture specimens of all of the game animals of Tanganyika" kept meat and guns out of the mouths and hands of natives. The imperial system reserved game for white travelers like Roosevelt or William, shooting and trapping for pleasure and profit. The native populations in the heart of the trapping region were farmers by necessity. In British colonies, laws restricted native hunting as well as access to guns and ammunition. Between the 1890s and 1930s, British colonial governments, in particular, instituted a series of restrictive game laws that enclosed prime game lands in the name of animal conservation and severally restricted the ability of native populations to hunt. Local peoples well understood that colonial conservation typically meant reserving large animals for the pleasure of white hunters. The Society for the Preservation of the Wild Fauna of the Empire, with intimate ties to British colonial administrators, lobbied successfully for the creation of game reserves from Southern to Eastern Africa, and even outside of the reserves, newly empowered game wardens enforced game regulations and the issuing of permits. At the same time, a stream of legislation stringently restricted the rights of native peoples to own guns or, for that matter, to hunt animals with nets or snares.[55] To hunt or capture big game required a permit, and these were open only to white colonists or visiting white tourists and trappers.[56]

Hungry mouths made tractable workers—at least until rations suffered. Then workers rebelled and colonial officials were forced to intercede. Out on the veldt, William realized why workers stayed in his camp despite its obvious discomforts. "What they really go for is meat," he concluded, and the "only time they feed well is when they are out on safari."[57] Like so many others who had hunted or trapped in Africa on safari, William ate and drank well, often enjoying a few of the culinary pleasures of home, carted from city to veldt by his many porters. Perhaps it was his own full stomach, but when he looked across his own table at his native workers, he saw only predators consumed with meat-lust. "There is no limit to the amount of meat a native can absorb," he sniffed. Twenty natives could consume an entire zebra in a single feast, he insisted. As long as supplies of meat were steady, his workers seemed willing to put themselves in the path of charging animals and stalking lions.[58]

The conflict began when William became desperate to capture a giraffe and cut the meat rations. In Tanganyika, at the behest of the local colonial officials, Sultan Chanzi Machafula supplied the hundreds of workers

William needed to track giraffes and other animals. The work was difficult and dangerous. Most of the workers holding nets spread out in a
thousand foot line across the hot terrain. The remainder beat the bushes
to force animals into the nets. The terrified animals took their toll. One
worker was badly gored and showed his injuries to the unsympathetic
William. According to the zoo director, the injured worker was simply a
native asking for "bakshish." Workers grew restive as the harvest season
approached. Then William sliced the meat rations. "We were given now
and then a portion of meat," complained Saidi bin Kasanda, "but there
were so many of us that it didn't go far."[59]

Soon, even promises of cash bonuses for the capture of giraffes were
not sufficient to quell discontent. William had promised each worker
enough money to pay his annual "hut tax," the local tax imposed by the
colonial government. In cash-poor areas, access to coin could smooth the
work of a trapping expedition. In this case, though, William had brought
with him to Africa ideas about how to speed up work that he had learned
from domestic American factories. On the veldt, he introduced piecework. He decided his workers would be paid in cash and meat only when
they captured an animal. He also paid additional bonuses to workers who
obeyed the expedition's strict rules. He gave an additional twenty cents to
those who stayed in camp, avoided beer, and did not "eat meat all day."[60]

Typically, trappers in the region had paid natives a flat wage of around
fifty cents per day in addition to meat and grain rations. Yet William's
workers complained that they received a mere sixty-nine cents for two
weeks of work with only the occasional meat ration. The "natives," concluded the British colonial district officer, "were naturally very dissatisfied."[61] With scant meat, piecework pay, and crops spoiling in the ground,
William's workers turned to colonial authorities far more powerful than
the sultan who had sent them on safari. In a strike out on the veldt, the
threat of violence lingered. Would native workers simply refuse to hunt, or
would they kill the already captured animals to get the meat they were
owed? Only Goss and Mann possessed firearms; would they fire them?
Tensions mounted. It was up to the colonial government to turn its back
on its white friends to save an expedition that had collapsed in recrimination and hunger on the veldt.

It wasn't the first time that in the challenging conditions of the African savanna an expedition spiraled from colonial order into near mutiny. Trappers were well aware that the people they employed were more knowledgeable and more able to withstand the rigors of tropical travel than an itinerant American trader. There was an obvious contradiction between trappers' beliefs in their own racial superiority and their dependence on locals, and because of this, workers were able to carve out a certain sphere of autonomy and profit. During the long journey of animals from jungle to zoo, native workers bore the responsibility for keeping valuable specimens alive, and they used this power to renegotiate the terms of their labor. As many as 200 drivers might be needed to transport animals across jungles and deserts to ports and waiting steamships, and each species needed its particular feed. Some animals could only be transported in heavy cages, often on human shoulders. On occasion, in the humid heat of the moment, an expedition simply refused to advance, and "the wild animal dealer suffers terribly from a pecuniary point of view." With the lives of already stressed animals in the balance, traders likely offered additional payment or extra rations, especially fresh meat.[62]

Workers understood as well that any animals inadvertently killed in the midst of capture instantly became food. William competed with his own workers over the fate of captured animals, and he never quite trusted his workers to privilege American zoos over their own meat hunger. "Whenever anything was captured," he told American reporters after a later African expedition in 1939, "it was a question whether it would survive long enough to be placed in a pen"—and not always because of the stresses of captivity. "[O]ur own hired hands," he complained, killed many of the "specimens" collected for the zoo.[63]

Here then was the key contradiction of the African animal trade— workers labored for fresh meat but were expected, ultimately, to keep animals alive. The line between life and death, meat and exhibit was dangerously blurred. The trader Henry Trefflich remembered his first animal-trapping expedition with his father who introduced him to the business. When Trefflich and his father caught up to their white hunters, things had gone terribly wrong. One of the hunters was prostrate with severe malaria. His native workers, sensing his death, feared for their pay. They turned the valuable live specimens into another sort of payment, killing most of the animals for meat. Trefflich and his father had already realized how

important meat was for native workers. Even as the Trefflichs raced to find
their white hunters, their porters and gun-bearers reminded them of their
agreement to shoot game for fresh meat. Ben Burbridge, who caught and
sold some of the very first gorillas on display in the United States, also
recalled the "meat chant" of his forty workers. Burbridge shot an antelope
and stepped back to watch its butchering. It was, for Burbridge, confirma-
tion of the similarity of tribes and feline predators. The men "fought and
elbowed each other, as smeared in blood and refuse, each strove to get the
larger share." One buck did not provide much meat for forty workers, and
soon there was only "a piece of hide and a few scattered bones."[64]

William, Burbridge, Trefflich, and so many other trappers and traders
seemed to relish their role as masters of their safaris, providing meat for
the "hungry mouths" of workers they derided as savage or primitive.
When their workers devoured antelopes or zebras, their hunger was, in
fact, a consequence of colonial restrictions on native access to game.
Forbidden to hunt, meat had become a sought-after scarcity. Yet behind
the laws, collectors and workers subtly struggled over whether animals
were exhibit or food. Natives in West Africa might have been amused by
the trapper J. L. Buck's desire to capture a pygmy hippo, but they certainly sensed an opportunity. Hippos and these villagers clashed when
hippos invaded and trampled rice patties. The trapper may have regarded
local peoples as a ready source of labor, but they considered him an
unwitting ally in the struggle between animals and people over valuable
riverbanks. A hippo removed from the river saved crops and offered a
chance for fresh meat. When a hippo blundered into a pitfall, the
American's vision of profit turned to disappointment when it became
clear the hippo had broken its leg. His native employees, however, were
delighted. The mortally injured hippo promised a rare feast. Like so many
other failed American trappers before him, J. L. stood to the side, cursing
his luck, the jungle, and its mosquitoes while his workers celebrated with
shared meat.[65]

Thus, William, in his frustrated quest for giraffes, had led a safari to the
edge of disaster, and it fell to colonial officers and magistrates to solve the
meat strike. Workers complained to colonial officials that they were
"detained" with miserly meat rations until they captured more giraffes.
The local regime, normally so willing to enable the animal trade, reminded

William that natives "be employed voluntarily only" and arbitrated the relations between the zoo director and his workers, once greased by meat. To Goss, they threatened to revoke his permits for this expedition and to deny him any future licenses. William realized the expedition had fallen apart. He fired his workers and hurried to crate his animals to return to Washington. He quietly purchased a pair of Sudanese giraffes—Hi-boy and Dot—from professional trappers and the colonial government in Sudan and returned to Washington in 1926 with a hard lesson in the work of the African animal business.[66]

When William disembarked from the *Silverash* in fall 1937 to buy giraffes, he must have relived memories of the 1926 mutiny in the open savanna. After all, he wasn't just browsing the African animal markets; he was avoiding yet another brewing struggle, this time pitting sailors against the Manns and the ship captain. Again, animals were dying, and again, William was suspicious.

Gaddi's Ark to America: Port Said, Atlantic Ocean, Washington

Layang Gaddi, native to Borneo, was living behind a hospital in Bangkok when William learned of his skills. The *Silverash* only had a few berths to spare for its return journey—an alligator, in fact, had been given its own room—and William had to choose his crew carefully. Gaddi was the perfect choice: an experienced forty-eight-year-old trapper, he could care for stressed animals in cramped cages across two oceans. Gaddi spoke perfect English and boasted a special touch with wild animals. When the boat arrived in Washington, this tribal with a knack for animal care caused as big a sensation in Washington as any of the *Silverash*'s animals.[67]

When he first boarded the ship, Gaddi's practiced eye recognized the telltale signs of a forced animal migration going bad. Near Port Sudan, the *Silverash*'s animals were already weakened and dying and William was dangerously ill. The ship's sailors, mostly Chinese seamen, were angry over the arduous labor needed to care for animals stressed beyond their normal endurance. The Manns, meanwhile, had become suspicious of the sailors over the loss of particular animals. Had dangerous animals been killed? Had others become food?

Tensions that simmered in the tropical heat erupted in open mutiny. As the crew in Port Sudan strained to load William's new purchases—four giraffes and two buffalo—they watched sailors on a nearby German ship enjoying liberty. In an "ugly frame of mind," the sailors confronted the ship's captain. A sailor attacked the captain with a jagged broken bottle. On a day in which the black leopard attacked his handler, all the small mammals seemed on the brink of death in the heat, and William fell into his cot with a threatening fever, a mutiny "seemed like a little too much" to Lucile. By Port Said, at the other end of the Suez Canal, several sailors were in shackles.[68]

By the time the ship crossed the Atlantic in late September 1937, Gaddi was exhausted and dirty and prize animals had perished. In the world of animal trading, it was a fairly normal trip. On any long voyage across oceans, animals escaped and quarreled with other animals, often their natural prey. Sometimes an unexpected storm washed animal cages from their moorings into the sea. On one trader's voyage, during an Indian

"Gaddi on the *Silverash*." (1937). In the hold of the *Silverash,* dwarfed by the piles of cages, Gaddi prepared food for future zoo specimens.

Ocean storm, a leopard escaped its partition and attacked a young elephant, beginning a battle that eventually engulfed an adult elephant and two other leopards. A leopard and the adult elephant died.[69] Animals were ubiquitous on ships sailing from Singapore, Calcutta, and Bangkok, and their deaths were frequent and devastating.[70]

Animal deaths could lead a collector to the brink of ruin. Traders often hired natives to mind the wild cargo, fully knowing their exotic appeal could add to the excitement at the ark's arrival. At sea, these skilled animal keepers struggled to keep animals alive, but they couldn't overcome the effects of cold, cramped space, and storms. Frank Buck had more connections and local influence than any other collector since the heady days of the Hagenbecks. Yet a single voyage in 1928 brought him home penniless when a typhoon swept his animal cargo into the tropical seas.[71] Such animal losses underscore the reality that, as resource extraction alone, the animal trade was risky business.

Traders entered a rapidly changing competition between locals, native nobility, colonial officials, plantation owners, and animals. Animals themselves were simultaneously prey, hunters, sporting game, menace, nuisance, and meat. And the land in which they lived was simultaneously wild jungle or veldt, hunting grounds, subsistence farms, or arable land fit for plantations. Animal capturing, for humans, was hard work. For animals, it was a frequently fatal struggle. Maybe like William's giraffes they died during or after capture. Others like Lucile's pet gibbon died at sea. Traders boasted about the contrast between their industry and the luxurious safari. The pleasure seeker enjoyed the safety of the high-powered rifle; the trader brought the animal back alive. Yet this, too, demanded killing, and a lot of it. The standard method of capturing animals, perfected by Hagenbeck agents, persisted among American traders and William's white hunters. Adult animals were too hard to catch and transport and impossible to domesticate. Instead, trappers killed adult animals to catch their young. To capture just two rhino calves—one for the Bronx and the other for Philadelphia in 1920—Frank Buck's Nepali hunters shot a dozen of Nepal's "resources."[72]

Zoos promised wild animals to visitors, but key to successful animal trading was domestication. Animals that continued to resist capture and confinement were much less likely to survive the rigors of the voyage from jungle to zoo. Traders sold animals that had accommodated themselves to capture. In other words, they had become, often through force, domesticated wild animals. Traders often kept animals in compounds for weeks after their capture even though they had to shoulder the substantial costs of feed. For antelopes, buffalo, and giraffes, this long stay in compounds, often in Singapore, prepared them for the trip ahead and life in the zoo. In fact, Buck suffered the most serious injury of his career not in the jungle but in his compound as he tried to salve the wounds of a frightened, orphaned tapir. The typically gentle animal suddenly became enraged and mauled his captor. Only the timely arrival of an assistant saved Buck from a less than romantic tropical death.[73]

On the *Silverash* Gaddi and Lucile realized that animals fought back— and were domesticated by traders—in ways unique to their species and their individual character. Some like a tiger or orangutan fought against confinement even through the ocean voyage, and they paid for their resistance with their lives. Animals charged the bars of the cages or refused food. Others, like Lucile's favorite gibbon, could touch the sympathy of their handlers and win an extra serving of food or an ounce of freedom outside their cage. Animal resistance had its immediate effects. With hints of unrepressed wildness, the snarling leopard or tiger became popular exhibits while placid animals seemed mundane. The cost of animals was, naturally, shaped partly by their wild abundance. Animals that ranged in enormous herds were often the cheapest on dealers' stock lists. Resistance also mattered in determining value. Animals, like the gangly giraffe, the muscular rhino, or the fanged tiger, that were harder to transport cost more regardless of their natural population. Yet some animals that rarely survived the journey or were simply too hard to capture did not make their way into zoo exhibits and did not figure in the way ordinary Americans imagined the natural world of the jungle. The animal trade acted as a kind of filter, determining the kinds of animals Americans believed populated jungles and veldts. They knew a great deal about tigers or lions because they were common in zoos. Even the most educated travelers, like Lucile, were astonished when carnivores, so visible in zoo enclosures, were so hard to glimpse in the wild.

Animal resistance mattered as well in determining another aspect of animal economic value. Animals that fought their capture with fang and claw, paradoxically, made their hunting worthwhile. While their successful delivery to zoos could promise healthy profits, such animals were also the most likely to perish en route. Yet the stories of their captures could be told, sold, and retold.

A "thick wool of fog" surrounded the *Silverash* when it docked in Boston, but the press and the dignitaries were waiting. NBC strung microphones over the hatches. A black leopard roared for the cameras. He also, unfortunately, swiped a gawking sightseer, and the microphones had to be spirited away lest they broadcast profanity. Again when the crated animals arrived by train in Washington, the press crowded around, amazed at the animals, repulsed by the "most gosh-awful odor," and entranced by Gaddi. "How he managed to survive that odor is beyond comprehension," read the reports. Despite the fatigue and the smells, Gaddi seemed fresh, posed for pictures, and answered questions in clipped English. The ship had traveled 12,854 miles and another 225-mile train "hop" to Washington. Then visitors could see the giraffes William purchased in Port Sudan. Gaddi, "the bespectacled descendant of Dyak head hunters in Borneo," was just as big a draw. Somewhere on the ark to America, Gaddi, a highly skilled animal keeper (the equal of William Blackbourne, the zoo's head keeper who would unload the animals in Washington), had become a curiosity whose glasses and loose clothes nestled humorously over his savage ancestry.[74]

Fittingly enough given how little he had done to capture his prizes, when the animals arrived in Washington by fall 1937, William wasn't there. The zoo director was dangerously ill and, Blackbourne assured, "trying to get some sleep."[75] Meanwhile, visitors flocked to the zoo, and from his sickbed, William was delighted. The arrival of the ark from Asia was a chance for visitors to imagine these animals and their Dyak handler wild in the jungle. In visitors' minds, the traps, instead of rusted excess baggage, were the tricks of the trade. The guns smelled of powder. The safari suits and pith helmets were torn and stained. William's illness wasn't a tropical fever but the scars of the dangerous animal game. Visitors were ready to believe that the animals on display had been captured in the lush jungle, not purchased in a shop. The many-hued birds and chattering monkeys were wild treasures, not faraway pets bartered before breakfast.

Like so many visitors, Lucile had "dreamed about Sumatra, read about it, and talked about it incessantly."[76] When they could, the Manns feasted on curry. Now, though, she had been to Asia. She had sipped cocktails in Singapore, traipsed through the jungle, tasted a durian fruit, bargained for animals, and then helped Gaddi keep them alive. Once the *Silverash* was in port, the animals safely in the zoo, and William finally recovered, Lucile tried to sell the story of her adventures south of Singapore. All those eager applicants destined to be left off the *Silverash* were ordinary Americans, thirsty for adventure. A handful of elites were rich enough to experience the thrill of the big game hunt, and ordinary sorts had eagerly devoured their published stories of animal expeditions, first in the American West and then on safari in Africa or shikar in India. Stories of animal trading substituted the trap for the gun and harsh conditions for chilled champagne on the veldt. The animal trade seemed that much more exciting and more accessible to ordinary zoo visitors than hunting tales. This was especially true when traders changed their accounts to describe jungle camps, not luxury hotels, and angry wild beasts, not domesticated specimens.

Animal traders, zoo directors, and Lucile all learned from those applicants' letters offering good food or clean shaves. If Americans could not join them on safari in person, they might instead buy their adventure stories. Lucile chose to tell her story "quietly and honestly"—and it could not sell. Each publisher returned the manuscript with a polite note of rejection. There was too much truth in a tale of unused traps, comfortable lodgings, and willing local animal traders. "Buying animals rather than capturing them first hand," noted one rejection, was both honest and novel but impossible to sell in a market glutted by adventure stories. In her telling, Gaddi wore working clothes: a blue beret, raincoat, dirty pants, and tennis shoes. NBC, though, announced he was wearing a "native Dyak costume." That exotic detail was what audiences wanted to hear.[77]

Gaddi answered a few radio questions but never had a chance to profit from describing his adventures. Lucile's publishing dreams, meanwhile, fell victim to the public desire for high adventure in the tropics. American traders, who in reality did scant little to put animals into cages, monopolized the profits from adventure stories. The countless thousands of laborers who worked to capture animals may have sometimes outwitted traders in subtle, local struggles over meat or game. Yet the harsh conditions of their work and the risks they took for American entertainment and education were obscured in the stories traders were able to sell to domestic audiences. Locals couldn't claim a share in the substantial profits from books, magazines, or films that chronicled the animal trade. Crucially, they also could not shape the roles they played in these stories. Native workers appeared as a blend of cowardly, loyal, primitive, savage, animalistic, venal, bloodthirsty, and, like Gaddi, the ancestors of headhunters, never as intelligent entrepreneurs or skillful animal keepers.

American readers wanted to believe the animals on display at the zoo had been captured, not purchased by daring Americans. The "bring-'em-back-alive school" of stories sold. The reading public craved animal collectors' "hair raising adventures," not cocktails on a veranda. In 1939 when the Manns prepared to depart, this time for Liberia, Frank Buck treated them to a farewell party. The press delighted in photos of the zoo director, his wife, and the great adventurer. In the end, though, it was Buck and his motto of "bring 'em back alive" that won the publishing wars. Buck, another rejection letter confided to Lucile, told his "fibs excitingly and the reading public has been vicariously thrilled by these fabricated exploits."[78]

The animal trade, as exciting as it might have been for Lucile and for ordinary Americans, was a risky business. Animals died, but, as Buck realized, stories lived forever. The real money in the "animal game" was in books, films, and more. In effect, animals were processed into fiction, just as raw rubber was transformed into tires or tin ore into hard metal. Packaged and sold, animal capture narratives became an increasingly important source of profit and a new source of competition for zoos. How could the staid zoo filled with dozing animals compete with the adventurer's jungle of fang and claw?

3

JUNGLELAND

The Money in Wildlife

HARRY cost almost $10,000 when he arrived in St. Louis in 1934, a fortune for a zoo to spend on a single animal in the midst of the Great Depression. Even so, Harry turned out to be a good investment. Local newspapers touted his arrival as a sign of the zoo's stature as the equal of the Philadelphia and Bronx Zoos, which boasted Peggy and Bessie, the only two Indian rhinos in America. In 1934, 3.5 million tourists visited the zoo, a million more than the previous year.[1] Until his death in 1961, Harry remained one of the zoo's most popular animal attractions— even as he acted much like any other rhinoceros in captivity. The Indian rhino, the placard on his enclosure informed visitors, is generally most active in the nights and early mornings when it feeds leisurely on hay. Harry dozed. He stumped ponderously to his mud pond. Even the zoo admitted the Indian rhino was "stupid, rather timorous."[2]

What, then, made Harry a star? For the masses of visitors in the 1930s, the allure of the beast was more important than its wild numbers. Harry promised animal rage—even if his daily life was one of quiet mastication and wallowing. He came to the zoo escorted by his captor, Frank Buck. George Vierheller, the St. Louis Zoo director, finally owned an animal that could mark his zoo's importance, and Buck had a wild story of adventure, danger, and tropical animal capture he could resell in everything from advertising copy to the big screen. "[F]rom the dealer's standpoint," Buck claimed, "this is the most valuable animal in the world today."[3]

"For the last eight or nine years," Buck boasted, "I have had a standing order from George for an Indian rhino." He had once almost delivered on his promise, but his captive died in Burma en route to a port. "You understand, I'm not selling this beast to the St. Louis Zoo" for profit, he proclaimed. Instead, the price tag merely covered his costs. Instead, Buck relished the opportunity to promote what had become, by the Great Depression, his principal occupation: jungle adventurer and film star.[4]

Harry's arrival provided Buck a rapt audience, and he seized the chance to spin yarns about tropical adventure and exotic natives. "[Y]ou can write it down," he proclaimed, "Frank Buck doesn't like big game hunting, or big-game hunters for that matter." Instead, he relied on the muscles and courage of his thirty workers. They dug a pit in the jungle mud and built a cage, lashed together with rattan vines. The calf fell into the pit, and his workers struggled to confine the panicking rhino. After eight hours of "the hardest labor I ever did in my life," Buck's workers dragged the rhino into captivity. His native employees fed his prize jackfruit leaves—"that's a kind of bread fruit." Back in Singapore, the sultan of Johore came to visit and feed the young calf. "I think he would have paid me pretty nearly any price to leave the animal with him, but . . . I had promised George."[5]

Confined in his concrete paddock in St. Louis, Harry offered visitors a lens onto the captivating world of the Asian jungle. Though an uncommonly placid animal, Harry could never escape the stories Buck told of his violent resistance to capture. Harry still recalled lush vegetation, superstitious natives, and the backbreaking struggle between native workers and Buck's prizes.

"Is there any money in it, Frank?" asked Edwin Hill, a noted film reporter. If Buck sold trophy animals like Harry simply to cover his costs, where was the money in animal trading? "I made a living and had a grand time," Buck recalled. But how? Harry was a living trophy, his capture embellished on film, newsprint, and memoir. The real money in the animal trade was in shaping animals—captured, often injured, and domesticated—into the wild beasts exciting to American audiences. The wildness of the animal, its ferocious attack or panicked defense, emerged from the moment of capture by humans—trappers and native workers.[6] Their resistance lived on as a marker of their essential animal character. "When an animal is trapped and begins to fight for his freedom," Buck explained, it "will become vicious and cruel" and visitors longed to see murderous rage, not somnolence. He described the "fiendishness" of the cornered cobra and the rage of a "man-eating tiger" that "throws down a native . . . and devours him."[7]

Buck sold more than animals. He made his real money selling stories about exciting tropical adventure, and such stories, a welcome relief from the stress of the Great Depression, helped shaped an image of the tropics as governed by savage laws of fang and claw yet profitable to the intrepid American. Zoos were just barely managing at the outset of the Great Depression. Even the St. Louis Zoo with its rapidly increasing attendance had declining numbers of species and specimens. Circuses, the other major outlet for Buck's animals, were faring even worse. Many circuses folded their tents or sold off their menageries. Animals didn't sell in hard times, but Buck thrived. "Of course the by-products of the game have paid—motion pictures, books, and so on." For Buck, the "so on" ran to a long list: celebrity endorsements, trading cards, comic books, memoirs, and World's Fair concessions. Admirers compared Buck to men like Charles Lindbergh who seemed to embody rising American global power. Frank Buck had become his own brand.

As a brand, Buck, the individual, disappeared into his fictional persona. So much so that he typically appeared in public in his trademark costume: pith helmet and khaki jungle suit. To this day, an older generation still calls this outfit—frankly as out of place in the humid Malayan peninsula as in New York socialite parties—a "Frank Buck suit." Buck was hardly the first animal trader to profit more from the stories he could sell than the animals themselves. Rather, he mastered a trick of the trade introduced a decade before, notably by Wynant Hubbard in southern Africa and Charles Mayer in Malaya (where Buck would do most of his collecting). Mayer, Hubbard, and Buck were self-consciously and self-styled literary and cinematic heroes who defined colonized lands for American audiences with little hope of international travel as destinations for natural adventure among wild animals and savage peoples. In their bluster and self-promotion, American animal collectors presented to

sedentary Americans an image of perfect empires: loyal assistants, dependent natives, valuable resources, and fascinating adventure.

The Animal Game and the Virgin Jungle: Cambridge, Rhodesia, Terengganu

Before there was Frank Buck, there was Wynant Hubbard and Charles
Mayer. Hubbard in Africa and the Mayer in Asia learned the same hard
lesson: the real money in animal trading was not in selling live animals to
zoos or circuses but in selling animal capture stories.

Wynant Hubbard and Margaret Carson Hubbard were married to
failure. Wynant was a football player at Harvard (with a reputation for
particularly rough and unsporting play). He followed other Ivy Leaguers,
like Theodore Roosevelt, on the path of natural adventure in European
colonies. Margaret had equal pedigree with a degree from Vassar.[8] After
leaving college just after World War I, the couple traveled to Northern
Quebec to work in the asbestos mines—the first in a series of disastrous
decisions. When they arrived, the mines were largely shut down in the
postwar dullness, and the area was rough with the awkward combination
of too many stranded but armed workers and too few jobs. Fired from his
job and even excommunicated by the locally powerful Catholic Church
(for reasons he never divulged), Wynant accepted a tentative offer in
South Africa to work in an asbestos mine. If he had chosen to work
with gold or diamonds, Wynant might not have ended up leading his
family, including two young babies, across South Africa and into
Portuguese West Africa in the fruitless effort to catch animals to sell to
American zoos.

By the time their steamer arrived in Cape Town, the asbestos mine had
closed, leaving Wynant with time on his hands like other white men,
including many Americans, who had arrived in colonies looking for
adventure and fortunes. Many like Wynant grasped at fantasies. He
entered into a partnership with a white farmer in Rhodesia who was eager
to turn his water-parched ranch into a depot for the animal trade. Through
trades with natives, the Hubbards built up a small zoo of their own.
Margaret assumed the day-to-day care of the animals, including a tame
cheetah—virtually all animal collectors in Africa seemed to keep a pet

cheetah. Wynant, meanwhile, headed off into the Zambesi Valley to track specimens.[9]

Wynant was a good tracker but a lousy businessman. He claimed to be able to identify and follow an animal merely from paw prints in the dust, yet the Hubbards never seemed to sell anything. Elephant hunting ended unsuccessfully, and hippos escaped before British authorities closed Northern Rhodesia to the animal trade. After freeing their animals still in captivity, the Hubbards and their sizable contingent of native workers trooped across the veldt into Portuguese territory and straight into floods, disease, and starvation. They could barely afford even second-class steamer tickets home. And somewhere along the way, their marriage dissolved.

Natives seemed to be on the only ones to profit. Carefully, subtly, and deliberately, natives converted the Hubbards from aspiring trappers into providers of meat and commodities. Wynant, fully aware of the precari- ousness of his own marriage and finances, craved affirmation of his hunting skill. Grasping at straws, if he couldn't bring animals back alive, at least he could face down—and kill—Africa's biggest beasts. Wynant rel- ished the praise of Kayingu, a local chief: "the white man is a great hunter." Hippos had arrived in the local watering hole, presenting an opportunity for meat and hides for trading, but in game-scarce Rhodesia, only white hunters had the right to bear arms and the permits to kill. The hippos, moreover, had made trips to collect water perilous. A few gifts of yams and eggs and a little flattery convinced the "white chief" of his skill. Armed with a rifle and a sense of importance, Wynant labored to capture a young hippo. The baby escaped, but its adult family died and hundreds of pounds of fat and tons of meat as well as hides and skulls entered the local trading economy. Wynant left only with a good story and his pride.[10]

At the cost of sycophancy, Kayingu circumvented imperial game laws. Wynant, though, had another explanation. Blandishments and obsequious- ness fed easily into his racism. Perhaps because of his business failures, Wynant relished his authority over native workers. He believed he had become a great hunter whose power over lions made him seem almost supernatural to natives. They treated him with respect and reverence, he insisted. Yet his workers also went out on strike. Equally, when faced with mutiny in the harsh conditions of their compound of wood huts and a

Art Department, "New York World."

WYNANT HUBBARD, MRS. HUBBARD, THEIR FAMILY, AND SOME
OTHERS.

"Wynant Hubbard, Mrs. Hubbard, Their Family, and Some Others" (ca. 1926).
Wynant Davis Hubbard, *Wild Animals: A White Man's Conquest of Jungle Beasts*
(New York: D. Appleton and Co., 1926).

faltering garden plot, Margaret resorted to a brutal wooden club perfected by Portuguese police. She ostentatiously hung the weapon over the doorway of their hut. This poisonous mixture of violent discipline and flattery helped Wynant understand his business failure as a "white man's conquest." Native workers stood at the front line between the trader and the animal, grasping ropes, constructing rattan cages, or pulling animals from traps. Yet traders alone held the guns. As a bull hippo charged Wynant's "poor fools," they seemed to Wynant like beggars and cowards, imploring him to shoot. In the heat of the moment, as they struggled to grasp the young hippo in crocodile-infested waters, he berated them as "black devils."[11]

Charles Mayer was only marginally more successful in building an animal business in Terengganu, one of British Malaya's quasi-independent sultanates (now a Malaysian state). Mayer was seventeen when he ran away from home with the Sells Brothers' Circus to become its elephant keeper. By 1883, he followed the circus circuit across the country and then the Pacific to Singapore. There, the circus replenished its menagerie and Mayer first encountered the city's animal markets.[12]

The circus moved on, but Mayer stayed, and for the next two decades, he was a struggling trapper in Singapore and neighboring Malaya. In 1898, he offered for sale to American zoos half-grown rhinos for a mere $1,000, tigers for $350, tapirs for $175, and orangutans a bargain at $250. Thirty-foot-long boas (longer than anything in zoo collections) cost $250. At the National Zoo, Superintendent Frank Baker was startled that Mayer's prices were so much lower than the more established German dealers like Louis Ruhe and the Hagenbecks. However, like many of the stock lists distributed by animal dealers, Mayer's lists were more fanciful hopes of what he might capture than of what he could really deliver. In fact, Mayer failed to build a consistent trade in the United States, and Baker never placed an order.[13]

Mayer had to find other ways of supporting his animal business. He even brought one of the first merry-go-rounds to the Far East. By the 1920s, Mayer, debilitated by "jungle fever"—malaria—returned from the jungle to the United States to sell serialized stories about the tropics. "[C]atching wild animals alive is a profession new to most Americans and they want to hear more." Mayer finally found a way to make a profit in the animal game. Over the next decade, Mayer produced two memoirs and numerous stories that described the "long chances in the animal dealer's game." He had read plenty of accounts of tropical travels, especially by British explorers describing plants, animals, and hunting in India and Malaya. Mayer also enjoyed American pulp fiction, those cheap melodramatic accounts of love, murder, and gangsters. He combined both genres to describe man-eating tigers, murderous pythons, and vengeful elephants. At the same time, he stocked his tales with human characters that expanded existing stereotypes. His "companions of the East" included "Mohammed Ariff, the crafty animal dealer; Ali, my faithful lieutenant; and, most of all, the Sultan of Terengganu."[14]

He befriended the sultan and became the first trader allowed to collect Terengganu's still abundant wildlife. Though the sultan barred "white men" from his still-independent country, Mayer was drawn by rumors of two large elephant herds that had crossed into Terengganu. Alone—with his Chinese manservant—he snuck into the country and waited patiently for a reception with the sultan. Confiding to his American audience, he was contemptuous of this petty prince. His "palace" was a half-finished two-story brick house, his coffee was bitter and muddy, and his reception room cramped and crowded. In front of the sultan, however, he was respectful and careful to give what he considered a "Malay answer" to questions about his desire to capture the elephant herds. The sultan seemed reassured as well that Mayer was not English, French, or Dutch, that is, not obviously a colonizer.

In the pages of *Boy's Life*, Mayer led readers into the Terengganu jungle, darkened by vines and creepers. The insects were voracious, and he soon fell ill with his first bout of malaria. Fortunately, he had impressed the sultan, who provided the crowds of workers needed to cut a jungle path, construct an enormous corral for the elephants, and beat the herd toward captivity. Their capture was a "wild nightmare," as confusing for Mayer and his men as for the elephants. The men drove the elephants into the corral in the middle of the night. The insects swarmed, and the elephants panicked. A giant bull charged and crushed a local priest who had just blessed the hunt. By dawn's light, sixty bellowing elephants were in the corral. The sultan selected the choicest elephants for his own domesticated herd. In return, he granted Mayer a monopoly on capturing animals in Terengganu. It was "virgin country, filled with animals."[15]

Americans relished Mayer's description of the jungle, and many probably believed his tales of ferocious animals and savage peoples. In his tall
tales, the tropical jungle was the stark opposite of the temperate forest
many Americans had come to describe as uplifting and peaceful. If
accounts by nineteenth-century travelers like John Muir ennobled the
temperate American forest, Mayer joined other twentieth-century
American animal traders in casting the tropical jungle as dark, lush,
deadly, and, in the hands of the shrewd businessman, profitable.[16]

In Mayer's telling, natives were sometimes fiercely loyal but other
times, as dangerous as the tigers he captured. Mayer was, above all, confident that he inspired both awe and fear among the natives. To his readers
back home (and just like Wynant), he boasted that natives regarded him as
a magic man who could best the fiercest beasts and the most vengeful
spirits. He was proud of his success: he was a new kind of manly American
hero who could turn jungle nature into profit. Far braver than the big
game hunter, that effete man of leisure who hid behind his wealth and
powerful rifle, the animal trapper faced down both natives and animals all
in search of wealth.

A good trapper, Mayer confided, needed to know as much about how
to manage natives as about the haunts of prize animals. "It is not difficult
to win the friendship of the natives," he said, "if you know how to treat
them." And he claimed to know better than anyone else. He learned their
language and customs, but his friendship with local nobles and his assistants was a business "stratagem." Friendship, for a chronically broke
animal trader, could be a form of trickery, a business confidence game he
proudly unleashed upon unsuspecting natives. Through his intimacy
with local villagers, he learned their "junglecraft"—the indigenous knowledge of the sounds, smells, and textures of the jungle.[17] Junglecraft was
knowledge of animals in the jungle and of the peoples who lived there. In
the hands of the American, junglecraft also became the ability to weave an
exciting story out of animal confrontations, failed business, and exploit-
ative labor.

Mayer often claimed to natives that he was a magician or a witch doctor
who could tame tigers. In 1924, he boasted to the Singapore newspapers,
"I told the whole village that I was a *pawang*, an exorcist. I assured them
that never since the day of my birth had I been molested by an evil spirit."[18]

He never seemed to doubt that natives believed in his magical powers. Walking through a village, Mayer cherished what he interpreted as the fear and respect of natives. "They were deeply awed."[19]

Was Mayer, in fact, a witch doctor in the eyes of locals? Was that awe perhaps also a stratagem? Mayer boasted to his American audience that he could manipulate natives from the sultan to villagers. Yet there are other ways of reading Mayer's stories of superstitious natives, effete nobles, and fearsome beasts. At the end of decades of struggle, Mayer had decamped feverish and broke for home to swallow quinine and retell his tales. Along the way, natives had succeeded in convincing Mayer to intercede in a very real competition between villagers and animals, especially elephants and tigers. Even the sultan, at second glance, seemed a more astute businessman than the American.

Mayer described the false flattery he offered his "friend" the sultan. Trickery certainly, but Mayer, in turn, readily accepted the adulation of natives and the sultan. The sultan appeared in Mayer's writing as a stock character drawn from many other British and American representations of native aristocracy. In Mayer's telling, the sultan was ignorant of modern technology and of the value of the natural resources of his domain. Because of his vanity, he was easily duped. Yet, from the sultan's perspective, he had struck a good bargain with Mayer. With only the outlay of labor, the sultan had rid his small country of two large herds of elephants that posed tangible risks to villages, farms, and roads. Moreover, he had acquired choice elephants that had substantial value for labor and transport and as evidence of power and wealth. The sultan himself took no personal risk. Mayer alone broke and trained the elephants. At the end of the hunt, with few profits, Mayer was dangerously ill.

Villagers equally profited from their seeming acceptance of Mayer's magic. They had persuaded Mayer to shoulder the very real risks of confronting tigers. Once, tempted by flattery, he faced down and shot an angry and badly injured cat. The villagers alone would not have had access to the sophisticated rifle Mayer carried. Without the American hunter, villagers would have confronted the tiger either with knives and spears or untrustworthy, antique guns. Yet it was the villagers who profited: the cat that had been preying on their livestock had been killed. The only price they paid was vocal acceptance of Mayer's status.[20]

The reality was that Malay peoples and Malay animals did, in fact, share the same space, and their proximity led to conflict in which both animals and peoples suffered. Tigers or leopards did occasionally prey on domesticated cattle and less frequently on people. People, in return, avoided jungle paths in the evenings or early mornings and hunted tigers that had ventured too close to villages or farms. Tigers and people developed mutual fear and, above all, sought distance. When animal collectors like Mayer arrived, however, they led natives directly into the tiger's den and provoked confrontations between animals and peoples fashioned by lifetimes of tense coexistence. Yet Mayer—the only one with the comfort of a gun—interpreted human reactions as superstition and animal responses as engrained viciousness. When Mayer watched his native workers spitting and cursing his dead tiger, he dismissed them as superstitious savages. In fact, he was witnessing rituals that had developed out of an intimacy between animals and peoples that the occasional jungle visitor, like Mayer, never could comprehend.

Mayer was particularly determined to "conquer" Terengganu's "Ghost Mountain," feared by locals for its evil spirits.[21] He brought his temperate readers along on the expedition into sensuous "tangled" tropics. The jungle was quiet at night, save for the cough of the tiger, the yowl of leopard, and the insistent drone of the mosquito. The dawn, however, was shattered by monkey's howls and birds' shrieks. The wet undergrowth awoke with leeches that grew fat on human blood. The smell of rotting, rank vegetation penetrated the camp at night. Meanwhile, death prowled constantly. A leopard leaped from above as Abdul, one of Mayer's local guides, hacked a path through the dense rattan. As Abdul screamed in terror, the cat ripped his jugular vein. Abdul died during the night, and Mayer struggled to convince his men the leopard was not "an evil spirit sent to destroy us."[22] "I can close my eyes now and bring it back," he wrote. So, too, could his readers.

At the mountain's top, Mayer recounted his natives' elation. As he stood to one side of the dancing circle, they danced, beat upon their tom-toms, and offered rice and herbs to the jungle god—everything his audience would expect from a savage celebration, including the watching American who refused to take part. The night was for primitive feasting; the day belonged to the modern animal businessman. His workers swatted mosquitos as they wove nets and hammered traps. Mayer and his men set fires in the jungle to drive the animals toward hidden nets. For the natives, "history began from the day" the fire "routed all the ghosts" and Mayer netted the richest catch of his career. With thirty-five native workers, he hauled two tigers, three leopards, a tapir, a rhino, and numerous other prizes to the coast for transport.[23] Still, even when Mayer stood at the Singapore docks watching his prizes sail away, he still could not be "sure of the success of my expedition." "[T]he animals I had yanked from the jungle might die before they reached their destination." At least, he had a story he could sell and resell.

As he transported his Ghost Mountain prizes to port, another tiger terrorizing a nearby village tempted him. Every morning the natives discovered its enormous pugmarks as well as missing ducks, chickens, and bullocks. Mayer laughed at the natives' "terror-stricken" response. "[W]ith true native instinct they decided that it must be a spirit." For Mayer, this was an opportunity to capture a particularly fine specimen—"there was always a market for them." Finally, after five nights of nervous waiting, Mayer and the villagers cornered the tiger in a hut. In his rage, the tiger knocked over a lit coconut oil lamp and perished in flaming agony. Like most American or European tiger hunters, Mayer insisted on measuring his catch. He boasted that it was the largest he had ever seen in Malaya but regretted he had not been able to catch this "perfect specimen" alive. The natives, though, were "overjoyed to know that the evil spirit was dead." Their fear gave way to an orgy of hatred for the tiger. They cursed it and spat on the animal's burned carcass—until Mayer insisted the cat be buried.[24]

The interests of trappers and locals, whether nobles or villagers, sometimes aligned. When they did, Hubbard or Mayer could enjoy sufficient and even brave workers. On other occasions, native desires trumped his business. Mayer lost this valuable tiger specimen partly because of an unfortunate fire and partly because native villagers were unwilling to put their livestock at risk simply to help Mayer capture the tiger alive.

Trappers were not the only victors in the jungle's struggle for survival and profit. Natives also understood the economic value of tigers. American traders like Mayer or Buck entered the jungle at a time when tigers already had real monetary value as well as spiritual importance to natives. Since the early nineteenth century, both Dutch and British officials in their Malay colonies had formalized a system of bounties for killed tigers. The British government in Singapore offered a bounty of 20 Straits dollars per tiger. By the 1930s, the bounty was raised to an astounding $100—about what a normal agricultural family might earn in a year.[25] Natives also invented numerous types of traps to protect livestock and villages from tigers. As early as the 1840s, Dutch and English colonial laws even mandated villages to build traps in areas infested by tigers. "Mr. Frank Buck," mused one Singapore newspaper, "was not the first man in Malaya to 'bring 'em back alive'"—neither was Charles Mayer. Sometimes locally captured tigers were simply killed; others were presented to local nobles for ritual purposes. Mayer, and later Buck, helped turned local practice into an export business.

As a businessman seeking his fortune selling tropical animals, Mayer failed. Later when the trapper became a writer, Mayer became a successful failure like Hubbard. Long before his writing was serialized in American newspapers and magazines, Mayer's vivid reports of elephant captures and tiger fights appeared in Singapore newspapers. Yet after he returned to the United States to convalesce and write, his colonial audience became suspicious. Mayer had presented several orangutan specimens to the Raffles Museum, Singapore's famed natural history museum, but it was the museum's director who realized Mayer's *Trapping Wild Animals in Malay Jungles* was an "amazing case of plagiarism." Mayer's personal memoirs had become an authoritative work on the natural history of the Malayan jungle. British colonial officials even included it in the Malaya Pavilion at the 1924 British Empire Exhibition, and the primatologist Robert Yerkes cited his descriptions of orangutans in his definitive *The Great Apes*. Yet Mayer's specifics about Malaya in fact were drawn from a less-known book written by G. P. Sanderson, a British official in charge of elephant catching in Mysore, India. Even personal observations, for example, on man-eating tigers were identical.[26] The accusations spread in Singapore but never in America. Americans still devoured his adventures. The animal trapper, dishonest, broke, and ill, had become a new American hero, far more exciting than a mere hunter. The "ordinary African game killing expedition seems tame compared with the adventures of this lariat-thrower of the jungles," boasted Wynant.[27]

The Strange Lure of the Tropics: Texas, Malaya

"I'm not a picture man," Buck protested in 1934. "Any films I make are incidental to the main purpose of my expedition which is to catch wild animals for the zoos."[28] It was one of many fibs Buck would tell over the course of his long career. By the 1930s, the animal trade was in tatters, and Buck was better known for blockbuster nature films than for prize catches. As one film critic wryly noted: "Buck decided to make some money so he quit collecting wild animals for zoos . . . and fooled around with a motion picture. Now he's sitting pretty."[29] As a trader, he had fared as well as could be expected: he built an animal compound in Singapore and later on Long Island. Dependent on assistants in Malaya and India, Buck imported many of the treasures of American zoo collections, including, of course, Harry. Domestic financial collapse, however, offered new opportunities.

The smash success of his *Bring 'Em Back Alive* (1932) was followed by *Wild Cargo* (1934), *Fang and Claw* (1934), and *Jungle Cavalcade* (1941). *Jacare* (1942) featured a fictional animal collector as a hero but depended on Buck's voice for the narration. By the late 1930s, Buck also began playing the animal collector hero in fictional adventure dramas. Clad in his safari suit and pith helmet, Buck played himself in *Tiger Fangs* (1943), a wartime drama about Nazi agents in the Malayan jungle. In *Africa Screams* (1949), an Abbott and Costello comedy, Buck again played himself. *Jungle Menace* (1937), a serial of fifteen installments, cast Buck as Frank Hardy—an American animal trader dressed, not surprisingly, in a safari suit and pith helmet.

Buck, more than any other animal trader, succeeded in turning the animal business into a popular culture industry. At least since Buck's childhood, Americans had delighted in stories of tropical adventure. They enjoyed accounts of safaris and hunting expeditions, like Theodore Roosevelt's *African Game Trails: An Account of the African Wanderings of an American Hunter-Naturalist* (1910). Pulp fiction books and magazine stories that described jungles as savage battlegrounds where predators stalked prey were even more popular. Novels like Edgar Rice Burroughs' best-selling *Tarzan of the Apes* placed stories of survival of the fittest into a distinctly colonial setting. Tarzan, after all, might have been the savage king of the apes, but he was also the lost Lord Greystoke. Typically in such stories, white adventurers and colonists represented the forces of civilization and progress. In *Tarzan,* apes and natives peoples were both savages and primitives.[30] The young Buck must also have enjoyed such stories. In later years, reflecting on his own jungle business, he lingered in gory detail on jungle violence and cast himself in the role of the manly, civilized white hero.

On the back of his film success, Buck's empire expanded. Comic books recounted his adventures, blending his own recollections with fantastical stories of plotting criminals, maniacal maharajahs, and savage animals. Playing cards chronicled his bouts with tigers and snakes and explained the habits of savage peoples. He offered his name and adventures to Corby's whiskey, Camel cigarettes, Armour meat, Goodrich tires, Dodge cars, and other products from toothpaste to guns. With Chicago's Century of Progress Exhibition (1934) and the New York World's Fair (1939–40), he entered the exhibition business, reconstructing a Malayan jungle camp on these fairs' midways.[31]

Whether on film or in cigarette ads, Buck never stepped out character. Buck emphasized his Texas birth (in 1884), a hardscrabble birthright that distinguished him from wealthy East Coast hunters, like Roosevelt, against whom he would often contrast his own manhood. "In the steamy jungles of Malaya, or on the heat-choked dusty plains of India—any of the primitive places," Buck wrote, "the measure of a man is judged by his abilities and not by his birth."[32] Like so many other animal collectors, he described a moment of youthful bravery that guided him irreversibly toward animal collecting—in his case, a capture of a copperhead snake. By the time he was seventeen, Buck was working as a Chicago bellboy and selling vaudeville songs. Armed with poker winnings, Buck left his drama critic wife and fled for Brazil where he collected a cargo of birds. Two decades later and nearly broke, he returned to show business to dramatize the animal trade.[33]

By the time Buck turned to film, jungle pictures had become staple fare for the biggest studios, driving one film critic to sigh, "Another animal picture! Heaven help!"[34] The same year, for example, that RKO launched *Buck's Bring 'Em Back Alive,* it also released *King Kong,* the fantasy picture of an animal collector returning to New York to display his prize.[35]

"It Takes Healthy Nerves for Frank Buck to Bring 'Em Back Alive!" (1933). One of many of the advertisements that featured Frank Buck and his exploits. This particular story of wrestling a tiger trapped in a pit—however fraudulent—would feature in his films and advertisements.

Other animal traders, like Don Taylor, also abandoned the jungle for the stage set. Taylor had spent fourteen years in Burma as an animal collector before the market collapsed in the Depression. By 1934, he mourned: "you can't give a tiger away nowadays or sell an elephant for enough to pay his fare from his native Burmese cane brakes." Taylor turned to film, serving as an advisor for the Warner Brothers' picture *Mandalay*.[36]

Buck repeated, and then embellished, his stories of animal capture to turn the mundane reality of the buying and selling of animals into adventures among wild animals and primitive peoples. As a trader, at most a transient visitor to the jungle, he relied on assistants, primarily Ali, a Moslem Malay, and Lal, a half-Bengali and half-Nepali Hindu. They provided multilingual help as he visited animal markets in cities like Singapore or Calcutta and smaller towns like Makassar and Palembang.[37]

Buck preferred natives over white workers who, he insisted, demanded and deserved different conditions. "It makes a big difference," Buck noted. "I have been so used to handling natives that sometimes we forget how a boy of our own race should be treated."[38] He had worked alongside some of his native employees for years, but he still viewed them through racial lenses. "I have always employed native boys," declared Buck, and he was convinced that as native Malays or Indians they needed only meager pay. "[T]he cost of native labor in the jungle countries is very cheap" with wants of little beyond "fish, rice, and sarongs." Yet they had a natural "knowledge of animals . . . and native customs." Natives' low wages, he argued, were merely subsidiary to their natural gathering and subsistence from the jungle.

Though his highest-paid boys earned no more than 6 to 12 Straits dollars a week (most earned less than half that), they were "loyalty itself." "I have watched some of them actually weep when they saw their tuan's ship pull out of the harbor at Singapore." His wages kept them in luxury, he argued, while he was around, but as natural parts of the jungle, raising tapioca, collecting beeswax, cutting rattan, and gathering fruit while he was in America.[39] In his stories, native jungle workers and assistants loyally helped him capture the animals that, truthfully, he had merely purchased in colonial cities.[40] He played, in some sense, the role of colonial master, and they were docile, if primitive, subjects. In the encounter of animal and trader, Buck performed the role of a manly hero. So, when Buck advertised that as a heavy smoker he chose Camels, he traded on the "thrills, excitement and real danger" of capturing "savage live cargo."[41] Animals were valuable wild beasts, and natives appeared in his stories as loyal, if primitive, colonial subjects.

Buck's "Number One Boy" Ali was the most prominent character in Buck's stories as the link between Buck—the modern trader—and the mystical Orient. In Buck's narration, Ali helped his master understand both exotic races and wild animal nature. As a translator, Ali spoke English and could converse in some of the various languages of Dutch and English colonial subjects in Malaya and Indonesia. Inscrutable yet loyal, Ali offered the key to "oriental" wealth. In broken English, he pledged his dedication. He was less a worker and more a servant, awed by his master's courage. In Buck's telling, Ali said, "Tuan, there big chance of success for man like you in my country . . . I work for you always—hard—and in future until I die."[42] According to Buck, Ali provided the local knowledge of "how to deal with these primitive people," and as a translator, he also recruited labor. These were the "boys" Buck relied on as he "really penetrated into the jungles and began to know the East." To know the East was "a complete course in the customs and habits of the people and the strange ways of wild animals in their native habitats," he said.[43]

Like the animals they helped collect, Buck's assistants appeared in his stories as specimens of their race and examples of the vast differences that separated Buck's American audiences from his Asian workers. If Ali was the key to the Malayan jungle, his Indian "servant . . . teacher" Lal represented mystical India. Buck described Lal as the perfect colonial subject featuring the best traits of British India's subjects. For Buck, Lal combined the fearless and loyal qualities of the Ghurka with the spiritualism of the "Hindu." He was the perfect specimen of "loyal help" without any of the laziness that condemned the typical "Hindu boy." "There was no salaaming" or demands for "bakshish," and his "eyes looked straight into mine, frankly, bravely, as if he had nothing in his whole Hindu world to conceal from anyone—even from his god Vishnu," said Buck.[44]

For the comics version of On Jungle Trails, Buck illustrated Ali's devotion as they struggled to feed an emaciated elephant calf. The elephant

"Too Young to Feed" (ca. 1934). Frank Buck's longtime assistant Ali feeds a young orphaned elephant with a bamboo tube. Buck stands, in his typical safari suit, outside the enclosure, looking on.

climbed on Ali's back while the standing Buck demonstrated his inge-nuity in crafting a bamboo feeding tube.[45] The imagery was easy to under-stand: Ali provided strength and loyalty, even when it placed him below the animal. Buck permitted Ali little role in devising captures and rebuked any unwillingness or resistance. "You do what you are told," Buck once snapped when Ali doubted Buck could shoot the branch from under a snarling clouded leopard.[46] Yet to demonstrate the servant's loyalty to his master, Buck recounted Ali saving his life from an angry tapir. When Lal lay sick in a San Francisco hospital, it was Buck's turn to sit by his assistant's bedside. "Don't worry about it, Lal," he reassured, "your sahib's here." After he paid for Lal's operation, he proudly recalled the wonder-ment of "white men" in India who could not understand his paternalism. "There are over three hundred million other natives in India, Buck," they told him. Whether Buck wanted to admit it or not, he needed his assis-tants alive. The bond of "servant and master" was the cornerstone of animal capture.[47]

"On Jungle Trails." In the comic book version of the same event, Buck claims the role of inventor of the elephant feeding tube. Ali, now pictured in exotic garb (not working clothes), is merely the elephant's support. Frank Buck, *Classics Illustrated: On Jungle Trails* (September 1957).

Fang and Claw: New York, Semang, Singapore

It was "just another street corner brawl between a couple of jungle rough-
necks." A tiger stalked a helpless young rhino. A black leopard locked its
fangs and claws on a python. A python slipped into a rank pond unaware
of the waiting crocodile. In 1934, Buck sat with his feet up on the desk
with his signature pith helmet incongruously on his head. The Rockefeller
Center dominated the skyline out the window as the jungle suit–clad
showman chuckled and defended his latest film. *Wild Cargo* brought
bloodthirsty battles to the big screen. A black leopard crushed by the
enormous python "is likely to horrify the more squeamish," noted one
reviewer.[48] Reporters began to wonder aloud if the scenes of mortal
combat were staged.

They were, of course, and virtually everyone knew it, but Buck was presenting a version of the tropical jungle Americans relished. Buck's jungle
was profitable yet primitive, exotic, and deadly. In Los Angeles, nearly
3,000 moviegoers sat back in the plush seats of the RKO Hillstreet theater
as Buck's voice greeted them: "Frank Buck's life work is to dare death. It is
his business to penetrate the darkest depths of poisonous jungles to procure rare and dangerous beasts which fill our circuses and zoos."[49]

The opening of a Frank Buck film was itself a newsworthy event. At the
Mayfair in New York where *Bring 'Em Back Alive* premiered, the theater's
marquee became a jungle of life-size papier-mâché tigers, buffalos, crocodiles, a python, and a black leopard. Hidden noisemakers filled the heart
of Manhattan with grunts, roars, and hisses. Crowds packed the street as
the film ran continuously. In Washington, record audiences demanded
that Keith's keep the film an extra week. In Los Angeles, the RKO
Hillstreet advertised that the sultan of Johore was coming to watch his
own jungle on screen. Even in Singapore, the film drew crowds—and
headlines—when it opened at the Capital.[50] "The spell which the production casts over its audiences is a thrilling thing to witness," marveled a Los
Angeles reviewer. The fight between a leopard and python particularly
impressed him. Film reviews focused especially on such jungle brawls.
Bring 'Em Back Alive is the "best animal picture ever made," praised an
Atlanta reviewer, and the battle between a thirty-foot python and a tiger
"is the most thrilling thing of its kind ever recorded."[51]

As the camera panned across the towering trees and massive palm fronds of the Malayan jungle, Buck's solemn voice admired its "beauty and solitude" yet warned that underneath the glorious canopy there was "danger that creeps. Death that crawls." It was a crude Darwinism, but audiences loved it. Buck had maintained all of the jungle violence of *Tarzan,* updated the action with staged fights, and added in a douse of businessman bravado. As Buck crossed into Johore to begin his trapping on screen, he "left man's laws a few miles out of Singapore. Out here was only the law of fang and claw. The grim code of survival of the fittest."[52] The jungle was "unchanged by evolution" and "prehistoric." This was a "savage, fierce country." The jungle's peoples, vegetation, and animals all clashed in a primitive struggle to survive.

The struggle for survival favored the white, American human entrepreneur. Animals, Buck lectured his domestic audience, were one of many products of the jungle, commercialized by empires, that had become everyday objects in American cities. "[T]he jungle of Asia has a commercial life," he declared, "that eventually touches everyone who lives in the world." The tires made in Akron or the shoes manufactured in Massachusetts came from Malayan rubber. The glue on the postage stamp, the tapioca pudding dessert, even chewing gum gave Americans daily tastes of the jungle. A wicker chair, gunnysack, hemp rope, and a plethora of fruits, oils, and spices brought the jungle into the American home. He looked into the jungle and saw adventure and profit: "The jungle takes life—but it gives life. It is generous to those who know it."[53]

Of course, Buck wasn't alone in the jungle filming primitive splendor and savage violence. In 1927, Ernest Schroedsack joined his longtime collaborator Merien C. Cooper in filming *Chang,* the story of Kru, a poor Thai farmer struggling against the depredations of jungle animals. From its opening credits, *Chang* transported its audience back in evolutionary time to meet primitive tribal peoples "who have never seen a movie camera." They voyaged into the jungle, "a great green threatening mass of vegetation." The "Jungle" even appeared in the credits as a sinister character intent on destroying Kru, his farm, and his domestic animals. The leopard and tiger were "the Jungle's stealthy, ferocious wild beasts." The tiger was especially villainous: "Tiger—the bully of the jungle . . . Cruel . . . bloodthirsty." The natives, locked in mortal combat with leopards and tigers, set traps to catch the predators and then to kill them with spears and antiquated guns. But the jungle enjoyed its revenge when angry elephants, those "dread destroyers," trampled Kru's house.[54] By 1933, Schroedsack and Cooper followed up on their success with *Chang* and the popularity of Buck's animal capture films with their epic *King Kong*. That film featured Carl Denham, a khaki suit–wearing animal trader who evoked, for many, Frank Buck.

Kru might struggle against the savage jungle. He could never profit from it. Buck, though, played the hardscrabble white American with a thirst for adventure and a desire for wealth. His films, and related stories and novels, offered a narrative of financial success entwined like a jungle vine around tropical adventure. In his deliberate placement of jungle exoticism into lowbrow films and pulp stories, Buck presented a jungle for the common man.

Like Schroedsack and Cooper, Buck filled his jungles with animal and human criminals. His personal twist was appointing himself—the animal trapper—as the jungle's police officer. Buck protected both animals and people and placed animal and human criminals in jails—zoos for tigers, leopards, and crocodiles but real prisons for his human villains. With its criminals and heroes, the tropical jungle merged with the urban jungle of the gangster film genre.[55] In describing the "sensational" *Bring 'Em Back Alive,* one film reviewer recalled the crimes of a tiger, the "leading racketeer" of the Sumatran jungle, feared even by his "gangster cousins, the leopard and cheetah" (little mattering, of course, that there were no cheetahs in Sumatra).[56]

By the end of the 1930s, Buck consummated the marriage between gangster pulp fiction and the jungle picture in the serialized *Jungle Menace* and the coauthored pulp novel *Tim Thompson in the Jungle*. Set in the jungles and plantations of Semang, Malaya, *Jungle Menace* used a tropical setting for a very American drama of smuggling and murder. Buck battled murderous tigers and a gang of rubber smugglers, dressed in the wool suits and fedoras typical of the urban jungle. Buck's safari suit, the lush foliage, and tigers provided tropical allure. Semang was a "melting pot"—a typical metaphor for American cities—of different races and people where the "striped gangsters of the jungle" matched the human "gangsters" in their viciousness.[57]

The American trapper, played by Buck, brought civilized law and order to the jungle through the animal trade. Spliced between animal brawls, Buck rescued native humans from top-of-the-food-chain predators. Once, his assistants slipped a noose around the neck of a thrashing crocodile while Buck tied its jaws shut. "No jungle native will have to fear him again," Buck narrated. American zoos, in this telling at least, were not merely customers for Buck's prizes; they were jails for the jungle's criminals. The message was clear; Buck wasn't trapping animals purely for profit or even to test his manly courage. Rather, when he played the role of policeman, he cast himself as a civilizer, earning a profit while bringing order and progress. This filmed capture of a man-eating tiger was, for Buck, just part of the trapping job of protection, paternalism, and profit. In the film's narrative, a tiger had killed a "coolie" on a Johore plantation before it tumbled into Buck's trap. With Buck's help, "natives were free to tap their rubber trees without fear of death and for the killer . . . he was committed to a zoo to serve a life term for murder."[58] Too fantastical, too superhuman to be real, the capture was pure fraud. The tiger was already dead when Buck leapt into the muddy pit to wrestle the big cat.

As he uncovered profit in his encounters with savage animals and peoples, he civilized the jungle. His movie jungle camp supposedly near Johore offered a romantic, idealized vision of the tropics—captured, profitable, colonized. Buck was living a jungle adventure not even its official colonial masters could enjoy. The "jungle man awakes" with the barked order for a breakfast of tropical fruit, a luxury "Queen Victoria wanted from her Indian empire, but never could have." The simple Texas businessman had surpassed the queen herself. In the background animal calls echoed: a leopard's "savage snarls" and birds' "gorgeous wild sounds."[59] Buck treated this abundance as a "jungle department store," albeit a dangerous one. Amid lurking perils, Buck carried on a brisk bargaining with "jungle people" for their prizes. Meanwhile, Buck directed his "boys" as they fed captive animals, ministered to the injuries of his "Number Two Boy," and, despite the terror of his assistants, returned an escaped python to its cage.[60]

The problem was: it just wasn't true. Even as crowds flocked to the movies, few fully believed Buck's promises of authenticity. He insisted *Wild Cargo* was a genuine record of the jungle and of the animals he captured, but accusations of faking followed him across the United States and back to Singapore. The epic battles featuring tigers, leopards, snakes, crocodiles, elephants, rhinos, and native humans seemed too violent, too fortuitous to be real.

By his own admission, Buck paid his native workers extra to identify clearings in the jungle where the lighting was good enough for filming. In one clearing, an excited native promised a giant curled python. Buck hired native villagers to beat the jungle to drive a leopard that had been preying on local dogs and goats toward the waiting python. He hoped for "a rare and unique fight, a fight few men—even those who have lived all their lives in the jungle—have ever seen." With the cameras running and the sun slowly slipping behind the jungle trees, the leopard let out a "savage scream." A huge coil of the snake wrapped around his neck. The leopard fought back with claws and fangs, tearing at the snake. Three minutes later Buck had "in the can" the animal fight that helped define the jungle and its exotic animals for American moviegoers. Few probably cared that the black leopard had lost his life as an unwilling actor.[61]

Armand Denis, the young director who traveled with Buck to Singapore to film *Wild Cargo*, later recalled his disappointment with Buck's methods. Once Denis admired Buck as one of the "heroes of America" and a "tough resolute adventurer with the tropical helmet." Then Buck scoffed at Denis's delight in filming the jungle: "I intend to stay at the Raffles Hotel," said Buck firmly. Denis's disillusionment deepened when Buck staged the epic fight with the coolie-killing tiger.[62] In fact, the tiger was a mangy, domesticated specimen purchased from a local Singapore shop, not a man-eater. The tiger died during the filming, drowning in the pit during a sudden downpour. Undaunted, Buck wrestled its corpse to capture the film's most talked-about scene.

Certainly, Denis's revelations were evidence of Buck's crass nature-faking. Yet staged fights reveal just as much about how Buck depicted tropical nature. Denis provided clear evidence, but nothing that would have shocked Buck's audience. For Buck, the tropics reached their commercial potential when they were manipulated into a cinematic wildness in which animals could be forced into actions unnatural to their real characters. "Animals don't normally fight to the death," Denis laughed when Buck proposed a fight between a tiger and an orangutan. "When I'm around they do," Buck replied, and one Atlanta reviewer was delighted. The battle in *Bring 'Em Back Alive* between a thirty-foot python and a tiger "is the most thrilling thing of its kind ever recorded."[63]

Quietly, and only to his friends, Buck admitted staging jungle violence with already dead animals. Ironically, protestations of genuine filming landed Buck in trouble with British censors who accused Buck of animal cruelty in filming *Bring 'Em Back Alive*. Buck was incensed. "This is an injustice to me," he complained in 1932 to C. Emerson Brown, the Philadelphia Zoo's director. After all, the only animal he killed during filming was a charging tiger and, even then, "only to save my life." That tiger corpse, he admitted, came in handy when he filmed the signature sequence of a tiger tracking a baby elephant.[64]

The Adventurer's Club: Ceylon, New York

In 1934 a young girl named Mary L. Shike probably looked forward to each new Frank Buck film. She was already a fan of his radio program. She even sent in a Pepsodent carton to receive her official Frank Buck's Adventurers' Club handbook. Mary used its adventure map to follow the "Bring 'Em Back Alive Man" during his Pepsodent-sponsored radio program. With Singapore at the bottom of the map, the Malayan peninsula covered much of the two-page spread. On the left-hand side squeezed Ceylon—and its elephants—and Nepal—and its rhinos. Rendered in a kid-friendly map, this was Buck's view of Asia. With the sole exception of Singapore—Frank Buck's Base Headquarters—all of southern and urban Asia had been returned to savage jungles, tribal peoples, and animals.[65] "Five nights every week," Buck spun radio yarns about the tropics that, for all their fiction, had become familiar adventures to zoo visitors, film audiences, and children like Mary who sent in Pepsodent cartons.

"I'll tell you why *Bring 'Em Back Alive* is a great picture," Buck once boasted to a room full of reporters. "I'm the only maker of an animal picture who knows animals." Yet to an admiring public, the jungle picture was as much about people as animals. Even in his early and pioneering African films, Hubbard interspersed images of wildly dancing natives with scenes of wild stalking lions. As *Chang* vividly demonstrated, jungle films portrayed the tropics as a primitive jungle where man and beast clashed in a struggle for survival. Native human and animal were all part of jungle nature, and each offered an exotic savagery. Animal traders, from Hubbard to Mayer to Buck, bought and sold animals, but their depictions of the wild jungle depended as much on the display of tribal people. Buck narrated native peoples as just another part of the wild jungle. "Environment," he declared, is "responsible for the wildness of wild animals" and there was "nothing tame about . . . their 'human' neighbors."[66] Tropical wildness, in its contrast to temperate wilderness, depended on the everyday intermingling of the primitive human, the dangerous animal, and abundant vegetation.[67] Zoos may have brought tropical animals into the heart of temperate cities and jungle films imported the sights and sounds of exotic animal battles, but zoo visitors and filmgoers came to believe that the tropical and temperate were worlds apart.

"White men seldom come" to central Malaya where the Sakai, a "race as old as Malaya itself," inhabited this primitive "impenetrable jungle." On screen while Sakai tribesmen threw spears and demonstrated blowguns, Buck's voice assured that these rare prizes were almost as hard to find as tigers. They, too, were the intrepid traders' prizes and, if he couldn't sell them to zoos, he could at least display them in their wild nature on film. Buck's 1934 picture book about his capture of elephants in Ceylon similarly illustrated tribal "Veddas" and other natives hard at work building a kraal—a massive elephant trap—deep in the jungle. One photo of a jungle dweller, part of a tribe of "genuine wild men," appeared twice. First, in close-up, he demonstrated how to shoot a bow with his legs. "They have long, bushy hair and beards. Like monkeys they can use their feet almost as well as their hands," Buck explained.[68] A few pages later, Buck, his hand close to his gun, towered over the squatting tribesman. Even the youngest reader would have understood the message: Buck was lucky enough, while capturing wild beasts, to encounter the strangest, wildest savages. In the search for profit in the jungle, he dominated humans and trapped animals.

From this particular capture, he selected a few elephants to sell to zoos, but the story of the capture itself was Buck's main source of profit. The story highlighted his book *Wild Cargo* and provided copy for a Camel cigarette ad. After the elephants were safely in the kraal, Buck savored his Camel. The ad had its double meanings. It spoke of the excitement of

tropical adventure, but the cigarette also marked the lines of race that separated Buck from his workers. The American cigarette was forbidden to
his native workers. Instead, the cigarette helped Buck, the manly hero,
enjoy a puff of civilization in the heart of the savage jungle. Pepsodent,
meanwhile, promised the "strong, healthy body" the jungle businessmen
required, highlighting the distinct needs of the white adventurer (who
needed clean teeth). All those products Buck endorsed helped define the
racial differences that divided Buck and his native workers.

On the back page of Mary's handbook, Buck appeared reading an ad
for Pepsodent in a Malay newspaper. The ad featured a beautiful white
woman and her shiny smile. Even in the jungle, the image suggested, Buck
was drawn to an alluring white woman. Native women, by contrast, were
just jungle inhabitants. Buck relied on his assistants and admired—and
maybe was jealous of—their skills in animal tracking, but he insisted on
maintaining the segregation of races. He was the white businessman pen-
etrating a tropical jungle in search of profit. On screen, he used images of
native women to accentuate the differences separating the white trader
from his native employees. The press guide for *Bring 'Em Back Alive*
offered titillating pictures of bare-breasted women. Like contemporaneous pictures in *National Geographic,* this nudity was described as
merely natural. On screen as if he were gazing at a naked ape, Buck was
deliberately oblivious to their beauty in contrast to the bewitched Ali.
Buck's sexual disinterest highlighted how distinctions among people, in
the context of the animal trade, took on the status of species and taxonomic hierarchy.[69]

Buck carefully maintained the segregation of human races, partly by
blurring the distinctions between native humans and tropical animals.
Both were, in his telling, fascinating creatures of the wild jungle. Animals,
of course, couldn't speak, even when they played gangster roles in Buck's
films. Less obviously, "picturesque" natives were just as quiet. Buck narrated his films, but his native assistants remained as silent as animals.
Buck described even the multilingual Ali as silent, more prone to animal
gestures ("a wave of one lithe brown hand, a toss of his little head . . .")
than language. In his writings, Buck's assistants speak only when their
words evidence loyalty, tradition, dependency, and natural, primitive
knowledge. And, if they spoke, it was always in broken English.[70]

"Bring 'Em Back Alive." This film advertisement
featured several of the clearly staged but
dramatic jungle brawls that headlined the
1932 film. A local woman, one of the exotic
attractions of the jungle and unashamed of her
nakedness, appears below Buck's portrait.

When Buck began planning his Jungleland concession at the New York World's Fair from 1939 to 1940, he knew that fans like Mary Shike wanted to see his loyal assistants as much as the animals. He couldn't stage animal fights in the fair's reconstructed jungle camp, but he certainly could show exotic humans.[71] Jungleland featured Indian monkeys, souvenir lassos, satay cooks, and his Malay "boys." This reconstructed jungle camp, built exclusively by Malay labor, evoked jungle adventure and the thrills of colonized peoples as ready "to announce tiffin" as to attack savagely. "Visitors," he promised, "will likely pinch themselves to make sure they are not in . . . the jungles of Malaya." Native assistants were as essential to the display of the jungle as tigers or luxuriant foliage. "Elephants appear with a native howdah loaded with people, rambling through the jungle as in India. It does not take much imagination to feel that at any moment . . . a native arrow may whistle past your ear," read the ads. Arrows and elephants notwithstanding, Malay workers did double duty at the fair. They were exciting displays and cheap labor who constructed Buck's midway concession.[72]

"This time," he boasted, "I brought the whole jungle back alive." Ali, Ahmed (his "Number Two Boy"), and four other Malays appeared as exotic curiosities plucked from European colonies—not as workers whose long working days might concern the unions that otherwise kept close watch over fair work. In Jungleland, Buck's exciting vision of the animal trade was on display and for sale—from trinkets to pith helmets. In a New York park, hard-bargaining merchants, recalcitrant employees, or resistant animals could not threaten Buck's domain. Those Chinese merchants whose stocks populated American zoos were absent, but Buck never mentioned them and no one complained.[73] As he surveyed his little empire, he felt on top of the world. "I was now Frank Buck, who had achieved fame in the jungles, in motion pictures . . . and in the show world," he recalled. Jungleland was Buck's great triumph as a manly American hero.

He didn't keep many souvenirs from his Jungleland extravaganza, though its success was a fittingly glorious ending for his autobiography. As the fair wound to its close, the nation's attention began to shift from jungle adventure to world war. Buck shuttered the exhibit, sent his workers home, and sold off the animals. Traveling back to his California

"Jungleland" (1940). Buck, the manly adventurer, placed himself as a key star of his Jungleland concession, surrounded by elephants and other exotic animal attractions.

mansion, he brought along a pair of candlesticks, props from the fair. Forged from cheap metal and plaster, the candlesticks provided a garish exoticism to Jungleland's displays. With their miniature white elephant tusks, they evoked faraway Asia. Perched on their base sits a primate wearing a vaguely Thai golden conical hat and its hands are tilted up in a *namaste* greeting. The hirsute primate is neither wholly monkey, ape, nor human. It's just a jungle native—exotic, fascinating, and on display. No wonder Buck kept it.

On the radio, at the fair, on the flickering movie screen, the animal trade became high adventure rather than mundane transactions. Buck spun his tales and told his lies to avid American consumers. Americans wanted to imagine colonized jungles just like those mapped in the Adventurers' Club handbook, rebuilt on a fair midway, or shrouded in Camel cigarette smoke. Here was an exciting and profitable empire: the natives were primitive yet obedient. The animals, even the most dangerous and the most villainous, fell to the trappers' bravery. The jungle at the fair, at the movies, and in Mary's handbook was lush with life and ripe for business.

Buck had perfected empire. On his radio program, Buck narrated a "savage, mysterious place" in which Harry's capture became a parable of

"Jungleland's Candlesticks" (1939–1940). These souvenirs kept by Buck from his Jungleland concession showed the primate partly as an exotic savage and partly as a performing ape.

adventure, risky business, and colonial order. Buck's story described his ingenuity, his managing of native labor, and the savage anger of the rhino resisting his capture. In "the rhino country," Buck assembled his " 'boys'," native workers "born and bred in the jungle who knew the trails and the lay of the land." The rhino, Buck told his radio audience, charged. "He charged again . . . ! More furious—more savage!" His eyes were red with fury. Even as he was forced into his bamboo cage, Harry continued to struggle, "mad, savage, angry, roaring." "Should any of you ever happen to be in St. Louis," he told Mary and the rest of his audience, "you can see this rhinoceros there in the zoo." If facing the charging Harry confirmed Buck's fearlessness, his brushing "twice a day with Pepsodent" and the animal's sale brought order to jungle savagery.

Across the midway, the Bronx Zoo, once a customer for Buck's first rhino, was struggling to compete with Buck's "complete show of the rarest animals in the world."[74] Buck, not the zoo, had won the concession to exhibit animals at the fair. Zoos, once customers, now were competitors in displaying animals—and they were losing out to midway concessions, adventure clubs, and movies. In the midst of the Depression, zoo officials knew visitors wanted more than just animals in cramped cages. They realized that if Mary was going to turn off the radio, skip the movies, and visit the zoo, then zoos needed to change.

4

THE MONKEYS' ISLAND

The New Deal Builds a Modern Zoo

I N CHICAGO, planners of the new Brookfield Zoo promised that a
towering island of fake stone would rise from an artificial pond.
Monkeys, crowded onto the island, would scamper, fight, and entertain.
Monkey islands were new to Chicago and to America in the midst of the
prosperity of 1927. With the idea of an economic collapse unimaginable,
the Brookfield Zoo promised, at last, to give Chicago a zoo to rival the
finest zoos in the country. The planned zoo depended not on the support
of local government but on the munificence of the city's moneyed elites.
Even the land for the new Brookfield Zoo was a gift from Edith Rockefeller
McCormick, one of the city's wealthiest. The new zoo looked to the past
as it sought new designs for the future; its centerpiece monkey island borrowed the hallmark of zoo design pioneered by the Hagenbecks, the
leading German animal traders. Decades before Chicago began construction, the Hagenbecks' Stellingen Zoo had featured a monkey island. The
Stellingen Zoo had been badly depleted during World War I, and the firm
was only beginning to recover by the time Chicago's elites were considering their new zoo.[1]

Just as the Brookfield Zoo was rising from open fields, the economy collapsed. Fortunes dissipated, unemployment soared, and suddenly monkey
islands proliferated in American zoos. During the Great Depression,

Americans visited zoos more than ever before. Zoos went from bust to boom as planners of the New Deal, a suite of federal programs designed to counteract the effects of the Depression, recognized zoos as the ideal recipients of urban relief. Jobs at zoos put unemployed laborers and even artists to work on projects that could excite urban populations with little money to spare. Nothing could advertise New Deal relief programs better than brand-new animal houses, open enclosures, and towering monkey islands.

Yet life wasn't easy on Depression-era monkey islands. Eager for little rivalries, brawls, and comic belly flops into moats, zoos crammed far too many monkeys onto their islands. Animals fought. Some died. Others drowned. When winter came, zookeepers herded monkeys into cramped indoor cages and mortality rates were, not surprisingly, high. At least one zoo even resorted to putting only male animals on display, but sex segregation only increased the violence. Public monkey mating was just too much to show visiting children, but monkey fights made for good tales, cute versions of animal battles visitors had grown accustomed to from the movies. To visitors, especially kids, fights over dominance on the crowded monkey islands seemed like games of "capture the castle." In the midst of the Great Depression, zoos presented heartwarming stories that turned monkeys into miniature children—only funnier, cuter, and with fur and tails. Of course, for the monkeys, squabbles were hardly cute.

Partly because of New Deal programs and partly because zoo officials were anxious to meet the competition from filmmakers like Frank Buck, zoos changed in the Great Depression. With them, Americans' relationships with the animals on the other side of the enclosure also changed. At the outset of the Great Depression, zoos knew evolution was desperately needed. Sure, animals were aging and cages were outdated, but zoo displays looked even worse when compared to the lush scenery and prowling beasts of jungle films. As they surveyed crumbling exhibits and measured dwindling attendance, zoo officials recognized they needed to compete with the popular culture of the animal trade. Like the prowling tiger, jungle adventurers were encroaching ever closer on zoos' territory. Buck offered radio programs, films, comics, pulp fiction, toys, and even midway concessions. He (and other jungle filmmakers) promised exciting stories and close encounters with jungle villains and exotic peoples. As zoos feverishly rebuilt their enclosures with federal dollars, they laid a new foundation for the future based on peanuts and candy. Kids, they invited, come to the zoo, laugh at the monkeys, and make friends with the animals. Let your imagination about the tropical world run wild, and despite the signs, please feed the animals.

Rusting and Rotting: Philadelphia, San Diego, Toledo

In 1932 in Philadelphia, a young boy offered to house the zoo's lions or tigers. The big cats, he promised, would find "a warm part in our hearts, as our family loves animals." His father was a butcher, so, surely, feeding the cats would be easy. Other children volunteered to board monkeys, raccoons, and rare birds—just until the zoo could get back on its feet.[2] They emptied their piggy banks to help pay for the zoo's "hungry hordes," and a lemonade stand donated its profits. Director C. Emerson Brown thought the children had "done fine." The zoo was $89.13 richer, but it needed another $100,000 to keep its animals fed. The city, facing its own financial straits, had recently halved its annual appropriation to the zoo. Doubling the admission charge merely reduced attendance by almost 100,000 people a year from a pre-crash total of almost half a million.[3] "We are lost," Brown warned. Gate receipts, in fact, had fallen off more than $12,000 in the first year of the Depression, about a fifth of the total budget deficit. Without city aid, the rarest animals would have to be loaned to other zoos or sold. Animals with "no market for them" would be killed.[4]

Zoos, like so much else in the nation, had gone from boom to bust. The 1920s had been a happy time for animal traders and for the zoos to which they sold animals. Existing zoos expanded their collections, and in both Chicago and Detroit, city boosters launched plans to replace cramped downtown zoos with sprawling parks at the cities' limits. Smaller cities, like Evansville, Indiana, and San Antonio, Texas, eager to prove their civic importance, planned to enlarge menageries into proper zoos. Regardless of city size, boosters imagined similar new designs for their zoos. Open enclosures became the hallmark of new plans while barred cages represented the stodgy seriousness of an older generation. Yet promised new zoos, even in Chicago and Detroit, were running out of time before the Depression.

Less than a decade separated dreams of grand new zoos and the real chance the nation's most famous zoos would close their gates forever. When the stock market crashed on October 29, 1929, a disastrous chain of events began that threatened the very future of American zoos. Some major donors lost fortunes; even those that hadn't suffered as badly in the stock market were now ever more reluctant to donate. By the time that young Philadelphia boy offered to house big cats, the unemployment rate had rocketed to 25 percent. Visitors just didn't have spare change to pay admission fees, especially if the exhibits seemed decrepit. In past hard times city governments had pitched in when private donations failed, but now they too suffered. With industrial production and tax receipts falling, city governments could no longer serve as zoos' last line of defense against collapse.[5]

In San Diego, only a quiet tax revolt stalled the sale of the zoo's entire animal collection. The brainchild of a local doctor, Harry Wegeforth, this privately held zoo depended on local philanthropy and attendance fees. The zoo was enormously popular, the finest attraction in a growing city. Yet by 1932, it was badly in debt and in arrears on its taxes. The projected animal sale pitted the state tax assessor against the San Diego city council, well aware of the zoo's local importance.

The Great Depression witnessed a unique form of resistance to eviction, but the strangest "rent strike" was at the San Diego Zoo. Often during a rent strike supporters prevented an eviction by moving furniture right back into the apartment of tenants no longer able or willing to pay their rent. Sometimes such actions were spontaneous sympathy; often they were actions organized by labor or radical organizations. At the San Diego Zoo, resistance to the sale of the zoo's animals was deliberate. The city of San Diego even dispatched police to the zoo to prevent state officials from removing any of the animals. "You can't make any mistake if you buy one of these fine animals," the tax assessor pleaded, as the zoo's animals were sent to auction. The audience remained steadfastly silent.[6] Even so, Wegeforth reluctantly put 600 animals up for sale just to feed the remaining 1,500. Keepers were laid off, and the zoo made plans to free its elk herd in the nearby mountains. "[S]ixteen years spent in making the San Diego Zoo one of the largest and best in the world, will be wiped out in a few months," Wegeforth worried.[7]

Without ever looking at their account books, visitors could sense zoos were rotting at the onset of the Great Depression. The animals appeared ragged. The buildings smelled. Their walls felt slimy, rotting to the touch. "[W]e had a terrible zoo," admitted Frank L. Skeldon, the Toledo Zoo director. "The plant was miserably inadequate, and animals were continually escaping from the place."[8] Repairs had been delayed so long that cages were collapsing, bars were rusted, and ceilings threatened to fall. The aging Philadelphia's Zoo and its animals were also "worn-out." The crisis was just as bad across the state in Pittsburgh where the local Humane Society threatened to close the Highland Park Zoo after newspapers printed pictures of mangy polar bears, a blind monkey, and a stork with a broken wing.[9] In St. Paul's Como Park Zoo, tropical animals had little protection from harsh winters. The zoo even admitted that of the around 100 monkeys in its cramped winter quarters, only five or six might survive.[10]

At first, help was hard to find. By the beginning of the Great Depression, many zoos' original patrons had already died. Ironically, their bequests in the 1920s had helped fuel the animal buying spree, but now they couldn't help failing zoos. Other elites were neither begging for relief nor selling apples in the streets, but their generosity to zoos waned. City governments could not fill the gap. In Philadelphia, it was the "duty of 'rich men'" to maintain their zoo, insisted Edwin R. Cox, the city council president. "When I see a child crying," he declared, "I am inclined to forget the zoo."[11] Unfortunately, the city's wealthy also forgot. Donations plummeted, and the zoological society's membership rolls shrank. "It's up to the city," concluded Williams Cadwalader, the society's president.[12] For $150, the zoo sold a pair of lion cubs to an Atlantic City boardwalk concession just to purchase feed for their parents.[13] Every month, the zoo's animals gobbled 5,400 pounds of fish and five tons of meat, three and half tons of cracked corn, and countless barrels of fresh vegetables. In hard times, it was a staggering burden.[14]

Hundreds of Worthy Unemployed: Philadelphia, Toledo

At first, raising admission fees seemed the only answer, but it was the wrong one. Higher charges merely brought the zoo closer to bankruptcy when Philadelphians simply stopped going to the zoo, now a costlier day out in hard times. Convinced by 1936 that only the public could save the zoo, the Philadelphia Zoo's management was ready for a big gamble. Zoo leaders offered to trade the authority they had jealously guarded for decades in a public campaign for a free, modern zoo. The zoo sought "no gifts." Rather, it incited a "mass response" that looked a great deal like one of the many union campaigns currently sweeping through American industry.[15] At workplaces throughout the city, organizers encouraged rank and file to sign petitions for the modern, free zoo. The campaign to save the zoo became a mass social movement. The Citizens Committee for a Free and Modern Zoo had its well-known and wealthy officers, but it also listed as its members "1,092 workers in commerce and industry" and "1,428 women workers and youth workers."[16]

"ENLIST for a FREE and MODERN ZOO!" called their posters, plastered on walls and distributed on the shop floors. In July 1936, the zoo opened its doors to demonstrate the masses wanted a free zoo. The working-class public "shouted" their demand for the free and modern zoo. A mere 255,000 visitors had attended the zoo in all of 1935. One hundred thousand more than that visited in just nine days when the zoo was free. Lines formed outside the animal houses and filed past a booth inviting visitors to a mass meeting. Children had to perch on parents' shoulders just to glimpse into the gorilla's cage.[17]

These cages had to disappear if the zoo was going to become modern and free. "Iron bars a prison make," the zoo said, but in a free "people's zoo" the animals roamed cage-less enclosures. The mass movement was going to save the zoo's finances by making it free for humans and open for animals. If the masses supported the zoo, pressure might build upon the city to help rebuild a modern zoo. Eschewing the austere rows of cages, the new zoo would let its animals roam in open enclosures that gave the "illusion of animals being free." An impassable but disguised moat would protect the visitor from the "lion or elephant, bear or tiger."[18] Visitors, in turn, were free to imagine that they were looking into the heart of the jungle.

The jungle was coming to Philadelphia through a "modern, free zoo." Philadelphia's campaign for a modern zoo revived elements of an older German zoo model of animal display. The cage, once hailed by American

"Enlist for a Free Modern Zoo." In 1936, crowds gathered for a free week at the Philadelphia Zoo as part of the zoo's effort to prove that the modern zoo should be free.

zoo directors as the best way of displaying animals as specimens for quiet contemplation, now seemed like a jail. The finest animals to visit were alive, walking, staring, and interacting, sometimes with visitors, sometimes with other animals. An animal's actions and character now seemed more important than its physical traits. Taxonomy disappeared in favor of an illusion of tropical nature open to the masses. The orderly array of animals in cages smelled (and in the case of crumbling zoos, the stench was overwhelming) too much like an older, elite zoology. The model of open enclosures and moats evoked the tropical adventure so popular in films.

The model of the Hagenbeck zoo, the open warehouse for animals on sale, influenced the relief of crumbling American zoos in ways unfathomable to the German businessmen. In Philadelphia, organizers noted the finest zoos in the country were publicly supported. St. Louis's zoo, for example, had been founded through the energies of urban elites like the brewer Adolphus Busch III, but it had remained free and tax-supported. In 1934, 2.5 million people visited the zoo, about three times the city's population. In Philadelphia, by contrast, attendance in the same year had plummeted to fewer than 200,000, a mere 12 percent of the city's popula-tion.[19] The Philadelphia Zoo's modernization plan was ambitious, but there was a problem. A cash-strapped city could not act alone. With the help of the federal government, however, ragged zoos could be rebuilt.

The mass movement nurtured in Philadelphia came to fruition first in Toledo. When zoo directors applauded the ability of federal unemploy-ment relief funds to create a modern zoo, they typically praised the Toledo Zoo. As a city, Toledo suffered badly in the Great Depression. A local auto factory closed, and the city's largest bank folded. In 1934, workers at the Electric Auto-Lite factory walked out in a strike that foretold the work-place upheaval of the Great Depression across the country. Strikers bat-tled the National Guard.[20] Like the rest of the city, the Toledo Zoo had suffered with the coming of the Depression. After the stock market crash, the zoo was reduced to paying its workers with city script, effectively worthless coupons issued by a bankrupt city. The situation was replicated elsewhere. Keepers and their animals bore the first brunt of the Depression. When the elite patrons of zoological societies offered little help, the deficit came out of keepers' pockets. In the Bronx, employees faced 8 percent salary reductions in 1932, but the savings were still not enough for the struggling zoo. Two years later, the zoo furloughed keepers without pay.[21] The Philadelphia Zoo cut its workforce and wages to a "minimum" at the same time it sold its lion cubs. Hourly workers went on half time while those on salary swallowed 10 percent cuts.[22]

In Toledo, as picket lines snaked around closed factories, a new zoo emerged literally on the foundations of a suffering industrial city. When the federal government stepped in, the Toledo Zoo soon hired new workers—1,300 of them, in fact. They excavated foundations, hauled stones, and constructed new decorations. Arthur Cox was slowly dying from silicosis contracted on a job that no longer existed. In his new job at the zoo he carved stone animals, like an elephant just the right height for clambering children. Even when jobs could be done better or faster with machines, the zoo chose human hands. Artisans reconstructed the zoo with open enclosures and hewed a monkey mountain from stones dredged from canals and from abandoned buildings. They recycled a rotting industrial city into a modern zoo by simulating the tropical wild. "Salvaged workmen worked with salvaged materials," boasted officials. A derelict railroad boxcar became the entrance ceiling. Stones and timbers came from railroad shops and radiators from a closed lumber company.[23]

Rusted buildings became airy houses and open enclosures. It was all part of a new grand bargain: zoos would pass out of the hands of older elites into the embrace of public governance. Supported by more than $1,400,000 in relief funds, "hundreds of worthy unemployed Toledo workers" used the historic remnants of downtown factories and buildings to transform "a formerly obscure zoo into the third or fourth largest zoological park in America." Attendance rose dramatically, and the city had found its new prime attraction. By 1937, 1.5 million visitors gazed at 1,865 animals.[24] With attendance higher than at the venerable Philadelphia Zoo, Toledo had become the national example of how to build a new "ideal" zoo. The modern zoo in Toledo that rose on a foundation of industrial detritus was less the ambition of a conservationist elite than a public institution, aimed especially at entertaining children.

In 1938, the mayor of Columbus, Ohio, along with members of a newly organized zoological society paid their way to Toledo and returned promising "elephants and gnus, orangutans and giraffes, tigers, leopards, and boa constructors [sic]" for "good old Columbus." Soon, seventy-five workers paid by the Works Projects Administration (WPA) labored on a large animal building and new hippo pools.[25] The WPA was the largest of a suite of New Deal agencies that provided relief work to unemployed workers. More than three million unemployed men and women went back to work rebuilding public buildings, roads, bridges, and zoos. In a first wave of zoo reconstruction, nineteen of the nation's ninety zoos hosted significant WPA projects.[26] In larger zoos, WPA workers salvaged collapsing buildings. In the Bronx, where most of the buildings dated to the original turn-of-the-century construction, the zoo applied for $800,000 in WPA funds to help replace a corroded aviary, a patched and tarred yak house, and a glass skylight in the bird house that threatened to collapse on visitors.[27]

In smaller cities like Scranton, Pennsylvania, New Deal relief workers expanded zoos that before the Great Depression could hardly have been considered more than local menageries. Children's donations helped the Scranton Zoo purchase an elephant, Tillie, in 1935. With WPA support, the zoo finally constructed a new paddock for Tillie, and she was released from the leg chains that bound her.[28] In New Orleans, the Audubon Zoo had suffered in the 1920s. Donations plummeted during the "black years after 1929." Benches collapsed. Cages rusted and the animals developed a "moth-eaten air as if they realized that none of their own particular eccentricities could any longer be ministered to," according to local reports. WPA workers demolished collapsing buildings in order to construct a new monkey island, flying cage, and large animal house.[29]

Back in 1924, when zoo directors gathered to discuss Chicago's plans for its new Brookfield Zoo, many "ridiculed" the idea of the "barless zoo." William T. Hornaday, recently retired from the Bronx Zoo, still resisted "those accursed barless dens." Even so, by the Great Depression, he recognized that "Zoological Park superintendents and directors . . . seem to have gone crazy over that accursed German-born idea." Zoo directors, eager for relief funds, sent their experts to Europe to examine how its zoos produced such realistic rock sculptures.[30] WPA architecture and its claim to modernity offered a sensual contrast with the smelly zoos of the past. "It's slum clearance of a kind," noted one observer.[31] At the National Zoo, open skylights and ventilation systems meant the rank odors of the lion house "will soon be a thing of the past." With New Deal support, the zoo promised an air conditioning system. "[A]nimals should be seen not smelled," declared William Mann.[32]

Instead of the dour biological ordering of specimens by genus and species, the WPA promised Little Rock "new-fangled fenceless enclosures." Nothing but "Arkansas fresh air" and rocky moats would separate visitors from deer, elk, camels, llamas, and zebras.[33] Visitors could see, not smell, animals in settings just real enough that visitors could imagine them prowling, pouncing, brawling, or hunting—a little like in those wildlife films. In San Antonio, WPA workers designed a new "African Panorama." The "WPA Army of Workmen" also brought the "modernization idea" to Buffalo. They dug moats around a new African veldt and swamp.[34] In Evansville, Indiana, the zoo with its dirt roads had rarely been accessible in winter or wet weather. With WPA labor, the zoo constructed an "African Veldt." Visitors in all weather could gaze across a barless lion enclosure at freely roaming African animals. The small zoo's collection had grown as well. Visitors could now enjoy baboons, camels, chimpanzees, an elephant, monkeys, a Galapagos tortoise, zebras, and a tiger.[35]

In San Francisco, the reconstruction of the Fleishhacker Zoo grew to become the largest nondefense WPA project in the far West. Begun in 1935, it employed an average of 500 men at the cost of $2 million. The work accomplished was staggering: ten new buildings and 150,000 square feet of animal enclosures decorated with 205,000 square feet of artificial rock.[36] Lions and tigers soon roamed "as apparently free as in their native haunts. In Jackson, Mississippi, zoo officials traveled to St. Louis and Cincinnati to view open enclosures. Soon, their newly reconstructed zoo received a honey bear, a lion, and a rhesus monkey. The zoo also began building a tiger house.[37]

In Detroit, the WPA estimated attendance would rise from two million to a record three million after its improvements. Detroit had begun construction on a large and city-supported zoo in the flush years of the 1920s to augment the miniscule Belle Isle zoo. By 1931, however, factories closed, city support dried up, and the zoo remained only one-third completed. "But the big zoo functioned" because of an emergency federal relief grant. As winter 1934 turned to spring, visitors enjoyed a new "African Swamp" flapping with flamingos. A Siberian tiger prowled its barless enclosure. Attendance at modernized zoos jumped dramatically. In 1937 in Cincinnati, a record 713,017 visitors came to see recently purchased animals housed in a new African veldt and barless grottos.[38] As new WPA projects opened, 1934 was the busiest year ever for American zoos. Twenty million visitors visited 130 zoos.[39] More than ever before, those visitors were kids. Children wanted to do more than look; they wanted to toss treats. Animals may have been moved to more naturalistic open enclosures and airy monkey islands, but they roamed "free" in a shower of peanuts.

All the Little Publics: Toledo, Detroit

At the Toledo Zoo, the keepers made the new workers hold the snakes. As the snakes were being cleaned and fed, they urinated on the hands of their frightened handlers. For the keepers who had been working with snakes for years, this was a good joke. For the recent hires, it was a rude, if harmless, welcome to the wild world of captive animals. Behind this hazing lay the new reality of American zoos: the Great Depression was their heyday. "It is a peculiar thing," the Toledo Zoo noted, "that through the depression the Zoological Park has realized a dream that probably would not have been achieved in the next 100 years."[40]

As factories closed and their very stones salvaged for zoo houses, the unemployed became zoo workers. At most zoos, WPA workers rapidly outnumbered permanent employees (even if most WPA workers were less immediately involved in animal care). "Most of the labor now employed," reported the Buffalo Zoo, was unskilled and paid by the WPA through the city's harsh winter.[41] Almost $1 million of relief money raised an "army of workmen"; 827 unemployed men turned an "old, dilapidated and unsightly place" into the city's most crowded attraction.[42] With all its construction, the National Zoo also needed new keepers. The zoo hoped to increase its permanent staff from 98 to 115. In Toledo, the zoo interviewed applicants eager to be keepers in the new birdhouse, but none had any experience in caring for tropical animals. With new shipments of animals on their way, the zoo was forced to turn the hiring of keepers over to the city manager and the civil service commission.[43]

In Philadelphia, the zoo had been willing to surrender authority to the city and the public. Instead, the WPA, more than any zoo director, realized the modern zoo. The WPA praised the work of 500 relief workers who turned the Detroit Zoo into "The Most Modern in the World." No bars or cages obscured the view of animals in "surroundings closely approximating their native haunts."[44] The lions "glare hungrily" at the crowds across a thirty-foot moat. WPA officials described the new displays as the "counterpart of their native jungle." Even mistakes could become WPA publicity. Hundreds of visitors in Detroit watched as Nero killed Menelik, a fellow lion. Sure, it was a tragedy, but to WPA officials their battle seemed worthy of a Frank Buck film.[45] In WPA hands, the zoo had become an "institution for the advancement of public pleasure." As science lessons retreated to the past, the WPA completed Cleveland's new monkey island—"the modern barless den type."[46]

Monkey islands, the WPA assured, "sooner or later will be made a part of every well equipped zoo."[47] For a few critics, the monkey islands that proliferated at smaller zoos became emblems of New Deal waste. Critics used the $65,000 monkey island in Little Rock, for example, to cast New Dealers as pampered monkeys. In a deriding cartoon, the Little Rock monkeys perched on two fake trees labeled *House* and *Senate*. "I'm against a capitalist government; what we need is shorter trees," declared one monkey.[48] More typically, monkey islands were advertisements for the value of New Deal projects. What could demonstrate the role of relief more effectively than animal entertainment? The worthy unemployed were back at work in the heart of the city crafting a new zoo for the amazement of children. When William Schmuhl, the WPA's Ohio director, praised the Toledo Zoo, he answered back to critics of the New Deal: "Were the WPA workers who erected this building . . . leaning on their shovels?" Committed workers had performed the "miracle" of modernization.[49] Stone blocks from the silted bottoms of canals had become a monkey island. In St. Paul, the WPA boasted about "one of the finest monkey islands in the country." With waterfalls tumbling into the moat, "it has become the center of attraction for people from many miles around."[50]

Behind the excitement of openings and new buildings, there was a quiet transformation. The influence of private, insular zoological societies waned. Even in the Bronx, with its well-established zoological society, keepers' salaries were now pegged to city parks wages. That change also transformed the relationships between animals, keepers, the public, and children. Keepers, once kept distant from visitors by military-style uniforms and constant chiding by zoo management, were now encouraged to tell stories about their animals, especially to children. Zoos increasingly cast keepers less as militarized guards than as friends to youthful visitors and to animals. In Milwaukee, the WPA trained new guides and older keepers in "animal lore." "The object," noted the local WPA supervisor, is "to gratify the normal interest every child has in animals."[51] In Memphis, 25,000 people visited for the opening of the monkey island. "There are no better children than those in Memphis," declared the city's mayor, "and they are entitled to a real zoo."[52] In New Orleans, the WPA boasted that "thousands of children" could enjoy a "tiger still violent enough in captivity to show a terrifying array of teeth."[53]

In Pittsburgh, the new monkey island crammed seventy-five monkeys onto a mountain seventy-five feet long and forty feet wide. These common rhesus monkeys were "lively and playful," and their "monkeyshines" delighted the city's children.[54] In Columbus, WPA writers introduced the monkeys and their new island. They "are greedy, curious, and jealous. They squeal, whimper, cry, croon, and whistle. They sit and brood, or tear about so madly." And they were just like us humans: "They sulk, they spank their children, they shout and throw things when they are angry. They are 'our poor relations.'" No wonder the children loved seeing them.[55] In Detroit, the WPA promised that 200 rhesus monkeys, which unlike other monkeys seemed to enjoy the water, would romp on the island and leap from miniature diving boards and boats. The new monkey island was a treat for "Mr. and Mrs. Public and all the little Publics."[56]

Begging for Peanuts: Cleveland, Tulsa

In 1936, keepers released the monkeys at the Cleveland Zoo at exactly three o'clock. Fifty thousand visitors already circled the empty island, and even more were trapped in a traffic jam outside the zoo gates. For days the city, spurred on by both zoo and WPA publicity, wondered what would happen when the monkeys were released. First, though, came the ceremonies. The WPA district director praised the monkey island, built entirely by WPA labor at the cost of $35,000. A representative from the May Company, one of the city's largest department stores, boasted about the 150 monkeys the store had donated and the new world they would inhabit. He described the new island, 100 by 150 feet wide and reaching 15 feet above the water line. Despite the island's modest size, the store pledged another 150 monkeys for the following year.[57]

"Then the action began." A tangle of arms, legs, and tails, the monkeys dashed—bewildered—to the top of the mountain. Some tried to run back inside their storage room, only to be shooed outdoors again by keepers with brooms. In shock and confusion, they huddled on the top of the island. They began inspecting the trees and waterfalls only after the "historical day for the zoo" ended. Despite their traumatic first day, those bewildered monkeys registered Cleveland as "one of the finest examples of modern zoo construction." Soon, other "zoo men" traveled to Cleveland to view the new monkey island. They wanted their own islands, and the WPA promised the money to build them.[58] Ironically, the Brookfield Zoo, the pioneer advocate for open enclosures, finally opened its own monkey island in 1936, after many smaller zoos had constructed theirs. In Chicago, relief workers sketched advertisements and sculpted animals, but the zoo ultimately constructed its island from its own limited funds.[59]

By the time Brookfield opened its island, the WPA had already brought monkey islands to Lansing, Omaha, Memphis, San Francisco, Detroit, St. Paul, Buffalo, New Orleans, Little Rock, and Pittsburgh. In Lansing, thousands of visitors gathered daily to watch more than forty monkeys leap from rocky ledges and trapeze bars and, occasionally, fall into the watery moat. Until they fixed the top of the "mountain," a few monkeys had also figured out how to leap across the moat for the "freedom" of the zoo grounds.[60] The new Detroit zoo featured an "ultra modern monkey island." When boosters first imagined a publicly supported zoo in Detroit, they had traveled to Germany to admire the Stellingen Zoo. Similarly, Pittsburgh, once a zoo of "undernourished, sickly exhibits," sought to replace rank cages with a "modern zoo," and it turned to a former Hagenbeck animal trainer as its new superintendent. Just as in Detroit, Arnold Schaumann's dream of a monkey island as the new zoo's central attraction waited for New Deal relief.[61]

When the monkeys arrived, visitors, reporters, and zoo officials turned fights and falls into children's humor. Such was the fate of Sammy Rhesus and Billy Ringtail. Tulsa opened its monkey island in 1932 in the first wave of New Deal zoo reconstruction. The monkeys crowding the island were immediately popular, not just with visitors but with newspaper readers who read the "humanizing" accounts of King Jumbo and his monkey subjects. The columnist George Ketcham, using the pen name "Sim Simian," provided updates on the day's wrestling, fighting, and biting. His stories were light and the tone was laughing, the perfect way of introducing the city, and especially its children, to the pleasures of the new, modern zoo. Far from wondering if the island's daily drama was evidence of stressful living, Sim Simian labeled a rhesus monkey named Sammy as "the island bad boy." Once Sammy had his tail pulled by Tim and Tiny, two other "pranksters." Sammy was soon "nursing a well-thumped skull." Billy Ringtail was the next victim. This time Tiny poked King Jumbo with a sharp stick and laid the blame on poor Billy—"Oh, oh, what a calamity." The "irate monarch" shoved Billy, "sore and bedraggled," down into the moat.[62]

Monkey Island, Mohawk Park, Tulsa, Okla.

"Monkey Island" (ca. 1942). Monkeys at the Tulsa Zoo crowded onto a typical New Deal–financed monkey island.

Monkeys like King Jumbo, Sammy, and Billy were ideal animals to help zoos adapt to their new juvenile audiences. Monkeys were familiar enough that they looked human but small enough to appear like dolls with hair. Their frequent fights on overcrowded monkey islands easily appeared as comic versions of childish human squabbles, not fierce struggles for dominance. When they cradled or nursed their young, monkey tenderness was just as easy to recognize.

Monkeys, once rare, were also now easily accessible in pet shops. In New York, Henry Trefflich sold a broad range of animals, but his specialty was monkeys, and he marketed them to both zoos and everyday pet lovers.[63] For many, the childlike monkeys were irresistible, even if they inevitably proved troublesome pets. The National Zoo was so inundated with letters asking for advice about caring for monkeys that by the end of the 1930s, it developed a form letter replying to those considering a monkey pet or to current owners now overwhelmed by the responsibility.[64]

Pets and comic characters, monkeys had become familiar. For zoos focusing ever more on attracting children, monkeys were the perfect mascot. Monkey islands could also be a way of depicting natural drama in ways that focused on the cute. Visitors, longing for cuteness, sought to interact with animals on display. Open enclosures encouraged visitors to gaze at animals, but monkey islands also invited children to toss food. When the "monkey business began" on Memphis's new island in 1936, the zoo celebrated by handing out 7,000 bags of peanuts. The nuts were destined more for the begging animals than for the mouths of visiting children.[65]

The intimacy of visitors and animals that had so disturbed the past generation of zoo directors was now the point of the zoo. When Jane and her school class went to the zoo, she and her friends headed first to the peanut vendor. Each bought a packet and, eager to feed the animals, they headed for the elephants. Sundra stretched her trunk nearer the children when the keeper gently reminded them that feeding the animals was just not allowed. Jane looked down at her peanuts—far too many for her to eat alone—and back up at the keeper. "Oh, see her beg," she said. "She wants them so much." "Well, give them to her," the keeper relented immediately. Sundra "knows you are a good friend." So it went for the rest of the day, as the children happily walked from elephants to monkeys. The keeper even made the baboon shake hands before John gave it his apple. Ned, never the best-behaved pupil, was busy admiring the monkeys. "If I could be an animal that lives in the Zoo," Ned wrote when he got back to school, "I'd be some kind of monkey, and act like a clown." In between jumping and swinging, "I'd beg for food and peanuts."[66]

John, Jane, and Ned were just characters in a children's book, one of many that turned a day at the zoo into child's play. As zoos became popular settings for children's books, they revealed how zoos had become havens for children who expected to interact with animals. Zoos and children's books together cast monkeys and elephants as animal friends, eager for tossed treats. They were no longer mere biological specimens. Feeding the animals was the highlight of young Ann and Bill's zoo visit. They, too, bought their peanuts first thing, but the spider monkeys were "already tired of peanuts." The orange from their lunch, though, was a special treat. Ann kept the fruit just out of reach of the monkey's fingers; she wanted to see its prehensile tail. They made the bears dance and beg before they threw peanuts and waited until Topsy the elephant curled back its trunk and opened its mouth. Then Ann, Bill, and all the other children "began throwing peanuts, candy, and popcorn at the great red opening." Bill missed twice but landed another peanut inside its mouth. At the monkey island, they watched a hundred monkeys chirping, shrieking, fighting, and cracking peanut shells. "These monkeys couldn't have that much fun back in Asia," a keeper assured them.[67]

Of course, zoos mostly maintained the fiction that animals were fed a specially controlled diet and shouldn't be fed popcorn, candy, or peanuts.[68] The WPA even launched a study of how best to feed captive animals. "Every zookeeper thinks he knows what to feed his animals," the WPA noticed, but animals still kept dying. They need "properly balanced rations," the study concluded.[69] Yet everyone from keepers to visitors to peanut vendors selling elephant-sized portions knew it was an indulgently enforced rule. "Any observer at the Park," noted the Bronx Zoo, "can see that the multiplicity of signs regarding the feeding of animals command little respect and are continually violated." The desire to feed the animals and watch them beg just seemed like a human trait.[70]

The kids wanted to feed the animals, and the Bronx Zoo wondered if the urge could be an educational (and moneymaking) opportunity. The zoo announced a junior membership, encouraging "boys and girls" to make the "safari" to the Bronx to touch, feed, ride, and learn. Junior members would get a free ride on the elephant or camel, a butterfish to toss to the seals, and a special packet of food to feed to whichever animal "he or she chooses." Come midweek, the zoo invited, so keepers will have enough time to answer questions.[71] Across the open enclosure, Jane and Bill yelled to Sultan, asleep on his island. Sultan seemed to hear his name and opened his eyes. Later from the howdah on top of the elephant's back, the kids could see the giraffes all the way across the zoo. And, in real life, kids in Chicago or Toledo could clamber around animal sculptures just their size.

Zoos welcomed children. Zoos and children's book authors alike insisted that the tossed peanuts, popcorn, and candy, far from being threats to animal health, were tokens of friendship between kids and animals. Honey Bunch, a darling little girl, loved collecting ants out by the back fence. She also helped search for a lost pet monkey. "The way to really learn about monkeys," her mother said, "is to go see them. You and I, Honey Bunch, will go and see monkeys." Wait until Honey Bunch could tell her next-door friend! "Am I going to the jungle, Mother?" No, she was going to the Bronx Zoo, which was every bit as good, promised her mother. At the zoo, she ran to ride the elephant and marveled at the exotic howdah. Then she wanted to watch the monkeys play. A trio of kittens who ran away from home to see the zoo in another story also wanted to see the monkeys. Little Osborne the elephant, with his indulgent mother, was always hungry for tossed peanuts and popcorn balls. He, like the little kittens, also wanted to visit the monkeys. As they walked around the zoo after hours, the kittens and the elephant chanced upon a wee spider monkey whose dream, above all, was to live on the monkey island. The kittens led their new animal friends to the monkey island. It was an awesome sight.[72]

In these new children's books, the animals were droll, childish, and constantly hungry for tossed treats. Except the ferocious tigers—they were still evil tempered, just as any child who had gone to the movies would expect. That intimacy, the connection of cute and child, was what zoos could do best. After all, zoos had new competition in the 1930s, even if their old nemesis—circuses—had fallen on tough times. No matter how realistic the enclosure or the appearance of freedom, zoos could never match the sheer breathtaking splendor of animals in plain combat in jungle films. Instead, zoos could entertain children and promise them the chance to feed the animals, but they were (mostly) bloodless.

On-screen, animals battled in mostly cleared, fenced areas—open enclosures of a cinematic sort. The irony was that however much movies promised to bring the viewer into the gloomy jungle, filmmakers had, in fact, mastered the art of the open enclosure, the appearance of liberty that allowed animals to walk free and filmmakers to shoot dramatic scenes.[73] Visitors had learned to love their animal illusions—when they went to the zoo and when they went to the movies later in the evening. They enjoyed the fleeting thrill of eyeing the glowering lion before spying the comfort of the separating moat. Zoos with their clever, modern illusions of liberty met their match—and master—in the staged fights and spectacular shots that graced jungle films.

The Hagenbeck design was perfected not in America but in Malaya by American filmmakers, and now the best of them, Frank Buck, was coming home to construct a monkey island. In 1939 he was now in direct competition with the Bronx Zoo, a former customer. The competition, in its own way, was an epic battle featuring the two titans of the animal business. The Bronx Zoo had disdained the monkey islands of lesser zoos, and now the outdated zoo was fighting for its very future against the most exciting monkey island ever constructed.[74]

Funsters of the Jungle: Queens, Malaya

Buck planned to build a monkey island in Jungleland, his reconstructed Malayan jungle camp at the New York World's Fair. He promised his island would dwarf any at zoos and to populate it with 1,000 primates, including rhesus monkeys, gibbons, and baboons. Buck even had to turn to Trefflich, his competitor in the animal business, simply to find enough monkeys for his attraction. Not everything went according to plan: monkeys escaped his Long Island compound, then from the completed island. Never one to miss a publicity stunt, Buck used an escape as another opportunity to capture the excitement of the animal game. He made sure his native hunters, not zookeepers or police officers, chased the escaped rhesus.[75]

How did the jungle adventurer and animal trader come to exhibit his animals and his loyal assistants at the New York World's Fair? In the shadow of Buck's towering monkey island, how did the fair ultimately transform the nearby Bronx Zoo? Buck, the Bronx Zoo, and the Hagenbeck's circus (one of the largest touring circuses in the United States, owned by the old German trading firm) all jostled to win the profitable chance to display animals at the New York World's Fair. For the Bronx Zoo, the fair was a grave challenge. Even as the zoo's finances remained precarious, the fair threatened to draw away visitors.

The rivalry among the circus, the animal trader, and the zoo was a microcosm of the larger competition about how to exhibit animals: at zoos behind bars, in open enclosures, on-screen, under the big top, or at traveling fair concessions. At the outset, the Hagenbeck's circus and Buck seemed to have the advantages. They both had been successful a few years before at the Chicago World's Fair. In Chicago, the new Brookfield Zoo, bruised by the onset of the Depression, had struggled to open in time. By the time the Chicago Century of Progress fair began in 1934, the Brookfield Zoo had completed only four of its seven buildings—and the

monkey island was not among them. Yet Buck's concession at the Chicago
fair featured its own island and visitors could enjoy monkey antics at the
fair, not the zoo. His Chicago monkey island was a replica volcano (complete with fake smoke) that towered seventy feet above the moat. Five
hundred rhesus monkeys fought, played, scrambled, and "just have lots
of fun." They even had diving boards and rocking chairs.[76] Buck promised something even grander in New York.

In New York, fair organizers hoped the Hagenbecks, with their animal
trading business and traveling circus, might construct a zoological exhibit
for the fair modeled after the Stellingen Zoo, complete with a monkey
island. Encouraged, the Hagenbecks and their American agents pushed
to win the concession. A Hagenbeck animal show, they promised, would
be the "most outstanding feature of the exhibition." The show would feature as many as 40 elephants and 100 lions, leopards, tigers, and other
cats. Alongside trained animals, the show would house a "Ceylonese
Village" featuring natives who would perform eight times a day. Many in
the fair's administration seemed convinced. There was certainly nothing
the Bronx Zoo could offer that could compete.[77]

However, only Buck could offer his jungle adventure. That—and
promises of a towering monkey island and his Malay hunters—was
enough to win the animal concession. Even before the fair began, Buck
constructed a compound on Long Island showcasing the kind of jungle
camp made famous in his film adventures. "Visitors," he promised, could
easily imagine themselves in "the jungles of Malaya, instead of highly civilized Long Island." As they brought film to life near the comforts of
home, Buck's camps trumped both the more staid zoo and even the
circus. As the World's Fair progressed from plans to construction, circuses, animal traders, and the Bronx Zoo submitted bids offering different
visions of animal entertainment. Circuses had fared poorly during the
Great Depression. When they couldn't afford to pay for new animals for
innovative acts, audiences declined and circuses, both large and small,
folded the big top and emptied their menageries. Buck, however, was
flourishing.

"When the circuses began to fold up," his publicity declared, "Frank
Buck's Camp was the logical successor to the Big Tents."[78] Buck displayed elephants, monkeys, and humans as the wildly strange curiosities

of the tropics. "All Alive!" he advertised, the "greatest animal show ever produced," with man-eating tigers, performing elephants, and strange, native hunters. Of course, there was also the towering monkey island, upon which "these funsters of the jungle roam at will."[79]

It was natural that Buck could outdo any zoo in the race for the biggest monkey island. Filmmakers, like Buck, in fact, had already mastered the naturalism of the barless enclosure by the time zoos were beginning their own New Deal–fueled reconstruction. Films actually presented much more realistic enclosures than zoos. Buck freely admitted his films were shot in Singapore enclosures, carefully groomed to allow enough sunlight for filming without compromising the sense of the lush jungle. The noted animal adventurers and filmmakers Martin and Osa Johnson equally acknowledged that they filmed gorillas in a stocked African enclosure. The talent for designing naturalistic enclosures had passed from the Hagenbeck zoo to American filmmakers working in Africa and Asia. During the Great Depression, American zoos were desperately trying to catch up.[80]

Even on-screen, Buck had turned his jungle camp into a kind of monkey island. In *Fang and Claw,* he cleverly offered a respite from the savage violence of animal battles with the comic relief of pet monkeys. As ominous music gave way to chirping flutes, two tame monkeys wrestled. Then the picture cut to what Buck called the "mass production" method of capturing rhesus monkeys with a net baited with treats of sweet potato, rice, and tapioca. The monkeys descended the trees, like goods moving on an assembly line. Then they were off to America and its monkey islands. Tempted by tropical fruits and vegetables, they would snack on peanuts in captivity.

The rebuilding Philadelphia Zoo bought some of Buck's captured monkeys and hired Buck to provide dramatic flair for the opening of its new monkey island in 1939. The monkeys' actual entrance onto the island turned from drama to farce when Buck introduced the monkeys in front of 3,050 visitors. The "Star-Spangled Banner" played, but only a dozen shocked and confused monkeys popped their heads out of a barrel. They had only been in the country for a week and had never even glimpsed their future home before they were "persuaded" (as local reporters joked) by a keeper. In a hailstorm of 5,000 peanuts, the monkeys spent the remainder of their confused first day confronting a slide, a watery moat, a seesaw, and, of course, each other. A monkey "horde" forced one poor monkey into the water. It was all part of an exciting spring holiday weekend; 25 human children, meanwhile, were "losted," according to the newspaper reports.[81]

It was clear that filmmakers, more than zoos, had mastered the art and drama of open enclosures. First on film, then at the fair, Buck trumped the ambitions of the new modern zoo. In *Bring 'Em Back Alive*, the camp pet monkey danced, swiped treats, and became bosom buddies with a bear cub. Darting up and down its log like an agile child, accompanied by jangly circus music, the monkey provided a welcome cuteness that complemented violent animal brawls. The monkeys on the Jungleland island played similar comic roles. In a shared publicity stunt, the jazz musician Tommy Dorsey tested Buck's monkeys' reactions to swing. Monkeys, the joke went, must surely enjoy music often derided as primitive. Delighted, Buck expanded this comic, if overtly racist, humor by sending his monkeys to perform jazz nightly on Broadway with Jimmy Durante.[82]

Buck played the adventurer's role to perfection. He paraded through the exhibition grounds, perched in a howdah with Johnny Weissmuller, the *Tarzan* film star, next to him and Ali, his loyal assistant, below him on the elephant's shoulders. He drew admiring crowds of children to Jungleland to pay their admission and purchase souvenir trinkets from buttons to lucky coins to postcards. The fair, like zoos, was inundated with offers of both freak animals and disabled humans. G. C. Allen offered his two-headed calf. Alfonso Carvaial promised a hermaphrodite horse from Bogota. Most strange of all was W. Franklin Dove's proposed unicorn display. Meanwhile, one father hoped the fair would relieve his "burden" and exhibit his baby boy with no arms and one leg.[83] The specter of the freak show would hover over the fair's exhibits. Buck promised rare, humorous, and spectacular animals and exotic Malay hunters. His advertisements, hardly biology lessons, were similar to those of the "Strange as It Seems" sideshow, which featured a "lion-faced Chinaman" and a "leopard-skinned" family.[84]

Jungleland attracted adult audiences, too, but its ideal customers were children, eager to spend on Frank Buck souvenirs. Buck's plans of catering to adult audiences didn't materialize quite as well. Buck advertised that he had "captured" Balinese dancing girls, an anthropological substitute for the "girl" shows discouraged by the fair's organizers. He knew they would be a "seductive sight," but the dancers never received their travel visas.[85] The fair also restrained Buck when he tried to exhibit a dugong, a sea mammal from the Indo-Pacific related to the manatee. Buck promised that the dugong was the origin of the mermaid—with all her sexual allure. The dugong, he hinted, has "natural breasts and sexual organs similar to a woman." To fair organizers, the exhibit seemed too close to a "grind show." [86]

Meanwhile, the local Bronx Zoo couldn't even display reptiles at the fair on its own doorstep. World's Fair organizers awarded the reptile concession to Clif Wilson, the veteran of the Chicago and Cleveland fairs and proprietor of the "Monster Reptile Show." Wilson promised to bring to New York the display that had served him well in an otherwise poorly attended Cleveland fair. On towering poles of bamboo that evoked dense tropical jungles, Wilson's banner promised "monsters, alive, alive."

Buck impinged on Wilson's monopoly when he tried to display deadly cobras and a giant python. Yet as the fair's director of concessions wryly noted, Buck was "probably the most insistent among our concessionaires on others violating his contractual [*sic*] rights." Buck even wrote to insist that the "Merrie England" concession cease its strangely conceived monkey show. He protested yet again when, in the second year of the fair, the Firestone exhibition enhanced its reconstruction of a Liberian rubber plantation with animals collected by William Mann and the National Zoo. And he complained about the New York Zoological Society and its efforts to bring the rarest and strangest animals from the Bronx Zoo to the Queens fairgrounds.

The Telegram to Mrs. Roosevelt: Queens, the Bronx

The Bronx Zoo was desperate for a place at the fair. The zoo was eager to pull itself into the new era of public animal entertainment, but in 1939 it was still ill prepared. Just as the fair opened up the competition for the animal concession, zoo officials were reeling from the disturbing findings of a self-study. "We are content to exhibit animals in completely artificial surroundings that are often painful to them and unaesthetic to the observer," the zoo admitted. The Bronx Zoo seemed staid and outdated even in comparison to competing zoos. Other zoos had benefited more from WPA largesse, and the Bronx Zoo still featured rows of small cages with bars and concrete floors. At a time when visitors were enthralled by jungle films, the zoo didn't even allow cameras into the grounds. "We present," the zoo mourned, "the same spectacle that the zoological parks presented to the public in [Queen] Victoria's day." How could visitors befriend the animals or imagine themselves as jungle adventurers in such drab surroundings? The animals might as well have been heads and horns on a wall. Our animals are "beautiful but dumb," the zoo admitted. Alone in the their cages, "they cannot tell their own story." By contrast with Buck's films, "we have done a very imperfect story for them."[87]

Maybe at the fair the zoo could invent humorous and exciting new stories. The zoo eventually won a limited World's Fair concession. It was permitted to display only fish and a single animal: Pandora, a rare Chinese panda. The zoo and its parent zoological society were full of disdain for Buck's concession, just along the midway. Jungleland's "tom-toms" were as crass as "garish neons of the Aquacade." Yet atop the zoo's building stood a sign advertising an animal show, not a "scientific institution." "Alive!" it declared, "the Panda Pandora." As the zoo's exhibit drew close to 300,000 paid visitors—still less than Jungleland—the zoo realized it had found new methods of exhibition that transformed animals from examples of "this or that species" into "characters in a serial story."

Fairfield Osborn, the New York Zoological Society's president, admitted past mistakes. Visitors, especially children, had come to the fair to see animals full of life and character rather than rare beasts identified only by their scientific names. They wanted more than the "entanglements of Latin derivatives." They wanted danger and a good laugh. Pandora, at least, could provide a rare curiosity. Pandora with her large head and still strange black and white coloring seemed like a lovable natural comedian. Of course, the zoo also realized that when it emphasized the rarity of an animal, like a panda, the zoo was approaching dangerously close to "the showman who displays a two-headed calf." The ghost of the freak show had reappeared.[88]

Thus, the zoological society president had become a sideshow hawker, enticing visitors with "the giant panda, prima donna de luxe, the ineffable, 'Pandora' herself." Early on one autumn morning, William Bridges, the zoo's publicity director, rejoiced in the pleadings of a young visitor who slid his quarter under the bars of the ticket seller's booth. He promised to come later in the day and wanted to make sure he bought his ticket before his spent all his money: "I want to see the panda tonight." Pandora was an ideal substitute for the monkeys on their island. When Bridges left the fair that night at 11 o'clock, this lad was "still whooping with laughter at Pandora's antics." Osborn, similarly, watched another young visitor outlasting the patience of his tired mother. He insisted on remaining for yet another demonstration of the electric eel. Osborn knew the eccentric panda and bizarre eel that daily lit lamps and rang bells were "lures" that might bring the "stranger" back to the zoo. Education about animal habitats and evolution somehow would follow. The panda, the zoo boasted, was "one sight that mustn't [*sic*] be missed."[89]

"There is no use in our housing a zoological Rosetta Stone," Osborn argued, "if no one comes into see it." Bridges delighted in another visitor, perhaps well acquainted with sideshow hoaxes, who insisted Pandora was really a man dressed in a panda suit. "Look," he told Bridges, "you can't tell me that a dumb animal watches a crowd and knows what makes them laugh, and then does it again to get another laugh." The eel, meanwhile, promised spectacular feats. In June 1940, to mark the second summer season, the eel matched the marvels of modern engineering. Eleanor Roosevelt, the first lady and a member of the New York Zoological Society, had long been recognized for the personal letters she received; none was stranger than the telegram the eel sent with its own power. It modestly wrote (with the help of zoo translators): "[T]his is the very first telegram I have ever transmitted in all my aquatic life and probably the first one, actually written by a fish."

The eel's electricity linked the zoo's displays to the modern technologies so prominently displayed outside of the midway. But the zoo's exhibition promised adventure in the heart of the primitive, tropical world. The zoo offered a chance to step back in "time" to the primitive tropics, even as General Motors' nearby Futurama display provided a glimpse of the American future. Visitors recognized the contrast between prehistoric wild adventure and Futurama: "[I]n a fair that celebrated the 'world of tomorrow' with bright lights, amplifiers, and colors," the zoo's exhibit offered a retreat to the wild with "a grateful sense of relief."

The New York World's Fair helped the zoo move away from the mere display of taxonomy. Welcome to the "Zoological Park of the future," trumpeted Osborn. Once animals were grouped "in more or less natural association," the zoo could help visitors imagine new stories. The fair helped the zoo realize that "it is arbitrary and unnatural to show all mammals in one place, all birds in another, all reptiles in a third. This is a purely human pattern of thought that does not exist anywhere else in the realm of nature."[90] The zoo's "natural live habitat group" at the fair stretched the confines of its concession agreement when it displayed tree kangaroos, a wallaby, a wombat, Galapagos tortoises, numerous birds, and a kangaroo in a shared enclosure. This simulation "to a certain degree" of the lush tropics told laughing stories of animal friendship and childish squabbles.

American zoos, despite open enclosures and WPA-funded monkey islands, still presented a sanitized nature, by necessity removed from the predatory relationships Buck's films placed at the center of their narratives. At the zoo and at its concession, there could be no "serious animosities," admitted Osborn. In their place, he spun dramas of love and friendship. A fawn, tortoise, and kangaroo forged "almost scandalous friendships." A magpie, however, ruled the zoo's fair exhibit like a "self-appointed dictator." As the fair concluded its second season, the zoo was already making plans to overhaul its Bronx enclosures into natural groupings of animals in highly managed habitats. There was a danger in rebuilding the zoo. In substituting habitats for bars and drama for taxonomy, the zoo risked equating exotic animals with the freaks and wonders of the midway. The zoo, however entertaining, should never become a "side-show," reminded Osborn. Recreation and education must become "twin sisters"—a hard balance to achieve.

Osborn knew zoos were playing catch-up to films. Films could tell stories, and Buck, at least at the fair, could build an exhibit that zoos, even with WPA money, could never match. WPA labor and money, though, meant zoos would be there when the fair closed and when the films left the theaters. Day after day, kids could come to zoo, grow up with the animals, and imagine themselves as friends, bound by ties of peanuts. Zoos could build exhibits in which visitors' imaginations could run wild. Visitors could take a "jungle expedition" to Frank Buck's Jungleland. Or they could visit the zoo and step out of the Depression into an exotic world of danger and tenderness.

Buy trinkets at the fair. Toss peanuts at the zoo. Zoos knew they were competing to make animals exciting and jungles alluring. Stories had an effect. Visitors may have felt close, even affectionate friendship toward the elephants they rode, the tigers they teased, or the monkeys they fed, but they felt far removed from tropical lands and peoples. Visitors could get to know the animals and laugh at their antics, but they could only gape open-mouthed at the strange Malay hunters. Maybe they could sample their strange satay. As zoos peddled stories of lovable animals and exotic lands, their vision of the modern zoo depended, ever more, on exciting adventures set in the primitive jungle. Leave the city behind, invited the Philadelphia Zoo. "[Y]ou" brush aside the dewy moss sheathing hewn rock to enter the lion steppe. "Abashed, you pause." The lion, awoken from his nap, rises to face you. "Shall you run?" Brief fear turns to plea-surable observation, made possible by the "illusion" of natural settings, confined by the hidden moat. Across from the ominous lion, the tigress nuzzles her playful cub. *"In truth, the Jungle has come to Philadelphia."*[91]

5

APING

African Animals on Zoo Stages

Jo Mendi played the RKO vaudeville circuit but found fame at the Detroit Zoo. The Great Depression hit Jo's livelihood hard, and by 1932, the chimp squatted in the basement of a Detroit dentist, no doubt causing the usual damage that happened when apes lived in human homes. Nearby, the zoo suffered. Gate receipts declined as Detroit's hard-pressed workers saved meager earnings. John Millen, the zoo director, purchased the chimp for $1,000 from his own savings, and Jo Mendi became a local star performing at the zoo and at local vaudeville houses. When Bob Hope came to Detroit to appear at the capacious Michigan Theatre, he discovered his name below Jo Mendi's on the marquee.[1]

From the beginning of the century, when the large apes began arriving in America, until after World War II, when they remained popular in zoos but endangered in the wild, African apes were different kinds of zoo specimens. Elephants offered rides and trunks in which to drop peanuts. Tigers hinted at the jungle's adventure, and vipers the jungle's dangers. The big apes, though, could wear tuxedos, play football, drink beer, sip tea, smoke cigarettes and cigars, peddle tricycles, drive cars or rocket ships, ride ponies, play the piano, and pile boxes to reach a treat. They hugged, mourned, sulked, bit, nursed, and smiled—or at least seemed to.

Apes, especially chimps like Jo Mendi, were popular entertainers even though they did relatively mundane human activities rather than feats of strength or dexterity. Americans first flocked to see apes aping humans on stage in vaudeville shows, the popular variety acts that toured grand theaters in the early decades of the twentieth century. During the Great Depression, chimpanzees moved to zoo stages where they would continue to perform through World War II and into the Cold War era.

Zoo visitors always had their favorite animals, but they forged different and special relationships with the big apes. They still do, even after chimp shows have become extinct, odious and unthinkable in today's zoos. Sometimes a gorilla or chimpanzee sits close to the glass. Visitors reach out, almost trying to touch a body that, hair aside, seems so much like their own. They giggle, with some embarrassment, at breasts and penises. They sigh at the tenderness of breast-feeding, and if the ape puts her hand on the glass, they rush to measure fingers and opposable thumbs. We are so obviously connected in genetics but always separated by moats and glass.

There's a reason why: it took zoo directors years and many tragic deaths to realize germs jump from human to ape. Tuberculosis and other diseases have a way of jumping across the species line. "Our entire collection is behind glass," noted C. Emerson Brown on the eve of the Great Depression, "which I believe is a great factor in keeping them healthy over long periods." The Philadelphia Zoo owned one of the nation's most comprehensive anthropoid ape collections, including seven chimpanzees and a rare gorilla. Electric fans freshened and filtered the air. There were two advantages: visitors didn't get an "offensive odor," and the apes didn't catch human colds or worse. All the precautions were needed because the lives of the big apes in zoos, especially gorillas, were often difficult and short. Even Brown seemed surprised his gorilla hadn't succumbed to illness—yet. At first the gorilla had a chimpanzee as a "playmate," but as he grew, he was becoming dangerous except to his keeper. The keeper still spent "considerable" time playing with him. Despite all that glass, the zoo believed the anthropoid apes craved the human touch. Maybe the gorillas needed a romp with their keepers? Maybe the chimpanzees wanted a hug? They looked so human that they must crave human affection.[2]

Apes remain, after all, "almost human"—but, crucially, which kinds of humans? Jo Mendi was one of many performing apes heralded as a missing link. He reminded his audience of the connection between chimpanzees, gorillas, and humans—especially *Homo sapiens* of African origin. If other animals in zoos still represented specimens of their genus and species, primates were hard-working entertainers. Monkeys scampered on their islands, and chimpanzees performed in stage shows. Gorillas, in their youth, acted the part of the adorable baby. Only in sullen adulthood were they isolated in their cages, physically removed from keepers they all too often injured. (The Asian orangutan, rare in zoos, was a very different kind of ape and not only in place of origin. Interesting but intractable, orangutans never excited audiences, scientists, and promoters quite as much as their African cousins.)

When Jo Mendi strode the zoo stage in his tuxedo, his act suggested that somehow humans had stepped hairless out of the ape's skin. Theater turned evolution, the shared heritage of ape and human, into something very real but funny. The emerging science of primatology used the same tricks and the same acts to measure the evolutionary distance between ape and human. When acts like Jo Mendi's migrated from the vaudeville stage to the zoo, they bridged the dwindling distance between entertainment and primatology.

When first chimps, and then gorillas in far fewer numbers, became regular attractions in America, they lived with humans, sucked baby bottles, and dressed in human clothes. They were important characters in the changing relationship between ordinary Americans, zoos, animals, and faraway peoples. Apes on zoo stages were not small humanoids; they were African animals forced into behaviors that because of their origins and the social context of their importation encouraged visitors to measure the differences of savagery and civilization separating white, black, and ape. It was all very funny for human audiences—at least until apes rebelled and some humans protested.

Old Time People: Equatorial Africa, Newport

The big apes fascinated Americans, but primates were still mysterious animals at the end of the nineteenth century. Americans had avidly read the accounts of adventurers in Africa, but their descriptions of gorillas seemed too remarkable to be believed. In 1856, the French-American adventurer Paul Du Chaillu set off to explore the jungles of equatorial

Africa where he observed gorillas. After his return, he became a popular lecturer in Great Britain and the United States. During his talks, he displayed gorilla skulls and haphazardly stuffed skins. Published in 1861, his *Explorations and Adventures in Equatorial Africa* introduced gorillas as both savage and dangerous. Du Chaillu described gorillas attacking his hunters and bending rifle barrels as easily as thin wire. Just as important, his illustrations pictured gorillas as distinctly human.[3] One notable drawing depicted a towering gorilla standing tall on two feet in a verdant jungle clearing. His human victim stretched dead on the jungle floor while the gorilla twisted the hunter's rifle in his hands. Long after apes donned human clothes and lived in human homes, Americans still marveled at gorillas' extraordinary strength and evil temper.

Even as Americans debated the veracity of Du Chaillu's descriptions, the big apes were still rare in America. When in 1878 the Cincinnati Zoo advertised an orangutan and a chimpanzee, such large apes remained infrequent zoo, circus, and menagerie attractions.[4] F. E. Kitzmiller was one of those eager to see great apes. He wondered if the apes on display near his home in Ohio in 1890 were gorillas. He wrote to Frank Baker, superintendent of the National Zoo, for an explanation. They were certainly not gorillas, Baker replied. The animals on display were likely the vastly dissimilar baboons. Kitzmiller, like so many others, just didn't know the difference.[5] The Asian orangutan and the African chimpanzee and gorilla were difficult to capture and even harder to sustain in captivity. "The gorilla sulks almost continuously and soon dies," mourned the Bronx Zoo. They knew from experience. The zoo had imported a female gorilla from Africa in 1912. It lived for only ten days.[6] In fact, some of the first "gorillas" displayed in the United States were probably the more resilient chimpanzees, deliberately mislabeled to attract public attention.

In 1921 Carl Akeley snapped photographs to document his breathtaking accounts of mountain gorillas. For many Americans, such images were the first glimpse of gorillas in their native habitats. Akeley, the hunter and taxidermist turned gorilla conservationist, first encountered gorilla nests in the Congo mountain jungles. In the misty underbrush, gorillas wove their dwellings from cracked bamboo. There was nothing to suggest gorillas might somehow take to the human home except that they looked like people.[7]

Yet decades after Akeley traveled to Africa, Americans still wanted to cuddle apes, raise them in their homes, and dress them like children. Sometimes they did all this in the privacy of their homes. Sometimes it was on stage. Sometimes it was science. And sometimes it was all three. The lines between primatology and performance blurred when humans tried to civilize apes through intimate connection with them. The chimp show at the zoo or on the vaudeville stage did not begin and end with feats of acrobatic skill but with the intimacy of the home, the hug, or shared food.

There was another intimate kind of living together that fascinated American audiences. They came to believe that African and ape lives were entwined and not just through biology. They shared the same jungles, Americans believed. They read about the larger apes long before they saw them in person, and when apes arrived in the United States, Americans expected to see savage beasts. In the early decades of American zoos, most of what Americans understood about the big apes was a poisonous combination of myth and racism. Like Du Chaillu's first accounts, explorers' descriptions and newspaper reports of the import of rare specimens stressed the entwined lives of African humans and apes. The "gorilla," one observer in 1915 insisted, sometimes "picks up a negro baby and carries it about." If the gorilla was confused, so, too, were natives. When George P. Goll, a scientist at the Philadelphia Museum, returned from Africa, he reported that natives called chimpanzees "old time people." If ape and African recognized each other, they must also share biological and evolutionary heritage, which is precisely what the Cotton States Exhibition, the 1895 world's fair in Atlanta, argued. Fair organizers proposed a display tracing a lineage from the chimpanzee through the Bushman to the American "Negro." The message was clear: chimpanzees were "negroes with long arms."[8]

In 1907, Consul—one of several famous performing chimpanzees who shared that name—hobnobbed with America's elites in their beach playground at Newport, Rhode Island. Clad in a top hat and tails, he smoked a cigarette and sipped his tea. Oliver Belmont, the scion of one of the nation's wealthiest families, loved his act, and so did many Americans.[9] But not all—African American critics were less amused by such dinner parties. If an African ape could dine at the table with elites, why not the human of African descent? The color line, they reasoned, had been exposed as farce.

"White man," wrote one critic, "be consistent on your color line question." This critic knew well that Belmont and others associated the African ape with African American humans. He had stumbled on the significance of the performing ape for the American color line. Chimpanzees were on the black side of the color line. The elite Belmonts, white audiences, and zoo directors all looked at apes and saw human racial difference. Consul's act delighted his Newport audience—and disturbed his African American critics—precisely because try as hard as he might, he wasn't human.[10] Yet as the Newport dinner party suggested, his audiences still recognized him as African. One other promoter boasted that if you looked at his gorilla from behind, "you could easily imagine it was a little negro kid." And that was part of the act. His gorilla lived for only a short while, but the promoter delighted in every racial slur spectators shouted. In fact, he bragged about them to William T. Hornaday.[11]

By the early 1900s, chimpanzees had become more regular attractions, but on vaudeville stages rather than in the nation's still-small zoos. If Americans viewed these apes, the animals were most likely wearing human clothing. They were African animals performing "civilized" tasks. Vaudeville promoted the confusion between African and ape. In New York, one impresario dressed up his newly arrived chimpanzee and brought it onto a train. The animal appeared "ludicrous" but "for all the world like an ordinary but very homely little colored girl." "If Darwin only knew," laughed the newspapers.[12] Americans were never permitted to forget that the animals that increasingly graced the stages of small and grand vaudeville theaters were African animals a lot like their human neighbors.

Monkey Mad: Equatorial Africa

In 1897, Joseph S. Edwards owned a popular ape show. He toured the country, playing different vaudeville theaters. His act was a formal dinner party with his "nearly human" apes all dressed for the occasion.[13] Just as the trade in chimpanzees from Africa to America began in the 1890s, vaudeville was at the height of its popularity. Performers—humans and primates—followed well-established circuits from city to city. Some performed in dusty, small theaters, but the most popular acts played expansive and garish halls constructed by noted promoters like Edward Albee and B. F. Keith. In Keith's New Theatre in Boston, the Palace in New York, and Chase's New Grand Theatre in Washington, DC, chimpanzees shared the stage with touring singers and comedians.

One circus and vaudeville veteran recalled that "hardly a night went by that a trained chimpanzee was not rattling over the rails to fill an engagement in one of the country's multitude of vaudeville emporiums. From the gleaming Palace to the dreariest window-sill (a vaudevillian's word for a shallow stage) there were chimps imitating men, and imitating each other imitating men imitating each other."[14] Vaudeville roots grounded ape shows in a long tradition of racist imitation. White vaudeville performers donned blackface to present comic caricatures alongside apes that imitated civilized manners but still reminded audiences of apes' African origins.

Apes like Prince Flora acted like civilized humans while white humans smeared their faces with burnt cork or shoe polish and donned wooly wigs to assume black characters. In a stage show of multiple imitations, blackface helped stress chimpanzees' African origins. Flora, like so many other chimpanzees, appeared alongside blackface performers. At the Palace, Flora "eats, drinks, smokes and does everything but talk."[15] When apes dressed like humans who dressed up like other humans, they confused species to make differences of race seem starker.[16] When humans sang a "jungle song" (as they were advertised) like "My Chimpanzee Queen," ape shows seemed like another form of blackface.[17]

"Just now Paris, London, and New York are . . . 'monkey mad,'" marveled one vaudeville reviewer in 1909.[18] That year, when Chase's announced its autumn attractions, simian stars garnered the banner notice even above human celebrities, like Carrie de Mar, who had recently graced continental stages. De Mar could sing, but "Hairy" Lady Betty could sew, ride a real bicycle up and down stairs, and roller skate. The very next winter, Peter the Great headlined the show. "Peter," his promoters claimed, was "born a monkey, made himself a man."[19] When Americans first viewed chimpanzees whether at the theater or at zoos, they were typically like Consul Jr., the African ape advertised as "civilized." He was a "native of Africa" but served at the table "fully dressed, including hat and gloves."[20] Vaudeville's racial humor highlighted apes' African origins to render aping a multidirectional imitation. The "reigning sensation" delighted his audiences with his love of fine port and cigars.[21]

Vaudeville ape shows reached their height in the 1920s as Joe Mendi joined the popular comic headliners "Smith & Dale" on a tour of the leading vaudeville stages. (The famous Joe Mendi would inspire the similarly-named Detroit Zoo star, Jo Mendi.)[22] Simian stars earned amounts that dazzled the most popular human performers. Joe Mendi was even advertised as the "$100,000" ape.[23] Civilization, though, was just an act. In front of critics and reporters, chimps could "enjoy" the human comforts of a first-class hotel and black servants, yet beyond the view of paying audiences or reporters, conditions were difficult and sometimes fatal. Charlie the First was a valuable performer, but he traveled from city to city out of sight in a railroad crate, and he suffocated.[24]

This was the kind of story promoters wanted to hide, not because of accusations of cruelty (no one seemed especially concerned about cruelty) but because Americans were enticed by stories of chimpanzees living with African humans, leading the high life in big-city hotel rooms, or enduring civilization in the homes of trainers and keepers. Apes in crates disrupted a good act. Instead, onstage, audiences could watch apes dressed in tuxedos or smoking cigars and laugh at the ridiculousness. The message was clear: it was silly to think the ape—or the African— could be truly civilized. When Charlie the First arrived in America in 1909, his promoters delighted in tall tales of his journey from the African wild to his audiences with European nobility. Charlie's promoters spun a typical story of discovering an ape confused by African blackness. As their expedition cut its way through the jungles of equatorial Africa, Charlie emerged with a gift of coconuts instead of fleeing like other animals. He had recognized his kinship with the "negro servants" carrying water and fuel, and he began to help them.[25] Too similar in race and species, so the stories said, chimps and African humans were confused. According to one traveler, Africans captured their ape neighbors by plying them with enough drink that they could no longer distinguish primate from human. Another observer insisted that a chimp, already captured and on display at a show, whistled only at dark-skinned women.[26]

Chimps, promoters boasted, lived in first-class hotels, but as babies they had grown up like African human babies. American audiences gasped at stories of African women raising chimps and gorillas as their own children. Consul's proprietor claimed he was reared alongside a human African baby and nursed on the same human breasts: "She was a low type of woman, even for an African savage." Vaudeville show promoters and later zoos recounted that African women reared chimps and gorillas on the same breast as their human babies. Sammy came to the St. Louis Zoo through the animal trader Warren Buck who purchased the young chimp from an African woman "nursing her own negro baby and the baby chimpanzee." She called both her "babies."[27] Mae Noell, who later owned the traveling chimp show "Noell's Ark," even published a photograph of an African woman nursing a chimp. Chimp and baby suckled just as easily from the same breasts.

Gorillas, too, suckled at human breasts. When Massa was an infant, he clung to his mother's long black hair while she led a "placid" existence in the African jungle foraging for fruits and bamboo shoots. Native gardens with their succulent plantains, however, were the fatal "undoing of the idyllic life of mother and infant gorilla." In the competition between humans and wild gorillas, Massa passed from gorilla to human breasts. Natives ambushed the raiding gorillas, plunging spears into those like Massa's mother who didn't flee fast enough. Massa's mother—or so the story went—ended up in the "oblivion of the cooking pot," the main course in a native feast that seemed especially cannibalistic because of Massa's fate. Natives knew Massa was worth more than a feast because white traders were always eager to buy these "little waifs." For the next six months, a native woman suckled the orphan gorilla "along with her own baby."[28] A pygmy woman similarly raised Bamboo (eventually destined for the Philadelphia Zoo). As he suckled at her breast, she "crooned over him as if he were a human infant."[29]

In America, breast-feeding gave way to human service. A "negro nurse," black servant, or African handler frequently catered to the whims of performing apes. Newspapers reported on the 1904 death of Consul through the lens of Harry Hall, the chimp's "negro" handler and "playmate." They were raised together. Maybe Hall truly did feel a loss. Perhaps he had learned to enjoy his ape companion. Certainly, he was now out of work. The news reports, though, stressed the similarity of African ape and African American. "He was just like me . . . except that I can talk and he could not," they quoted Hall.[30] Descriptions of chimps staying in luxury hotels and served by black servants were standard fare in newspaper reports on upcoming vaudeville acts. Juan and his "wife," chuckled one reporter, ordered persimmons and bananas from late-night room service and commanded their black servant to pack their bags. Again the message was clear: once they had become trained to act civilized, chimps no longer lived with Africans or suckled at their breasts. They ordered them about, like masters or employers.[31]

Peter the Great: New York, Cuba

As the Bronx Zoo's director, Hornaday went to see Peter the Great's 1910 vaudeville show and came away convinced his evening's entertainment was the frontier of science. Hornaday watched intently as Peter "ate with a fork, using his fingers as 'pusher,'" about what a human child might be able to do. After lighting and enjoying a cigarette, Peter unbuttoned his overcoat and unlaced his shoes.[32] Back at the zoo, Hornaday adopted the very same vaudeville act to test the intelligence of his captive apes. He taught them to eat at a table, drink from a beer bottle, spit, smoke, and ride a tricycle.[33] Hornaday, impressed by Peter the Great's performance, soon had his own chimpanzee star at the Bronx Zoo. "Consul and Peter are in vaudeville. Baldy is only an exhibit at the zoo," noted the zoo's guide. Yet Baldy was as "clever" as his show business rivals, insisted his keepers after they taught him to roller skate. "With a little more training he would make some fellow's fortune on a vaudeville circuit," boasted one keeper.[34] Aping at the zoo entertained visitors, but Hornaday argued that it also educated. The formally dressed chimps and orangutans seated around a table at the Bronx Zoo illustrated "their wonderful likeness to man."[35]

Meanwhile, Peter's owner gleefully offered his prize for very public scientific scrutiny. Lightner Witmer, the director of the University of Pennsylvania laboratory of psychology, tested the chimp in his clinic, normally devoted to the study of "defective children." After administering the same series of tests given to such children, Witmer anointed Peter a mental "missing link" between human and ape. It was perfect advertising copy. Back on stage, Peter the Great's show promised to reproduce "the tests that convinced scientists a man's brain is in this Chimpanzee's form."[36]

Scientists, zoo officials, and ape show promoters discovered mutual interests and shared methods. At the zoo, lab, and theater, they trained chimps to measure the connection of apes to those they deemed human inferiors—whether "defective children" or Africans. Through his tests, Hornaday ranked the African apes in their proximity to different human races. He felt that the gorilla was nearest the human in physique, but the chimpanzee approached the civilized human child or the savage adult in intelligence. Place the chimp's head on a gorilla's body and the "missing link" between primates and the lower humans appeared, he said. Hornaday was convinced that a vast difference separated the highest human races from the lowest and that the "highest animals" were "intellectually higher than the lowest men." After Hornaday attended the ape show, he mused about how complicated the idea of Darwinian evolution truly was, especially with an added douse of racism. "I would rather descend from a clean, capable and bright-minded genus of apes," he wrote, "than any unclean, ignorant and repulsive race of the genus Homo. . . . There are millions of members of the human race who are more loathsome and repulsive than wild apes."[37]

The flirtation between aping entertainment and science that began on the vaudeville circuit continued in a private Cuban castle. Madam Rosalià Abreu, the wealthy daughter of a Cuban sugar planter, had assembled the finest collection of apes in the world by the 1920s.[38] Stage performers, zoo officials, and primatologists all clamored to visit her fourteen chimpanzees, a trio of orangutans, and numerous smaller primates.[39] Later, upon her death in 1930, American zoos augmented their collections through the dispersal of her pets. Her chimpanzees, however, were sent to an important admirer, Robert Yerkes, the Yale University primatologist.

In 1916, Abreu had invited Yerkes to visit her apes, promising they "behave very correctly at the table."[40] Yerkes, however, was still working for the United States Army, overseeing the testing of millions of draftees during World War I. After the war, eugenicists cited his tests as tangible evidence that certain races, especially immigrants and African Americans, exhibited low levels of intelligence. Once released from military duty,

Yerkes applied the intelligence testing of human races to primatology partly through his study of Abreu's apes. Among Abreu's chimps, Yerkes found the same diversity of intelligence he had uncovered in his wartime tests. He even believed that chimps were divided into different races based on skin color. Some chimps were "as white as the light-Skinned Caucasian," he said. Others were "as black as the African Negro," and skin color mattered. Lita, for example, was "a typical white-face," friendly, fearless, aggressive, and playful—an ideal specimen for future primatologists. The "dark brown" Malapulga was ill-natured and "rather stupid."[41]

Yerkes lived with the chimps he studied: Prince Chim and Panzee. He trained his Chim and Panzee to eat with spoons and sit at a table; they were particularly fond of powdered milk.[42] Yerkes was especially attached to Chim, and he believed the chimp returned that love. Whatever the affection he felt toward his chimps, he also knew he could attempt experiments on them that would be impossible on humans. The results, he insisted, could be reapplied to human problems.[43] Training chimps to accentuate their seeming humanity could "increase . . . our knowledge of the facts and laws of life as will enable us more wisely and effectively to regulate or control individual, social, and racial existence," concluded Yerkes. Vaudeville had made trained chimps objects of delight, and he was impressed. He hoped science would eventually catch up with "variety shows."[44] Yerkes, like Abreu and Hornaday, admired ape shows. After all, just as he was completing his studies of Chim and Panzee, a trained African chimp in Tennessee had turned the American debate over evolution into the best ape act of all.

Joe Mendi, the Missing Link: Dayton, West Africa

In 1925, "[c]ivilization is at stake," read the telegram. "Have missing link, wire instructions." The famed lawyer Clarence Darrow was in Dayton, Tennessee, defending John Thomas Scopes, a local teacher, against the charge of teaching evolutionary theory in defiance of state law. Telegrams like this one promised exhibits of strange people and trained chimps that proved the anthropoid origins of man. One particularly famous chimp, Joe Mendi, and his promoter soon arrived in Dayton, eager for the publicity of an evolution trial. They turned the "Scopes monkey trial" into vaudeville theatre. Outside the courthouse, Joe Mendi held court, pipe in hand.[45]

Aping, in its comic glee, had double meanings. When he paraded outside the courthouse wearing a formal coat and tails, Joe Mendi reminded trial attendees and reporters of the biological resemblance between human and ape. Yet the ridiculousness of this ape with a taste for tobacco and fancy dress also testified to the inferiority of those supposedly below civilization. Either way, the performing ape had become a well-dressed part of the debate about human evolution. As the trial progressed in Tennessee, visitors flocked to the National Museum in Washington, DC, for an exhibit on evolution. Models of gorillas, chimpanzees, and the Piltdown man (a fossilized skull of an early humanoid, later proved to be a hoax) shared space with skulls of contemporary Africans. In St. Louis, visitors crowded around the zoo's chimp enclosures, reenacting the trial's debates.[46]

Joe Mendi's appearance in Dayton was a publicity coup for his promoter, but he was hardly the first chimp hailed as the "missing link" between apes and Africans.[47] Consul, noticed one admirer of his 1895 show, "has a large mouth and his lips protrude just as the lips of other natives of Northwest Africa do."[48] Vaudeville shows like that of Jack the Chimpanzee in 1921 advertised the "thrilling story of the evolution of man" and promised missing links not simply between man and apes but between apes and Africans.[49] Joe Mendi's promoters also publicized his African roots. By the time he left Dayton, Joe Mendi, "vaudeville's greatest exponent of the 'missing link,'" had become a star.[50] "Direct from the Evolution Trial at Dayton," he appeared at B. F. Keith's in Boston advertised as the "nearest approach to a Human." It had taken Joe Mendi eighteen months to move from the African jungle where he was trapped to the Keith-Albee vaudeville circuit. On stage, his promoters described his show as an experiment in bringing up an African ape in the manners of civilization. Yet even in his name, his promoters reminded viewers of his African origins. He was named for the West African Mendi tribe, rumored to be cannibals.[51]

Around the same time as the Scopes trial and Joe Mendi's surprise cameo, chimpanzees were becoming more common in American zoos. In Milwaukee's Washington Park Zoo, for example, visitors could enjoy Chilo and Bobo sipping tea each afternoon at four o'clock.[52] Chimps were different kind of zoo animals. Elephants might be for riding and tigers for admiring, but chimps came to zoos to perform. Vaudeville shaped the unique way zoos presented their apes, even as traveling shows faced hard times during the Great Depression. Theaters closed or switched their offerings entirely to movies. A few ape promoters, like Reuben Castang, managed the transition to film. By the late 1920s, his apes starred in a series of comic shorts, including a satirical chimp version of *Uncle Tom's Cabin* that further blurred the lines between human blackface and chimp aping.[53] Other apes, like Jo Mendi in Detroit, moved from vaudeville to zoo stages.

Depression-era zoos, as they increasingly courted visitors of all ages with promises of entertainment instead of dour pedagogy, provided an ideal home for animal acts. Chimp shows proved moneymakers for zoos suffering bad times. The Bronx Zoo even offered a bonus to its keepers if they succeeded in teaching their animals profitable tricks, and by 1933 the zoo had found a worthy successor to Baldy. Buddy ate at a table, drank his milk from a cup, and played a small piano for visitors.[54]

As chimp acts moved from stage to zoo, they still promoted stories of chimp training as civilizing. Purchased in Africa, Skippy began his career with Gertrude Lintz's Gorilla Village ape show before becoming the Lincoln Park Zoo's "gentleman chimpanzee." He ate his meals at a table with a knife and fork and wrote daily letters—however, indecipherable—with a pencil.[55] In St. Louis, the zoo welcomed Prince Koko and Lady Jemima. Lady Jemima's name alone evoking not only the breakfast food icon of a smiling African American cook but also a popular blackface star named Aunt Jemima associated her with minstrelsy and blackface. The zoo assured visitors that the two apes had been found living among human "natives" in the Congo. Their savage birthright made their "surprising aptitude for the finer attributes of civilization" all the more remarkable. The zoo boasted they ate the "food of gourmands, and with the cutlery of civilization."[56]

With prizes like Koko and Jemima, the St. Louis Zoo boasted the nation's most popular show.[57] In the midst of the Depression, the zoo admitted that its ever more elaborate chimpanzee show represented the "only means of stimulating interest" at a time when the zoo had few funds to purchase new animals.[58] A chimp theater, though, proved a good investment. One of the largest performance spaces in the city before

World War II, the St. Louis Zoo's new chimp arena could pack in 4,000 visitors. During the summer months as many as 100,000 people a week viewed the shows, cementing director George Vierheller's reputation as a showman. As the shows grew in complexity during and after World War II, attendance continued to rise. By 1950, audiences topped 2.75 million. The zoo's chimp shows retained their remarkable popularity until 1982, when—despite public resistance—they were finally canceled.

"Jitter, St. Louis Zoo" (ca. 1943). Jitter, one of the stars of the St. Louis Zoo's famous chimp show, performed in a zoot suit, a clothing fashion associated with African Americans.

For chimps, growing popularity meant an extended performance schedule and ever more complex acts. Sammy and Billy had become stars in St. Louis for their boxing act. Soon, they were boxing six times a day until the zoo realized they needed a break because they were becoming punch-drunk. Off-season, star chimps trained arduously and, despite the new arena, lived in such squalid conditions that the zoo was surprised when they did not succumb to colds or fever over the winter. The making of the animal celebrity, its intense and often violent training, happened behind closed cages. The display of close intimate relations with humans, by contrast, occurred in the glare of the public eye.

The Strange Case of Mr. Moke: Congo, St. Louis

Born in the Belgian Congo and civilized in Florida, Mr. Moke arrived in St. Louis in 1959 when he was just three years old, destined to be the zoo's next great ape star. Then he was chimp-napped.[59] Robert Tomarchin, the Florida animal trainer who sold the chimp to the zoo, had misgivings. At least he left a note in Moke's zoo cage when he snuck back into the zoo at night. "Mr. Moke is a sub-human child in need of all the love and care due him," Tomarchin explained.[60] Tomarchin had landed in the "biggest bit of legal monkey business since the Scopes Trial." As the FBI joined in a nationwide chimp-hunt, media portrayed the "love" that linked Moke and his former owner. They ate, slept, swam, showered, and dined out together. "Bob and Moke lived for each other," said their friend Gertrude Lintz.[61]

Finally tracked down and on trial, Tomarchin could only protest that Moke needed him. The sale was a mistake. The court fight became a contest over whom the chimp loved more: the zoo and its keepers or his first trainer. When Tomarchin claimed Moke would "grieve himself to death," the zoo countered that Moke sought the comfort of Mike Kostial, the zoo's chimp trainer. Kostial was "second to none in the animal's affections," Vierheller testified.[62]

Yet how could a court, a trainer, and a zoo measure chimp love? The evidence of affection lay in Moke's anthropoid acts, not in any effort to document the chimpanzee's more natural behavior. Tomarchin claimed Moke shared his dream of living together on their "own little island." As

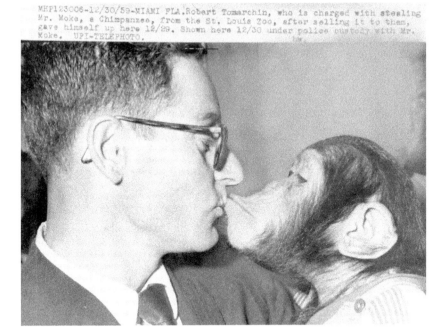

"Robert Tomarchin and Mr. Moke" (1959). In a photo taken after Tomarchin chimp-napped Moke from the St. Louis Zoo, the kiss, a typical part of the stage act, became a measure of who should rightfully care for the chimpanzee.

proof, he offered evidence of their intimate domesticity. The standard acts of the vaudeville stage replayed as domestic harmony when Moke and Tomarchin shared dinner. Moke preferred his steaks, pork, and chicken "well-done," said Tomarchin (even if chimpanzees are mostly vegetarian).[63] But some said the white, human Tomarchin was living too intimately with his African ape.

All that closeness seemed "pretty abnormal." To some who watched the trial, Tomarchin's "love" for his chimp appeared almost perversely sexual. It was one thing for African women to suckle a chimp; they shared proximity, noted observers. It was abnormal, though, for the white Tomarchin and Moke to share a bed. Tomarchin might have craved his chimp's affection, but it was still intimacy with a profit motive: Moke was Tomarchin's only livelihood. He had appeared on Miami television and on *The Ed Sullivan Show*. He even gave a special performance, sporting a gun and cigar, for Fidel Castro. After the chimp-napping, Tomarchin rented out Moke (under a false name) for the filming of *The Bellboy* (1960). Jerry Lewis might not have known that his chimpanzee costar was a fugitive.[64]

After a jury trial and a last effort by Tomarchin to flee the country with Moke, the star chimpanzee returned to the zoo. Both ape and human avoided bars. Tomarchin surrendered and was placed on parole. He sold off his last remaining ape—a young gorilla—to pay his debts.[65] Moke, meanwhile, skipped the cages of the primate house and began a whole new human relationship. He went to live with Kostial and his family, much to the delight of local reporters eager to document the life of the famous chimp. The local papers offered a lavishly illustrated "intimate study" of a chimp living as a child in the lap of white suburban domesticity. In Kostial's St. Louis home, Moke sat at the table, watched television, and raided the refrigerator for cookies. The only difference between Moke and the "human child" was his after-dinner cigarette.[66]

In a sense, Moke's performing career began long before he stepped into the zoo's arena. Audiences delighted in seeing the young ape living like a human babe. The intimacy of trainer, zoo director, or promoter with their chimps was crucial in justifying claims to apes as "missing links." Other animal acts, from elephants standing on their hind legs to tigers jumping through hoops, were easy enough to force with the whip. Yet chimp shows were designed differently. When they smoked cigarettes or walked trapezes, chimpanzees acted like humans, and most important, they were supposed to be enjoying it.

The kiss, grin, hug, shared drink, and domestic bliss all became part of chimp shows. Extending the performance far beyond the actual stage to include the suburban house helped highlight affection as the key motivation for ape performance. White keepers and their families had learned to love African animals. Vierheller boasted about his friendship with chimpanzees and other apes. Nero the Great was "my young friend," and Cookie stuck out his tongue: "Hi there, boss!" Occasionally, Vierheller shared his cigar with the gorillas. Chimps were supposed to be willing and cheerful consenting performers.[67] Especially because chimp shows represented a primatology experiment in civilizing, not simply training, signs of chimp pleasure were particularly important. This was why

Tomarchin and the zoo traded accusations of cruelty and beating. The zoo, for example, claimed Tomarchin had knocked out Moke's two front teeth.

Even those zoo directors like Belle Benchley at the San Diego Zoo who decided not to create their own chimp shows insisted that chimps enjoyed performing. "No one could doubt their happiness," Benchley recalled after watching the St. Louis apes.[68] Chimps' physiognomy was familiar enough to imagine their gestures and behavior as human-like pleasure, but not so human that their difficult show business tasks could be described as hard labor. Shows were meant to be natural, not exploitation or animal labor.

Moke's broken front teeth tell a very different story. Vierheller's accounts hint that while some chimps acceded to their training and relished its rewards, others resisted. Even popular performers had to be restrained from returning to more natural traits. Vierheller and Nero made daily trips to the refreshment stand after the show for a bottle of soda, but Nero had to be reminded to walk, human-like, on two feet. Kostial knew more than anyone about how to make chimps act like men. For reporters and their cameras, he gazed, like a proud papa, at Moke sucking a bottle. Yet Kostial also bore the scars of chimp resistance. "There is nothing more dangerous per pound than an angry chimpanzee," he said. They banged their heads, screamed, and attacked. "I had sixteen holes once in my arm from a chimp attack," he admitted. He knew several other chimp trainers with crippled hands and disfigured knees, permanently scarred by their work with apes. Performing chimps also subverted their shows in more subtle ways. Kostial believed chimps understood that trainers would hesitate to discipline them physically in front of an audience so they learned to become "cheaters," shortening and simplifying their acts while on stage.[69]

From vaudeville to zoos, chimp acts showcased behaviors that were anthropoid but were, in reality, coerced. Hornaday, for example, remarked that Peter the Great was "afraid of his owner." Even if he seemed to perform effortlessly, violence lingered. "Where he did not do a thing promptly he was told to do it and once the owner pulled his ear roughly," Hornaday noticed. Chimp trainers used a variety of techniques to wring tricks from distrustful animals. Trainers claimed to depend on kind words and kisses but still resorted to beating sticks. Some even used the "hot-shot," a pole that delivered a substantial electric shock. Lintz terrified her chimps with something she called the "bogeyman," a carved African head mounted on a stick. Merely the sight of the stick, she insisted, caused her chimps to hide under their blankets.

In Detroit, Millen always presented his relationship to Jo Mendi as one of tender camaraderie; the chimp even died "holding his paw in my hand," said Millen. Publicity described Jo Mendi as a joyous performer: "Once his audience shows the proper appreciation, he works furiously skating wildly about the stage."[70] In truth, Jo Mendi resisted his labor violently. His trainer Theodore Schroeder often appeared with bandaged hands from Jo Mendi's bites. Some at the zoo even admitted Millen fueled Jo Mendi's performances with cognac. The second Jo Mendi was just as violent. Once, the show had to be stopped mid-act when the chimp drew blood from his keeper. The ape soon retired.

Bamboo's Temper: St. Louis, Philadelphia

The first gorilla in St. Louis wore a dunce cap. The chimp shows were well established long before Yonnah arrived at the zoo in 1931.[71] Vierheller yearned for a new attraction for his ever-evolving spectacular. Yet like others before him who had tried to train gorillas, Vierheller failed. Gorillas just refused to perform. At least in their youth, some chimps could be cajoled, caressed, or cuffed onto the stage. Gorilla babies could be dressed into human clothes, but they just wouldn't act. So, Yonnah earned his dunce cap, the theatrical recognition of Yerkes's belief that gorillas were just not as close to the human as the chimp. Chimps, with their ability to ape civilization, sat at the top of this hierarchy while gorillas, "slow of intellect" and the apes that seemed most to resemble African humans, sulked below chimps.[72]

Carr Ferguson had trained Sammy and Billy to box, and their act was a hit. "[T]his act is without doubt the cleverest I have ever seen," admired a zoo donor. Ferguson was also in charge of Yonnah and her chimp play-mate, and he had a special fondness for Yonnah. "[S]he returns his affec-tion many times over," but Yonnah wasn't meant for the stage. "I've pushed that gorilla 700 miles in an effort to make her learn to ride," but

Yonnah preferred to "sit and look funny," Ferguson admitted. Yonnah was on display only for a few hours every day and disappeared into a downstairs, sky lit cage whenever she was unwell or suffered from the "usual bowel trouble."[73]

Gorillas never made particularly good stage acts, but in their remarkable likeness to the human, their bulk that hinted of ferocity, and the temptation to raise them as humans, gorillas were always a zoo favorite. As late as the 1930s, explorers still sold stories of gorillas leading planned attacks on humans. It was all "hooey," but gorillas were rare and still seemed savage.[74] Benchley enjoyed watching people watching gorillas. "I am surprised," she said, "at the people who come back day after day to watch them." They spent their days in front of the gorilla display.[75]

As they aged, captive gorillas became surly and dangerous. Lintz raised her gorilla Massa in her own home and weaned him from the savage breast and onto the civilized bottle. Then she displayed Massa at her Gorilla Village. She trained Massa (thinking he was female) to wear dresses and to help her mop the floors—until the day in 1935 Lintz accidentally spilled a pail of water on him. The startled Massa attacked and bit. A quick-thinking friend saved Lintz by stunning the gorilla with a cast-iron skillet.[76] Massa then went to live in the Philadelphia Zoo where the older gorilla Bamboo was beginning to turn on his keeper.

As a baby, Bamboo also cuddled in his keeper's arms and, like any human baby, squealed and pulled on pant legs until he was lifted up again.[77] By adulthood, Bamboo lived an angry life behind strengthened bars after two dangerous attacks on his keepers. In 1947 he sent one keeper to the hospital after a "savage beating." Bamboo escaped his cage and grabbed his keeper, battering him against the iron bars and concrete floor. Keepers struggled to force him back into his cage. They tried flinging boiling water and even scaring him with large snakes. Eventually, they used a fire hose to corner the gorilla and free the mauled keeper. A few months later, Bamboo snapped the arms of another keeper who came too close to his cage.[78] Bamboo in Philadelphia and Bushman in Chicago forgot about the keepers who had romped with them when they were young. As a young gorilla, Bushman had roamed the zoo's grounds with his keeper. Occasionally, the two played games of football, at least until Bushman grew up and became dangerous.

Partly because of gorillas' continuing rarity and partly because of their uncanny resemblance to humans—and, for so many Americans, to African humans—they were prizes so many zoos wanted and so few could acquire or keep alive. After the death of his short-lived specimen, Hornaday suspected zoos simply couldn't keep gorillas alive. So, when the explorers and filmmakers Martin and Osa Johnson returned to New York with three mountain gorillas for sale, zoo directors, circus owners, conservationists, and the general public were enthralled. Who could afford the gorillas, and would they live long enough to repay their cost? And, most important, who would care for them?

Congorilla: Congo, Harlem

During the Great Depression, Belle Benchley, George Vierheller, and Martin and Osa Johnson all dreamed of gorillas. Benchley, newly promoted to lead the San Diego Zoo, wanted gorillas, preferably a pair. Maybe they would mate, but at least they would draw visitors to her financially struggling zoo. She wanted gorillas for another, intensely personal reason. She wanted to reach out and touch the gorillas, run her fingers through their wiry hair, and turn the division of species into ties of affection. Vierheller wanted gorillas for more commercial reasons. His chimp shows had already transformed the St. Louis Zoo into a showman's palace; a gorilla pair would be its crown jewel and star attraction—even if they would never consent to perform vaudeville tricks.

The Johnsons had already made names for themselves as the nation's best nature filmmakers with *Simba* (1928) when they approached the American Museum of Natural History with an exciting idea. They wanted to make a documentary that followed Carl Akeley's footsteps into Africa and to record authentic jungle sounds. They would trap gorillas to bring them back alive, and they would find and film the illusive pygmies. They promised "our old friend the giraffe" and the "roar of the lion," but footage of the filmmakers living alongside the "almost human" was the film's real draw. The racism of the Johnson's "adventures among the big apes and little people of Central Africa" linked primate science, ape shows, and animal profit; it would soon spark controversy back home in New York. Bowing to the museum's authority, the Belgian colonial

government granted the Johnsons permission to capture a single mountain gorilla, but they came back with three—the first storm that would
engulf the *Congorilla* expedition.

Congorilla (1932) was "swell showmanship," according to a reviewer.[79]
The Johnsons trekked from the coast to the Kenyan highlands and eventually into the misty Congolese mountain jungles. Along the way, they
filmed lions gorging on a zebra and goaded elephants and rhinos into
charging. Then they entered the Ituri forest where (if they could find
them) they planned to live among the pygmies for several months. In the
humidity, their camera equipment rusted and film moldered, but they
captured, and narrated for their audience back home, the intimate
moments of pygmy life: their dances, a marriage, and, above all, their
physical oddity. In front of a curious crowd, the Johnsons measured
pygmy men and chuckled at their shortness. With their cameras pausing
on female breasts, the Johnsons stressed the pygmies' primitive nakedness. The gorilla was also naked in the jungle.

To visit the gorillas' and pygmies' jungle retreats, they argued, was to
step back into the mist of the primitive primate past—both human and
animal. Pygmies, for many Americans, seemed like another missing link,
anthropoid but not quite human. Almost thirty years before, the 1904 St.
Louis World's Fair had demonstrated that link when it displayed a group
of Congolese pygmies, including Ota Benga. By 1906, around the same
time Akeley first visited the pygmies' forest, Benga moved to New York
and into the Bronx Zoo's primate house. Hornaday was thrilled with his
new addition. Here was a missing link and an important part of his rare
ape collection. As one newspaper poet wrote: " 'Mid companions we provide him, / Apes, gorillas, chimpanzees, / He's content!'"[80] The racial
message was impossible to miss, and Benga's display angered African
American critics, especially religious leaders wary of evolutionary arguments about the proximity of ape and African. "Only prejudice against
the Negro race," a local pastor said, could explain the display of a human
in the ape house.[81]

The real trouble began for Hornaday when Benga proved too popular
and, simultaneously, too controversial. Critics protested the overt racism
of an African man on display and keepers complained about the chaos
when 40,000 people crowded the zoo to see the "wild man from Africa."

Some poked him in the ribs. Others tripped him, and still other visitors laughed and jeered. Despite the abuse, reporters asked him how he liked America and its zoos. Even after he was released from the primate house to wander the zoo, newspapers mused about his natty dress. Like a performing chimp or a dressed gorilla, the pygmy in "a suit of white duck and canvas shoes" seemed newsworthy. "Enough! Enough!" Hornaday yelled at reporters. "I have had enough of Ota Benga."[82]

Benga soon left the zoo, but Americans never did lose their fascination with "this grotesque people" or their belief that pygmies were as savage as gorillas but clearly kin to the African American. At least this is what Osa tried to suggest as she played a gramophone record of the "black bottom" dance for bemused pygmies. As they bobbed and weaved to jazz beats, the pygmy seemed to Osa to be just like the "Negro of Harlem."[83]

Leaving the pygmies' hidden villages, the Johnsons next sought gorillas. They crept through the gorillas' tunnels of crushed underbrush. The expedition found gorilla nests and knew gorillas surrounded them. They just hadn't seen them yet. So the Johnsons began laying traps. They captured two babies and purchased a third from nearby natives. At first, the infants resisted their captivity. They fought and bit, but eventually, like so many other animals caught for transport to zoos, they seemed reconciled.

Uniquely, though, the gorilla babies seemed to recognize their new dependency on their human captors and played roles in the vaudeville-like scene the Johnsons used to conclude their film. Just like the old-time vaudeville act, the young gorillas sat at a table and played with the utensils. They climbed onto the plates and tried to drink beer straight from the bottle. Good Americans, the narrator joked, the Johnsons themselves were strict prohibitionists. By 1931, the Johnsons came home to promote their film, and their gorillas were perfect props.[84] Alone in America, the gorillas, far from their own families, needed companions. The first two seemed closely bonded and would stay together despite the controversy to come. The third was given a chimpanzee as a stepsibling, and the two romped and played like the biological cousins they were.

All three also had humans to care for them, not as surrogate mothers but as fellow Africans. In Nairobi, the Johnsons had hired two natives, Manuel and Diosaner, to handle the animals and brought the two African men to America. (The newspapers were never quite sure about their real names and called them a range of different names.) The Johnsons expected the two handlers would also become exotic attractions just like Ota Benga had years before. Even as the Johnsons were sailing for the United States in July, domestic newspapers carried breathless stories that the explorers were returning not just with captured animals but with two native boys.

The Johnsons surely remembered the tension that surrounded Ota Benga's display a quarter century before, but the temptation to turn animal handlers into attractions was too great for born impresarios. Native handlers frequently accompanied African and Asian animals and just as often sparked local interest. Gaddi, for example, had returned from Asia with William and Lucile Mann's menagerie. Manuel and Diosaner were destined for a very different American trip that began, predictably, when the Johnsons decided to house them—or display them, more accurately—alongside the gorillas in the Central Park Zoo in the heart of New York. Back in Africa, the three captive gorillas often fell asleep in the Johnsons' bed, but the two men had never slept in a bed before, Martin joked. So, he housed them on the floor of the empty elephant house.

Martin later offered his own synopsis of the controversy that followed. His facts: he returned with two animal handlers and tried to show them the sites of New York, but the two "boys" were unimpressed. Soon, though, they began to get involved with bad elements, especially Communists, and (perhaps) even began drinking. Suddenly, they began making demands about the numbers of hours of work, but Martin refused. To him, the work was easy anyway, and the boys spent their days in the zoo eating ice cream and later in an apartment with real beds. Then he sent them home, a sour ending to an African safari saga.

The crisis was much broader than Martin admitted. The Johnsons hoped their gorilla prizes would sell, adding to their film's profits. They quickly realized that selling gorillas in the midst of the Depression wasn't easy, especially if those gorillas were, in fact, contraband. The Johnsons had a trapping permit from the Belgian colonial government to collect only a single gorilla, but they came back with three and the Belgians, as well as conservationists at home, were furious. Martin was earning a reputation less as a naturalist than a businessman. Frank Buck endorsed Camel cigarettes. Martin advertised Lucky Strikes. "I believe we may consider this gentleman as legitimately 'out after the money,'" scoffed one critic.[85]

The Belgian Minister of Colonies was angry, and the American Museum of Natural History and Mary Jobe Akeley, Carl's widow, whose authority had help the Johnsons win their solitary permit, were embarrassed.[86] Mary was particularly incensed at the "purposelessness of capturing young gorillas" and bringing them to the United States where they would almost certainly soon die.[87] Obviously, the "bootleg" gorillas had come to America for good, permit or not. The gorillas had been spirited out of the jungle—gorilla-napped, in fact. "It is a sad thought," said Mary, "that anyone would wish to make use of these animals who are 'almost human' for a matter of personal gain."[88]

Yet rumors spread that the gorillas were destined for a circus. Richard Sparks, a self-admitted gorilla enthusiast and friend of Mary Jobe Akeley and Robert Yerkes, was devastated. Circus lives were short ones for gorillas, and it "also means that the mountain gorilla will be classed by the American people as a side-show freak."[89] Some wanted the gorillas to go to Yerkes and his laboratory, but Yerkes or his benefactors could never meet the price the Johnsons wanted. Sparks mourned, "My thought is that with our country filled with poor millionaires it is going to be darn difficult to raise anything like the price Johnson asks for."[90]

Sparks visited the gorillas at the Central Park Zoo and chatted briefly with their handlers. He was worried about the gorillas' futures, not about the fate of Africans on display. Yet while the Johnsons toured the United States delivering lectures to packed audiences about their Congo adventures, another crisis was unfolding, this time over whether Manuel and Diosaner were human workers or ape spectacles. Sparks went to see the gorillas; many others went to gape at the Africans. A keeper at the Central Park Zoo even admitted: "[T]he native boys have stolen the show from the animals."[91] Reporters and visitors delighted in the oddity of two Africans in the city. The elevated railway and the Empire State Building (which had not terrified King Kong) frightened the "congenial little boys." Electric lights, "one of civilization's greatest miracles—to these boys," fascinated them.[92] Yet this chuckling story of savagery in civilization soon crumbled. According to Martin, the boys "went Harlem on him and he shipped them home."[93]

Martin had, in fact, taken the handlers to tour Harlem. Posturing for reporters, the Johnsons repeated the same gag they had used with pygmies, a gramophone, and the Black Bottom. They took the African gorilla companions to tour Harlem with its jazz clubs and dance halls, and Martin imagined laughing publicity about African savages encountering "people of their own color." He had jumbled up the lives of Africans, African Americans, and gorillas for zoo and film audiences to enjoy.[94] The problem was that with fresh memories of Ota Benga, African Americans were ready to protest the display of African humans as their missing link to anthropoid apes. Suppose scientists, wrote one exasperated African American critic, "should discover that the gorilla is far closer to the Caucasian than to the Negro?" Could gorilla studies ever escape "racial preconceptions"?[95]

Harlem was ready to welcome the two African men, but in ways the Johnsons never expected. Clergy had led the opposition to Ota Benga's exhibition. This time, pan-Africanist activists in Harlem opposed the zoo display of African humans. Such anticolonial activists argued that histories of slavery and colonization (not chimps or gorillas) linked African Americans and Africans. New York activists turned the Johnsons' savage spectacle into a labor struggle. After Martin's tour of Harlem, a group of angry Harlem women marched on the Central Park Zoo, and the pan-Africanist Native African Union of America appealed to the British Consul. The handlers, after all, were British imperial subjects. Later, the union provided a lecture stage in Harlem for the two Africans to speak.[96] The irony of their handlers competing with them on the lecture circuit could hardly have been lost on the Johnsons. In protesting against the "inhuman treatment" of the handlers "housed in the Central Park Zoo among animals," the union distanced ape from African. The handlers had found powerful allies and soon demanded "things essential in the Negro's life today:" union wages of $5 a day (instead of the $4 a month wage the Johnsons had promised), the eight-hour day, and hotel accommodations.[97]

The union wage and a real bed meant more than just fair pay for the hard work of animal care. They confirmed these two men were workers, not missing links. One curious African American woman regarded "these men" as an all-white crowd at the zoo tortured them. She was incensed. The handlers, she said, were taunted in precisely the same way as zoo apes. The crowd bought the handlers cheap lemonade and "tried to put a show on, laughing childishly about every misunderstanding."[98] Some visitors, like the Johnsons themselves, tried to find the comedy in Africans who had "gone Harlem." Perhaps the handlers had simply imbibed intoxicating "large doses" of "Western civilization."[99] They chuckled at newspaper reports that, alongside demands for union wages, the two handlers had also asked for ice cream, a car, and a radio. "It doesn't take long to Americanize even savages in such matters," sniffed an observer.[100]

The Johnsons were worried. Their apes still had no buyers, and their handlers had rejected the zoo enclosure for the union hall. The solitary gorilla, Okero, and his chimpanzee companion headed to the National Zoo, and Vierheller desperately wanted the other two apes. He even considered showcasing the ape pair in a paid exhibit, but the St. Louis Zoo board balked. The warm weather season for the ape shows had already passed, and purchasing the gorillas now was risky business. There was every chance they wouldn't last the winter. After all, Okero died shortly after reaching Washington.[101] The Johnsons were desperate to sell the gorilla pair—fast and at reduced price—and they needed to send their handlers home. As protests against the display of the African men grew around the zoo, the Johnsons begged the police to keep order outside the human enclosure. "A little civilization . . . is a dangerous thing," mused one observer, especially if it was the "civilization of Harlem delegates" and "race relatives."[102]

Martin rushed to return the "boys" to Africa. Publically, he declared it was all part of a promise to their chief. Armed with cheap souvenirs, they were ready to return to nature, to the "native veldts where living is less complex than in the white man's civilization," said Martin.[103] Black activists in Harlem had turned to the politics of pan-Africanism to reinforce distinctions between human workers and gorillas, and they hated *Congorilla*. For an African American reviewer, *Congorilla* was an "abortion" of a film.[104]

While their original handlers returned to Africa, the Johnsons finally found a buyer for the two remaining gorillas, Ngagi and Mbongo. For Benchley and the cash-strapped San Diego Zoo, the $11,000 price was still a struggle to meet. The Johnsons, though, had hoped for $20,000. They probably never earned "very much over the actual cost."[105] By the fall of 1931, the gorillas sat on a train to San Diego passing from the care of Africans to the trained hands of San Diego's zookeepers and its female zoo director. They still remained dependent on humans for food, health, and affection.

Ngagi and Mbongo, the Johnsons' Black Babies: San Diego

Benchley could never resist touching her anthropoid apes. They were, she admitted, "the most adorable babies which hold out their little hands." She believed that chimpanzees "always turn to me," but her true favorites were the gorillas.[106] Though she never liked chimpanzee shows, she dreamed of a pair of gorillas for the San Diego Zoo. Chimps in human clothes unnerved her. Apes aping humans revealed "the hideous likeness and pitiful differences between man and beast."[107] That feeling never left her even when she finally had a chance to add gorillas to the zoo's collection. When, finally, she saw her first gorilla and he reached out to touch her hand, she felt at the end of a long evolution that began with the "wild gorillas of the jungles."[108]

"[T]hese great black hairy men" had come into her life.[109] American gorilla conservationists were relieved the gorilla pair had found a haven in San Diego. Of course, their first choice would have been to send the gorillas to Yerkes's laboratory or, at least, to keep them on the East Coast, nearer to leading primatologists. Yet San Diego offered something special and unique: a woman leading the zoo. Manuel and Diosaner handled the gorilla, but Benchley could care for them as a white mother cared for her children. "I believe these animals strongly desire 'the feminine touch,'" claimed Sparks. "All the specimens I have seen are quickly responsive to affection."[110] Gorillas might somehow sense a woman's—a mother's—touch. Benchley agreed. When she first met her two gorilla babies, they "came up and handled my clothes." "I expect to spend a great deal of time" with the gorillas, she admitted, "largely for my own pleasure."[111]

She began a daily diary recording the gorillas' adaptation to California living. In particular, she registered the apes' growing comfort with her presence as well as her own disappointment when she confirmed what the two African handlers already suspected: both apes were "papas," and there would be no gorilla babies. The two apes were new in the zoo, and even after the traditional Labor Day end of the summer season, the crowds still packed the zoo to glimpse the two babes. In the evening after the throngs departed, the gorillas appeared happier and Benchley rushed to their cage. Mbongo lay on his belly on his favorite eucalyptus stump. "Some way," she wrote, "they seem very childishly dear to me."[112]

"I am trying definitely to make friends with them," Benchley said. As much as she worried about their loneliness, she also craved an "intimate closeness with the Johnson's Black Babies." A few months after their arrival, the two young apes remained shy, but Benchley believed they tried to attract her attention when she passed their cage. Then one day Benchley was elated. Mbongo came to the wire of the cage and plucked grapes right from her fingers.[113] She tried another time with raisins, and this time Ngagi pressed his finger into her palm. He ran off scowling, though, when she closed her hand.[114] Ngagi seemed especially affectionate, and soon Benchley could "put my hands on her almost at will."[115] She came to believe that without a woman's affection gorillas could never survive in the zoo. They would simply waste away with loneliness.

They didn't die of loneliness, but they did die young. Ngagi died in 1944, two years after Mbongo. Benchley was particularly affected by his death from Valley Fever, or, more scientifically, *Coccidioides immintis*, a local fungus native to the American Southwest. She suspected he caught the disease from an infected clod of mud hidden inside his hay. Valley Fever mimicked tuberculosis, and Mbongo wasted away in front of Benchley's eyes. She reached through the wire to the dying ape to "lay my hand on the huge black one." Briefly, their eyes met, and she realized, as he moved "the cold hand on which mine rested," that she wouldn't see him again. Certainly, his death was a loss to science, but he was also "a grand entertainer" and "a rare friend indeed."[116]

As much as she loved touching her gorilla babes and fervently believed they longed for her affection, she could only feel them through the wires of their cage. Before the babes arrived, she planned their cages carefully, and she was confident they felt secure. As she looked through the wires and into the gorillas' eyes, she realized that the zoo's job was to understand how a gorilla family would live in the wild. Only then could the zoo properly exhibit gorillas. Yet Benchley still wondered if gorillas could be civilized to live like humans. She once jotted down in her note-

"Weighing the Gorillas" (1940). Belle Benchley reaches through the fence to touch her gorillas during their weighing.

book complicated ideas about the development of the brain in young apes. She was trying to solve a simple challenge: "how a gorilla might be changed into a man." She often pondered that question as she sat watching Ngagi and Mbongo, especially when they came to the wires of the cage and looked "curiously and affectionately into my eyes."[117] Even as gorillas became more common in America, some still wanted to change them into men.

Bobo, Jean, Bill, and Raymond: Anacortes

Six years after Ngagi's death, Benchley took stock of America's gorilla population. She estimated there were "twenty-three or four gorillas"—more than ever before.[118] A year later and just two decades after Ngagi and Mbongo arrived in San Diego, Bill Lowman was entranced by chimpanzees, and like a surprising number of Americans, he wanted to raise a primate in his own home. Near to his parents' Anacortes, Washington, home, a keeper at the Seattle zoo hinted that a young hunter in Columbus, Ohio, might just have a chimp for sale. Once in Ohio, though, Bill discovered a gorilla, not a chimp, lying in the bassinet. Bill wasn't particularly wealthy, and $4,000 was a lot of money. In fact, he was deeply in debt, but a gorilla baby was hard to resist. A few days later, Bobo was his and ready for his long drive west to Washington, a present for Raymond and Jean Lowman, Bill's parents.[119]

By the 1950s, the Simian Society of America united Americans who wanted to rear monkeys or apes in their own homes. "Toys are not enough to satisfy" a pet primate, the Society told its members. "It needs YOU," even when it shows its teeth in attack. It's just "like the child." The society was excited for the Lowmans to join, especially since they had become the parents to a gorilla "man-child." At home in Anacortes, Bobo enjoyed "every civilized advantage," even a velvet security pillow.[120] Word spread that a gorilla, still rare in America, was living in a comfortable but small human house. Bobo became a local celebrity. Still an infant, he donned overalls, posed for photos as a Mobil gas station attendant, and earned his family a $12 fee.[121] A half century after ape shows had become popular, the public still loved the idea of an ape living in a human home and wearing clothes. At nights, Bobo crept into Bill's bed, and during the day when he craved affection, he went to Jean. She bathed him and oiled his chest.

At home, Jean, Raymond, and Bill struggled to bend the gorilla's nature to the confines of their house. Like Edwards and Joe, Schroeder and Jo Mendi, Tomarchin and Moke, Lintz and Bamboo, the Lowmans and Bobo forged close and dependent ties across the lines of species. Once Jean and Lintz talked about how they learned to love their gorillas. Maybe, they thought, it was because they were so "nearly like a human child." After raising apes in her home, Lintz wondered "where the animal leaves off and the human elements begin." A trail of destruction, though, also marked the species line.[122]

"Dear child" or not, the Lowmans soon had to admit Bobo wasn't growing up civilized.[123] Ripped from the wild, he ended up in human women's arms, first black African, then white American. Breast-fed first by a native woman, Bobo learned to drink from a bottle in the "civilized world." For some apes, that civilizing ended in violence and tragedy. Captain Jiggs was one of Lintz's favorite chimps. She raised him as an "experiment in civilization." He wore the nattiest of suits and lived with Lintz inside her Brooklyn home. She touted Jiggs, among all her "babies," as evidence of apes' ability to adopt human civilization. She boasted to scientists that someday he would learn to talk. She even took to the radio in 1938 to try to prove apes could converse with humans. That same year, though, Jiggs reverted to the jungle. His "rampage" began when a rat entered his cage and he escaped. He refused to be cornered. After five hours, Lintz implored the police: "shoot him. Are you afraid?" Jiggs died from police bullets; Gertrude Lintz fainted.[124]

In the case of Bobo, the civilizing experiment failed in the litter of broken lamps and stolen Ovaltine. When Jean put up new curtains and a fresh towel in the bathroom, Bobo tore them in two and, for good measure, tossed the toothbrush in the toilet. In Bill's room, he flung phonographs, and in Raymond's, he smashed the light, threw the radio, and upturned the television. In the kitchen, Bobo swiped the Ovaltine, an 89-cent complete loss.[125] Jean began tallying the cost of raising an ape as a human, maybe out of exasperation that civilization just didn't take. In one month, Bobo shattered a couple of panes of glass and walked straight through a closed French door. He chewed the toothpaste tube, smashed the aftershave lotion bottle, ripped three books, ruined a dollar-a-day's worth of food, and cracked $20 worth of dishes.[126] Unlike Jiggs, he never did attack his human parents. Despite all the tensions and confusion, he seemed to feel affection for his new parents. Years later, he still whined pitifully when he saw his mother—Jean, that is.

By 1953 when Bobo was three years old, the damage got worse, and soon the Lowmans searched for a new home for their gorilla child. Bill considered taking the ape on the road as part of a show. "I could have made a million bucks easily," he boasted. Instead, the Lowmans sold him to Seattle's Woodland Park Zoo, and for $5,500 the zoo had a new star attraction. The crowds came, and the zoo soon earned enough to afford a new primate house.[127] Not everyone was ready to abandon the civilizing experiment. "*How can* you people be so *heartless* after having raised" Bobo from a baby and now sending him off to the zoo, read one angry letter. Would the gorilla grieve himself to death? "Sam and Lola" were also disappointed. "Lola and I have had loving care for over 11 years" and still hadn't been sent to the zoo. They were pet kinkajous.[128]

The zoo hoped Bobo would breed with Iffy, his cage mate. He never did, but whenever the Lowmans visited the zoo, he recognized them with whines and a shaking of his cage bars. Bobo seemed forever caught between apes and humans. When Bobo died, the newspapers called him a "bachelor" gorilla, but in reality, he was an American ape, a subspecies of sorts: ape but almost human. Bobo, the American ape, was African in origin, gorilla in genetics, but human in his family. His habitat was first the human home and then the enclosure, estranged from other gorillas.

The American desire for both affection and entertainment across biological barriers had changed Bobo—and so many other apes—forever. "Bobo was a complete ham, a lovable showoff who liked people," remembered one observer.[129] Along the way, apes like Bobo forged relationships with the African women who suckled them, the white American women who gave them bottles, scientists who tested them, impresarios who trained them, and keepers who cared for them. Bobo began his human life with an anonymous African woman before passing into the arms and home of the Lowmans. John Nichols, his keeper, was his last human companion.[130]

Bobo, and other African apes, reminded Americans how they divided their species into races. Nichols, for his part, knew he patrolled the line dividing apes from humans, protecting Bobo from humans intent on interaction. Visitors tossed food, measured their hands against Bobo's on the glass, and made ape noises. Nichols realized that the thick glass protected visitors from Bobo and not the other way around. Despite the hard work of keeping animals and people apart, Nichols grew close to his gorilla. He had learned to care: "[Bobo's] table manners were atrocious," he remembered. "But Bobo was a friend of mine."[131]

6

DON'T FEED THE KEEPERS

The Labor and Care of Zookeepers

THE ANIMALS were lonely at the Bronx Zoo in spring 1961. The elephants longed for peanuts. Frisky monkeys waited for an audience for their antics. The keepers were on strike, but newspaper reporters wrote headlines for color pieces rather than the business pages. If assembly lines grinding to a halt evoked economic insecurity and memories of Depression-era class conflict, a zoo strike could be good fun, an opportunity to reflect on the tender ties that linked animals, keepers, and visitors, especially children.

Virtually every major American zoo faced or narrowly averted a strike from the 1940s to the 1970s, climaxes to organizing campaigns that stretched back to before World War II. Such strikes had two distinct contexts. The strikes were part of a wave of organizing that roiled municipal and other public sector workplaces. Zoos remain, today, a largely unionized industry and a big one at that. The zoo industry employs about as many people as Chrysler. Zoo strikes also occurred over the longer history of changing relationships between visitors, zoo administrators, keepers, and their animals. Ironically, although zoo strikes confirmed the place of keepers as intimate human companions for captive animals, their local intuition and experience did not necessarily translate into a global role in animal conservation. That role would fall to the celebrity zoo directors, scientists, and even public relations experts who straddled the fine line between biological science and popular education. Where did care end and conservation begin?

Once, when I took a break from reading in the Philadelphia Zoo's archives, I watched a keeper caring for an okapi. The zoo creates close relations between animals and their humans. She brushed the shy forest animal, murmuring words, easily drowned out by the tired toddler standing next to me. As she brushed in one direction, the okapi darted its long purple tongue over the keeper, perhaps in mutual grooming. Nearby was the species survival sign touting the zoo's place in saving these rare, shy jungle grazers. What was the connection between quiet words between a keeper and a rare animal and its conservation?

Zookeeping was a very different occupation in 1970 than it was in 1900. The hard work of animal care has changed as zoos have evolved. The changing goals of zoos and the growing imperatives of species conservation all transformed the workaday life of keepers. The challenges of managing visitors and animals—both unpredictable in their own ways—have, however, always confronted keepers. Saving endangered species demands human labor and lots of it: feeding, then shoveling manure, husbandry, prenatal care, and, frequently, hand-rearing. While it is easy to imagine zoo managers chuckling at the humor directed at the picket lines snaking around a zoo, these strikes showed just how important the keeper was in the daily life of the zoo and the animals that lived there. The zookeeper was the mediator between administrators (increasingly removed from the exigencies of everyday care), visitors (bound by ever-growing sets of rules on their behavior), and animals (the last precious examples of their species).

A Safe and Suitable Man: Washington, New York

George E. Crane called the president in 1908. He was a good employee, though he had "many whims and vagaries," noted Frank Baker, the National Zoo's superintendent. Among them, perhaps, was a tendency to complain about his pay directly to the "Executive Office."[1] In the still-young zoo, a constant stream of keepers' petitions frustrated Baker. Still, the zoo director admitted, he could understand why the keepers grumbled and Crane phoned the White House. Animal keeping was "arduous, confining, and often unpleasant," and even worse, Baker recognized, it paid poorly. Among the major zoos, the Bronx Zoo's keepers earned the highest pay at about $70 per month in 1905, despite the "frightfully high" cost of New York living. In Philadelphia in the summer, keepers worked eleven-hour days with a half day off during the week for $50 a month. They earned wages and worked hours equivalent to railroad workers, but the train engineer never faced a surly elephant.[2]

Around the same time Crane called the president, M. C. Hunter submitted his resignation from the National Zoo. "I have tried to do my duty," he wrote, but facing family sickness and his own ill health, he simply could not endure the long hours. "I shall be compelled to look for work at once."[3] Even though federal eight-hour-day laws bound the National Zoo from Monday to Friday, weekend work had no legal limits. "You have strictly complied with this requirement so far as the hours on weekdays are concerned," the keepers complained to Baker in 1906. However, they also worked nine- to ten-hour days throughout the weekend. They petitioned for two leisurely Sundays a month.[4]

Without the "Sabbath" free and respectable wages, zoos could hardly expect to attract the "suitable man," workers complained to Baker. "Married men with homes, and settled habits, who are likely to devote their lives to this work," they reminded, "have been found to be the most valuable and trustworthy employees." Sobriety, marriage, and a dose of physical strength represented the perfect alchemy for the work of caring for restless and caged animals. The zoo had its animal dangers, keepers warned, and only they—and their sober habits—separated curious, if ill-mannered, visitors from fascinating, if rebellious, animals. "A careless or incompetent" keeper might lead to loss of life of animals or visitors. "These men should in no sense be classed as unskilled laborers."[5]

Zoo directors had other ideas about what made for an ideal keeper. William T. Hornaday always imagined the zookeeper in uniform as part of a "force," and he demanded the same demeanor as a soldier on duty. They must not "lounge on guard-rails, at any time. When on duty they should stand erect, and present an alert appearance," Hornaday admonished his apparently sluggard employees. Even a loose button could drive Hornaday to circulate an impatient memo to his employees.[6] When the United

States entered World War I, the already militarized Bronx Zoo was prepared. Under the shadow of newly raised towering flagpoles, the zoo converted part of the lion house into a Red Cross branch. Keepers, already in
uniform, were armed and mobilized into a military corps—with zoo managers as officers. "Of all the Home Defense companies of this City,"
Hornaday boasted, "ours is the only one actually armed."[7]

As the war came to a close, the United States entered a new era of
upheaval. A strike wave in 1919 devolved into a Red Scare that targeted
political radicals, especially immigrants. In 1916, Madison Grant, still the
chairman of the Executive Committee of the New York Zoological Society,
published his alarming *Passing of the Great Race,* warning that "inferior"
immigrant racial stock was overwhelming the superior "Nordic" races.
The fervently anti-Bolshevik Hornaday enthusiastically joined the anti-
immigrant fervor Grant had helped stoke. Hornaday demanded all immigrant employees show their citizenship papers directly to him. Only citizens could work at the zoo, he declared, and preference would be given to
honorably discharged veterans.[8] He always liked to think of his keepers as
loyal soldiers in a conservation army.

The problem was that trim uniforms weren't well suited for caring for
unpredictable animals. Zookeepers lamented "our willing but sometimes
weary shoulders." This, then, was the essential contradiction of zoo work.
On the one hand, keepers recognized theirs was a caring job, akin to
motherhood. The keeper must display care similar to the "gentleness,
consideration and judgment due those who are helpless in the human
family." Keepers knew their job was "special work" that required not only
"utmost kindness and precaution" but also "quickness of action, love of
animals, and intelligent judgment." On the other hand, keepers dressed in
military-style uniforms for work that demanded the discipline and muscles of the soldier.[9]

When keepers donned their uniforms and entered the enclosures to
care for "valuable specimens of rare and costly animals," they insisted on
their sobriety, skill, patience, and knowledge, a set of traits they believed
marked zookeeping as a manly skill. There were certain skills they brought
to the zoo as "suitable" men. Their code of manliness balanced physical
exertion, knowledge, and temperate nature. Keepers weren't alone among
workers in the early decades of the twentieth century to highlight a manly

code when proclaiming their labor as a skilled trade. Trades as diverse as machinists and printers all linked physical strength to sober habits in codes that marked their work as manly and skilled. The difference was that zoo work was also caring work.[10]

Hornaday, Baker, and virtually every other zoo director shared ideas about what made the safe and suitable manly keeper. "The best man is the man who does the most to promote good-fellowship, and keep the building and ALL its animals up to a high standard of health, cleanliness and good order," declared Hornaday. That, and a respect for the uniform he wore.[11] Zoo managers demanded safe, sober, and strong men. Yet these same attributes had dramatically different meanings for keepers than they did for superintendents like Baker. For keepers, these were the terms of manliness so crucial to their sense of themselves as skilled workers. For Baker, they were simply the necessary requirements for work in "the muddy conditions of the inclosures [sic] which are barren of grass."[12] When Baker watched keepers laboring outside his office window, his mind turned to cattle stock handlers. Keepers, he believed, cared for animals but merely by following his orders. After all, keepers of the pre–World War II years boasted no formal training. Rather, they often rose from the ranks of unskilled laborers who tended the grounds, constructed buildings, repaired damage, and hauled manure.

In the gap between skilled work and dung shoveling, between sober men and soldiers, keepers and managers clashed. Frustrated keepers turned to collective action to demand the wages and conditions they felt due skilled, manly workers. In the Bronx, Philadelphia, and Boston they submitted petitions. In Washington they turned to the central city union, and just as Crane had done five years before, they appealed to the president himself. "In many instances the safety of the crowds," they declared, "depends upon the vigilance, care, and bravery of these men." To keep these men laboring past the point of "rational endurance" put visitors at risk. Only the keeper, after all, stood between the eager visitor and the angry wild animal.[13]

Baker remained unconvinced. To be sure, he admitted, animals knew their keepers best, but the zoo director made final decisions about their care. The arbiters of animal life, directors jealously guarded their right to fire workers and control their hours—it was only about what was best for their captive specimens. "We are very strict with our workers," Hornaday said. "Those that are not satisfactory, we do not hesitate to discharge."[14]

Codes of manly labor helped keepers define animal care as skilled employment worthy of healthy wages. Although their wallets were still thin, such codes were the rules they set for themselves based on the claim they knew how to care for animals as individuals, not specimens. In Philadelphia in 1902, they reminded Superintendent Arthur Brown of their "continued faithful service" when they asked for salary increases. After all, they had families.[15] Codes of manly work were a source of pride, but they weren't necessarily a reflection of keepers' daily routines on the other side of the enclosure. Quietly, when directors sought to impose their vision of order on the zoo workplace, some keepers and other zoo employees quit, rebelled, unbuttoned their uniforms, lazed, or just ignored the regulations. So in 1918, Hornaday fumed when keepers did " 'the guide act' for visitors." For tips, visitors might get questions answered or, even better, special access to the animals. Hornaday was just as infuriated when workers "gossiped," lounged on the guardrails, or left "offensive matter . . . in cages after it has been deposited."[16] After Prohibition, Bronx Zoo keepers began fermenting wine in the dark corners of cages and houses. The new director W. R. Blair sent around another angry circular.[17]

Crane wrote the president, Hunter resigned, and others chafed in their military uniforms. Still, many other men sought work at zoos and for many different reasons. J. W. Buchanan was one of them and, by all accounts, "a suitable man." In 1925 he was living in a modest Alexandria, Virginia, hotel room ("hot and cold running water in every room") desperate for a position at the National Zoo. He had worked for more than three years at the Philadelphia Zoo, balancing the seven-day week with caring for his ill wife. He left voluntarily and rejected offers from the Cincinnati and Bronx Zoos. His wife now recovered, he wanted to return to zoo work. "The care, feeding, and making clean and comfortable the lives of wild animals has been a labor of love instead of task," he wrote.[18] Arthur Clark also knew it would be hard to find a job at the National Zoo, but he was eager for more experience with the kinds of rare birds and mammals only a big zoo owned.[19]

For Max Spitz, by contrast, a place at the zoo was more a sign of desperation than of his love of animals. Among the National Zoo's employees in its first decades only a handful arrived with previous experience in animal care, usually at smaller zoos or circuses. Spitz was more typical. He was a veteran, no longer capable of doing a soldier's work. Perhaps, though, he could wear the zoo's uniform instead of spending "the remainder of his days in idleness at the Soldiers' Home."[20] Especially at the National Zoo, zookeeping kept injured or discharged soldiers in familiar uniforms and on the federal payroll.[21] Zookeepers followed circuitous routes to the zoo. However, when they arrived, they tended to stay. J. C. Mayer came to the National Zoo from the circus; by 1908, he had worked at the zoo for fourteen years. J. E. Dean had left a job as an attendant at the insane asylum for the zoo where he worked for more than six years.[22]

In the early years of American zoos, Spitz, Mayer, Dean, and thousands of others cared for animals within the strictures of the militarized vision of zoo directors and their own tough standards of what defined a suitable man. Sometimes when tough conditions worsened, keepers petitioned collectively to demand what skilled workers deserved. Other times, they fermented wine or otherwise bristled against a rule-bound workplace. Rules and resistance together helped shape the zoo workplace. So, too, did the intimate and dependent relations between animals and their keepers. The lines between affection, care, frustration, and anger were easy to cross, and both animals and people bore the scars.

A Reliable and Trusty Keeper: New York, Chicago

Things were not going well for one particular keeper. At work, the animals were restless; he hadn't brought their food on time. An air of rebellion had settled over the cages and enclosures. The keeper's attention drifted from dangerous animals to his collapsing domestic life. The code of manly labor that from the perspective of keepers had governed their work was unraveling.

It was, though, simply a play and a children's one at that, written and performed by the Federal Theatre Project, the New Deal agency supporting unemployed theatre professionals. The "Revue at the Zoo," published and performed between 1936 and 1939, reflected the New Deal's growing engagement with the world of zoos as well as zoos' growing orientation toward children. The "Revue" captured something very real about the intimacy of keeper and animal in the first decades of American zoos. A morality play, it pictured the zoo for the tenderness and tensions of interspecies care rather than dry taxonomy. Its spare stage set featured moated walls, depicting the style of zoo design the New Deal vigorously promoted. In their modern enclosures, animals debated why their food had not yet arrived. Encouraged by Fox—the play's villain—the animals conspired to replace their keeper. Only the timely arrival of a child saved the beleaguered keeper. The young visitor confessed to his animal friends that their keeper's child needed an operation the struggling keeper could not afford. Instead of revolt, the animals began a charity for their keeper.[23]

When zookeepers worked so hard and for declining wages during the Great Depression to keep animals alive and fed, those captive animals ceased to be specimens. They became individuals alternatively to love, fear, hate, and cherish. Unlike museum guards, zookeepers' charges had characters of their own. Some were individual traits; not all elephants were alike. From the perspective of the administration building, animals were still just specimens of their species. Rhinos were surly. Tigers were fierce. Snakes were poisonous. Closer to the cage, for William Blackburne, the National Zoo's head keeper, scientific nomenclature dissolved into individual temperament. Keepers saw the zoo for its daily drama rather than its taxonomy. "There are those who are ingrates and those who respond with devotion to kindness," said Blackburne. "Some are cruel and . . . some are the best-natured imaginable."[24]

In the close quarters of cages and enclosures, keepers and their animals developed a cross-species companionship that like any relationship had its moments of both affection and anger. So it was, then, that Raymond Payne sought the help of his congressman. He had been fired from the National Zoo where he had worked as an assistant keeper of the monkeys. Blackburne distrusted Payne; he was hardly "a reliable and trusty keeper." "Not too careful about telling the truth," Payne seemed careless and impatient with his charges. Once, "I found him abusing a monkey with an iron scraper," Blackburne reported. A banana would have enticed the monkey back into its cage. Instead, the animal suffered a loose tooth and refused to eat for several days. Blackburne, who often used his position as head keeper to advocate for zoo workers, was in this case swift to recommend Payne's firing. Payne had breached the manly code Blackburne so cherished.[25]

The National Zoo was quick to fire Payne, and even a congressman couldn't help. When animals caused harm, though, it paid injured workers up to thirty days' wages. Like most major zoos, the National Zoo promised surgical and medical aid to its keepers. The Philadelphia Zoo was more generous when it offered three months' pay to injured workers, and in the Bronx as early as 1907 Hornaday even drew up a "memorandum of understanding" recognizing the dangers of animal keeping. Any work with "wild animals in captivity" is necessarily perilous. The keepers' responsibility was to avoid injury to animals, visitors, and themselves.[26] It was a dry, contractual warning about the reality of caring work along the species line.

The Bronx Zoo warned against "those who are careless in the presence of animals, or who exercise defective judgment when liable to attack." Yet whatever the judgment of keepers, animals did attack. A de brazza monkey suddenly became enraged as Harry Dean washed his cage. Dean rushed to the hospital with a two-inch "severe gash" from the monkey's teeth.[27] Dean was luckier than a keeper at the Central Park Zoo who confronted an African rhino named Smiles. Smiles had come to the zoo after years of mistreatment in the cramped cages of a circus menagerie. Finally at home in New York, she remained suspicious of humans, charging the bars of her cage at the slightest approach from keepers. To enter and clean her cage, the keeper had first to rope Smiles. It worked well enough until the day she burst her bindings. "The beast rushed with lowered head and horn tilted forward to run its keeper through." Other keepers came running, armed with pitchforks and iron bars.[28]

Elephants were a particular danger to animal keepers. At the St. Louis Zoo, a female went "insane" and tried to kill her keeper. She tore off David Frede's ear and smashed him against the walls of her cage. Strychnine— enough to kill 200 men—failed to euthanize the elephant, and she was finally killed by rifle shots.[29] Male elephants could be even more dangerous. Males in "must," that is, sexual maturity, could become particularly violent, keepers insisted. The "erstwhile gentle, willing friend" could suffer "frightful paroxysms of rage."[30] George "Slim" Lewis came to the Brookfield Zoo after a career training elephants in circuses. In Chicago, the zoo needed an expert to care for Ziggy, an elephant too dangerous for a circus life. Lewis had worked with herds of elephants and succeeded in forging a "friendship" with each individual—except Ziggy. Zoo officials, despite Lewis's misgivings, had decided to breed Ziggy. Lewis knew it would lead to trouble; zoos just didn't understand enough about elephant's fertile periods or mating patterns. The closest the zoo came was a false pregnancy in one of the female elephants, but Lewis realized he had lost control of his resentful and now sexually frustrated elephant. He worried trying to breed Ziggy would lead to violence. "But," he said, "I was a hired hand and did not have final say in such matters."[31]

Ziggy and Lewis managed an uneasy truce until April 26, 1941, the day Ziggy turned and charged. Before Lewis could flee, Ziggy had pinned his keeper's arm to the dusty ground of the elephant yard. He stabbed with his tusks and tried to gnaw Lewis's leg. Lucky for Lewis, though, Ziggy's final stab buried his tusk in the dirt, giving Lewis crucial seconds to scramble to safety. Ziggy spent almost the next three decades enchained inside the elephant house.[32]

Ziggy was in chains, and Lewis felt defeated. Other keepers could care for the zoo's other elephants, but Lewis was "no longer master" of his elephant.

Ziggy and George "Slim" Lewis (1941). A visitor snapped this dramatic photo of Ziggy during his attack on his keeper George "Slim" Lewis.

Even when he drifted west to work in Seattle's Woodland Park Zoo, Lewis still remembered Ziggy. Despite their struggles, Lewis loved that elephant "as a man would love a wayward son or brother." When the zoo finally let Ziggy wander outside in 1970, Lewis was there to lead him into the yard.[33] Ziggy was an angry animal, never reconciled to captivity, but Lewis knew the vagaries of his care had unquestionably made things worse.

Animals like Ziggy were dependent on their keepers, but decisions about their care ultimately rested on directors and zoo managers. For Lewis, the division of zoo labor between keepers and directors almost cost him his life. In New York, it was fatal for a mountain sheep and tragic for his keeper. During a cold rainstorm, the animal caught cold and died, after eight years in captivity. His keeper was censured for forgetting to bring the animal into shelter. "This rare and valuable animal perishing miserably . . . will haunt him and sting his conscience for the remainder of his life." To add to the prick of conscience, Hornaday sent the keeper a picture of the late animal and demanded he hang it in his office.[34]

Backsliding against the Rules: New York

Hornaday was frustrated. He insisted that rules and regulations were easy to follow and uniforms just as simple to maintain—but they rarely were. On the other side of the enclosure, the sometimes tense, occasionally tender relations between keeper and animal turned military order into everyday disorder. Hornaday was supremely confident that he knew what was best for the zoo and, therefore, for the animals: strict military discipline and total loyalty. The orderly hierarchy of nature, though, was frayed by the reality of the animal care and the fragility of animal life, a daily pressure Hornaday, in the comforts of the administration building, rarely experienced in person.

In 1922, Hornaday railed against "backsliding against the rules." He warned against poor cleaning of cages and lateness. He was particularly angry if employees stopped work fifteen minutes before the final whistle. Hornaday decided to dock workers a full day's vacation for each offense. "If the old custom continues," he warned, "we will then have to consider the withdrawal of the vacation privileges now granted." His successor W. Reid Blair was just as incensed about carelessness with uniforms.

Workers stripped down to undershirts, and they wore dirty, torn uniforms, especially when they scrubbed out the cages. "With soap at 5¢ a cake, there is no reason why every man should not appear . . . in a clean suit of working clothes," Blair fumed in 1936.[35]

From the very beginnings of American zoos until the 1930s, zoo officials and workers defended different models of what made for an ideal keeper. The former sought military order, as if to mirror natural order. The latter praised a manly code of behavior. In some ways, they each ended up at much the same point: zoo work, however much it was caring labor, demanded sobriety and physical bravery, and neither side doubted it was men's work.[36] The Depression changed the working life of keepers with the speed in which aging zoos became modern New Deal zoos. Quasi-militarized keepers (along with other zoo workers tending the grounds or fixing cages) found themselves through relief programs as essentially public employees, akin to other municipal or parks employees. By the end of the New Deal, the New York City parks department directly paid most Bronx Zoo employees. The remaining few, paid privately by the New York Zoological Society (NYZS), pocketed wages set at exactly the same scale. Once a private society that enjoyed the largesse of urban elites, the NYZS was now entangled with city government.

In 1942, NYZS President Fairfield Osborn tried to remind his now essentially public employees that "the zoo is a special place."[37] On the one hand, Osborn was gently asking his employees to accept their low wages. On the other, he was acknowledging the new reality: thousands of new workers had arrived on the other side of the enclosure with little experience in animal care. With the wave of New Deal Zoo construction, zoo workers had arrived so quickly and with so little prior experience that they knew little of those codes cherished by an earlier generation.

Zoo managers often described the New Deal years of federal relief largesse as a golden age for American zoos, yet they still bemoaned their diminishing power. Managers confronted workers who eagerly aligned themselves with a newly powerful labor movement. In the 1930s, zoo workers joined a wave of union activism, galvanized by a new federation of unions, the Congress of Industrial Organizations (CIO). In 1937, the nation as a whole was convulsed by a series of strikes, including, most dramatically, a sit-down strike at a Chevrolet auto factory in Flint,

Michigan. A few months later, on a busy summer Sunday in the Bronx, zoo workers walked off the job for the first time. Amid the wave of strikes in heavy industry, the trouble in the Bronx never made it into the papers, and the strike ended quickly. The union, though, had arrived at the zoo, and the struggle over collective bargaining and animal care would continue for the next thirty years, slowed, but certainly not ended, by World War II. At the Bronx Zoo, union leaders derided Osborn as a patrician from "another age which most of America has outlived."[38]

By 1941, the majority of the Bronx Zoo's workers signed cards to form a union local affiliated with the CIO. "Even if it had to do with the method of feeding peanuts or of meat to the carnivores," Osborn responded, "we were at all times prepared to hear any complaint."[39] Except zoo officials had no patience for unions. The patrician elites who had once fostered zoos were just as frustrated. "I hope you are not having very much trouble along these lines," sympathized one modestly understated letter of support for the zoo's administration.[40]

The Bronx Zoo met the labor trouble with the same strategies of combating unions perfected in factories. The zoo mixed enticements—salary improvements—with threats of firing. Two weeks after employees wrote Osborn announcing their union local, the zoo quietly sponsored an employee's committee (in effect, the kind of company union despised by Depression-era labor activists). Osborn warned managers when setting up this committee not to talk with employees about pay raises—none would be given. Yet workers—union activists probably—attended the committee's meetings, sabotaging them into a platform to complain about unfair firings and forced overtime. Managers responded by distributing Christmas turkeys. Early in the new year, the employee's committee was forced to cancel its meetings.[41]

Behind the sympathetic veil of Christmas turkeys and the false promise of a workers' committee, the zoo identified and punished individual union activists. Union leaders complained that they were arbitrarily transferred, faced salary cuts, or were forced to work seven days a week. There were "numerous other indignities."[42] Joseph Hannaberry, the zoo decided, "has the welfare of the CIO at heart rather than the Society." Zoo officials wanted him fired.[43] William Zeillinger, "reputed to be the power behind Hannaberry," had worked at the zoo as a cleaner and laborer since 1939, and employees elected him a union shop steward. Soon after, he was brought up on charges and fined three days' wages for being away from his post. Zoo managers quietly evaluated and punished its workers, especially those identified as particularly active in union meetings.[44] Frank Crowley was placed on probation for an unexcused absence; his child had measles. Crowley was "rather poor material . . . inclined to be a trouble maker and agitator." Carl Reinicke had worked at the zoo since 1928. After he became a vocal leader in the organizing drive, managers described him as "never a first class worker." For the rest of the workforce, the zoo now promised a wage increase.[45]

As the nation mobilized for war, conflict at the Bronx Zoo reached an impasse. "It looks to me as if we have now come to the stage that any further communications—from us at least—would not have any point," wrote Osborn to the union in the summer.[46] The union, meanwhile, reminded Osborn that "human beings are a special kind of animals." "This little local fight," the union promised, "can be stopped right now. The New York Zoological Society, in recognition of the distinction between its human employees and its animal charges, can stop butting in to the union activities engaged in by the men." By spring 1942, the strangest attraction at the zoo was not the exotic animals in cages but pickets outside the gates. "Don't Feed the Keepers," mused one observer.[47]

Many of the zoo's workers, however, soon exchanged one uniform for another. Mickie Wolfe had taken care of tahrs, monkeys, and even the zoo's rare panda. He was also the union's president. Now he was a soldier, hoping to join the cavalry because he "just naturally likes animals." He left behind a message, perhaps in case he never came home. "Boys," he wrote, "keep up the fight to organize the Zoo. Make it a better place for both men and animals."[48]

Like workers in many other American industries, zoo workers chaffed against wage restrictions during World War II. Though prices inflated, zoo wages stagnated throughout the war, and Osborn recognized the hibernating union had awakened. By V-E day, "those rumblings you hear when you visit the Bronx Zoo may not come from the lions or the bears" but from zoo workers. Even Osborn admitted keepers were hard-pressed for money.[49] The union returned to the zoo. Wartime rumblings foretold a long labor war that would eventually close the zoo gates in a 1961 strike.

The Lonely Animals: New York, St. Louis

The strike began in the Bronx on the normally busy Easter weekend. The zoo remained closed for most of the spring season, the climax of intense postwar labor conflict on the working side of the enclosure. Reporters rushed uptown to cover the nation's strangest picketers. Zoo workers, members of Local 1501, District Council 37, brought their animals. Boas wrapped themselves around picketers' shoulders, and monkeys grabbed at picket signs. Children squealed in delight to see the animals even though the zoo gates were padlocked. It was the climax of a tense struggle between those who managed the zoo and those who cared for its animals.

The Bronx Zoo strike was the longest of the many strikes that shuttered zoos from the 1940s until the 1970s. After a spontaneous strike in 1943, the St. Louis Zoo advised its workers that they were forbidden to alter their uniforms in any way, especially with union buttons. The St. Louis Zoo endured two more strikes: a three-day strike in 1959 and again in 1961. The Highland Park Zoo in Pittsburgh had its first postwar strike in 1951 when zoo workers reported for work but refused to prepare animal food or clean out cages. In San Diego organizing began around the same time and culminated in a 1965 strike. The zoo hurried to offer an eight-hour day, a 3 percent annual wage increase, and increased vacation time in order to reopen the zoo before the July 4 weekend.[50] The following year, the Brookfield and Cleveland Zoos faced strikes and organizing also began in Philadelphia. Strikes would soon close that zoo as well. In Baltimore, zoo workers joined other municipal employees on strike in 1974. "Man, they care more about the animals in the cages," a picketer told reporters," than [about] the people who work here." He altered his uniform with a button in his hat that read, simply, "Zoo Power."[51]

Zoo power took a long time to win. The long march to the 1961 Bronx Zoo strike began all the way back during World War II. In 1944, James Reilly, the assistant head keeper of mammals, had worked at the Bronx Zoo for fifteen years. With the war nearing its end, he worked eight hours a day, six days a week during the winter and even more in the summer. Like others in New York's cultural institutions, Reilly earned less and worked more than other municipal workers. By September, organizing at the zoo joined drives at all of the city's museums. The New York Botanical

Garden, nearby in the Bronx, was soon "100% union."[52] Osborn, worried about the possibility of a strike, decided to prepare a "frank talk on the whole Union business" with each department. If open persuasion didn't work, the zoo also decided to "keep our ears to the ground." Secretly, the zoo hired detective Raymond Gill to infiltrate the union, and his reports were alarming. The union, he warned, was preparing a new petition, this time directed to the mayor himself. For an institution that depended on the sympathetic stories of cuddly animals and adorable babies to encourage visitors and persuade donors, the publicity could be disastrous.[53]

The zoo soon had a new employee: Walter Gordon Merritt, the nation's best-known (and, from the perspective of organized labor, most feared) management lawyer. Merritt had represented management for more than half a century before the zoo hired him. Most famously, he had represented Danbury, Connecticut, hat-making firms in their landmark 1913 case against a union boycott. In Danbury, Merritt had pioneered the practice of suing union members for losses incurred by a business during a strike or boycott. Now, in the Bronx, the aging lawyer helped develop a strategy that put animals at the center of a dispute between the zoo and its union not over pay or conditions but over the responsibility of care. If workers went on strike, who would take care of the animals?

Merritt devised the strategy because he knew the Bronx Zoo was inching toward a climactic strike. In a preview, the union walked out for four days in 1957, but the union itself assumed control of animal care.[54] In the aftermath of this skirmish, the fight over wages and hours dissolved into a struggle over animal care. The captive animal had become, the union protested, a "pawn in an anti-union endeavor."[55] For the zoo officials and their philanthropic supporters, there could be "no selfish goal" in zoo work. The affection joining animals and their keepers might alienate workers from the union—or so zoo managers hoped. They warned keepers the union "could" force its members to abandon their posts. "The Society is quite sure that those who actually feed and care for our living creatures would not on their own volition engage in such action," the NYZS jabbed at the union.[56] With few exceptions, the zoo's donors were delighted with a labor policy that cast striking as "deliberate cruelty to animals."[57] "I refuse to believe that our keepers would desert their charges if ordered to," wrote one donor. Catherine Huntington even volunteered to take keepers' places if they walked out on strike. "I could grate eggs, cut up fruit, etc.," but she could not cope with snakes, she admitted.[58]

Who would care for the lonely animals, even the snakes? Huntington might slice fruit, but she was hardly going to shovel animal droppings. Osborn, equally, wasn't going to walk from the administration building and into the lion cages. The union held a trump card in the relationships that tied keepers to their charges. The union even promised it would care for animals in the case of strike on the condition the zoo would remain closed in the event of the strike and not attempt to reopen using strike-breakers. For the moment, the zoo refused.[59]

By early spring 1961 the strike loomed, and the union and the zoo each presented competing understandings of care. If zoo managers described union solidarity as incompatible with the ties that bound keeper to animal, union leaders accused the NYZS of caring for captive animals but disregarding humans. Even *The Economist,* normally a strong voice of management, took the union's side. "The Zoological Society which is 62 years old, still shows traces of the Theodore Roosevelt era in which it was founded," the magazine noted. The zoo boasted the most modern methods of caring for its animals but "seems to be less enlightened in handling its relations with its employees." Zoo managers, Hornaday's protégés, took "unselfish devotion for granted."[60]

The zoo went to court to try to prevent its employees from striking. After a court-brokered agreement that ensured workers would maintain care for animals in a closed zoo (precisely what the union had already offered), the strike began on April 2, 1961, an Easter weekend and one of the busiest days of the year. A bitter seven-week struggle had begun. The union encouraged the city to withhold its budget allocations to the zoo. The zoo responded with weekly letters to its employees, appealing to the tender ties between animals and keepers. "[W[ould you want to leave old 'Sudana' without proper care, allow those new Nyala youngsters or young 'Ookie' to suffer?"[61]

The union and the zoo jostled to claim the mantle of animal care. "Care for animals but dignity for employees," read the picket signs carried by strikers and their animals. The novelty of slithering picketers brought

reporters scrambling to the Bronx. A pet African rock python named
Count Butchie draped easily over his owner's shoulders. A black spider
monkey held onto his handler's head, and Bonnet macaque monkeys
shrieked at the sight of snakes. Monkeys and snakes on the picket lines—
ironically, many of the keepers actually involved in the care of primates
and reptiles were still at work behind the zoo's closed gates.[62] Typically,
laborers, assistant keepers, or janitors walked the picket lines. Newspapers
reported that even picketers seemed uncomfortable with their slithering
comrades.[63]

The strike dragged on into the spring season, the zoo's most profitable
time. By the end of April, Mayor Robert F. Wagner Jr. tried—unsuccessfully—to restart negotiations. For reporters, eager for light news, a zoo
strike was, much like the zoo itself, children's entertainment that provided
a welcome break from the stresses of urban life. Newspapers delighted in
telling stories about the animals' loneliness. In the elephant house, Phoebe
and Peter, a pair of hippopotamuses, remained submerged in their pool.
Sudana the elephant swayed gently. As the "silence of the jungle" reigned

"Care for Animals. Dignity for Employees" (1961). Union rank and file and some of
their animals picket outside the Bronx Zoo.

at the zoo, animals mourned the loss of tossed treats. The Bronx Zoo elephants "settle into the strike's popcornless daze," joked one reporter.[64] The squirrels even took to following the skeleton staff around the zoo, desperate for dropped food. Oka the gorilla seemed disappointed with no audience while the monkeys acted "morose."[65]

It was a similar story wherever zoo workers walked the picket lines. In Pittsburgh, coverage of the strike focused much more on the response of animals than on union demands for a one-week vacation during the busy summer season. Without their morning feeding, elephants trumpeted and pounded on their cage doors. Polar bears rattled their bars. The zoo superintendent quickly shuttered the zoo gates, and supervisors hurried to feed the hungriest and loudest.[66] The animals in St. Louis were just as lonely when their humans went out on strike in 1959. The lions paced impatiently. Moby Dick, the sea elephant, yawned. Marius Crowe was a zoo policeman who patrolled the eerie and empty park. "The monkeys like people around," he told reporters, "they don't play as much now."[67] Duke, the star of the zoo's chimp show, helped diminish keepers' claims to skilled labor. Some of the zoo's famed chimpanzee trainers—who were not part of the union's proposed bargaining unit—fed the animals. Mike Kostial, better known in St. Louis as the lead chimpanzee trainer, took his turn cleaning the cages and brought along his animal charges. Unwitting strikebreakers, the chimpanzees posed for pictures carrying brushes, pitchforks, and wader boots. Initial news reports that detailed the union's grievances and management's response quickly gave way to a narrative about chimpanzee replacement workers. Even chimps could shovel manure.[68]

Animal antics could also tell a different story. During the Bronx Zoo's serious strike turned into light news, the "biggest comedian," Jimmy the Chimpanzee—and the reporter who covered his antics—helped confirm the intimacy binding keeper and charge. Whenever a keeper entered the quiet ape house, the chimpanzee "dances with joy."[69] Mickey Quinn, an ape keeper who had worked at the zoo since he was twelve, realized that in childish stories of lonely animals, striking workers could show the importance of the keeper's touch. He walked through the quiet ape house with a reporter. It was a personal tour, and Quinn shared how he—and he alone—knew how the strike affected animals "he regards as almost human." "[L]ook at these," he insisted, "how their lips are pursed. Look at how tense they are."[70]

"Hey Kids, Zoo Strike is Over!" read the headlines. Well into the warm weather, the Bronx Zoo reopened on May 19.[71] "It was a day of reunion for the human and animal kingdoms," and the animals were delighted—at least according to the newspapers. "The seals knew it, the birds knew it, the lions knew it, the great apes knew it, the gnus knew it, and the elephants never forgot it for an instant." The elephants, after all, had missed out on two million peanuts.[72] Both sides could claim victory. The union had won its place at the zoo, but Osborn was delighted that even behind locked doors, keepers had kept on caring. As he boasted to his donors, the agreement on animal care "may well have its influence upon other such agreements entered into by other institutions, including hospitals, where the protection of life is involved." In fact, the zoo model of a shuttered workplace and caring strikers was used in a series of nurses' strikes beginning in the 1960s at nearby New York hospitals.[73]

Zoo workers had won union contracts built around the unique relationships they had with their animals. They alone did the caring work. They swept dust off the elephants' backs. They fed the snakes and big cats. They mucked out the enclosures and explained to children about the animals. When animals sought human comfort, they turned to keepers. And when they were angry, they vented their fury on those who carried their food and shoveled their feces. But in the 1960s and beyond, as zoos remade themselves ever more into centers for animal breeding and havens for vanishing species, there was another conflict brewing. Did animal care also mean animal conservation?

Keepers and Conservation: San Diego, Oklahoma City

Zoo managers met in secret. Every year they gathered for an annual conference. They listened to reports, swapped stories and animals at cocktail receptions, and toured (and praised) the hosting zoo. After a weekend of professional hospitality, the American Zoo and Aquarium Association (AAZPA) published its conference proceedings—except when its was talking about unions. John McKew began working at the Bronx Zoo as its personnel manager after the 1961 strike ended, and he brought bad news to the association's conference: unions were "here to stay at zoos." By the end of the 1970s, the landscape of the zoo industry had changed. About 50 percent of the 9,000 employees at the nation's 166 zoos were unionized.

When McKew was first hired, zoo administrators across the country braced for more strikes. Zoos had been "thrust into the world of industrial conflict," and he understood why. In 1962, an executive order from President John F. Kennedy extended collective bargaining rights to federal employees, and numerous states followed suit for other public employees. A wave of organizing among municipal workers followed, and zoo workers were especially poorly paid.[74] In fact, McKew remembered, their salaries were linked to those of pet shop attendants or farm workers.

The strike wave crested over American zoos in the decades after World War II. Yet strikes at the Brookfield and San Diego Zoos were the last major zoo strikes. What had changed? McKew had bargained with the union at the Bronx Zoo, and he judged that the union had "matured." Unions had realized the "animal collection would have to be alive at the end of the strike in order to have jobs to return to." McKew remained dedicated to a labor policy that still assumed the "love of animals" was natural to zoo managers but exotic to unions with little regard for "living collections."[75]

Behind the headlines of striking zoo workers and lonely animals, the nature of the zoo work was shifting again. As in the 1930s, the workforce was changing as the public mission of zoos was transformed. By the late 1960s, zoos emphasized their roles in animal conservation and scientific breeding. Children were no longer encouraged to drop peanuts into the trunks of begging elephants. The days of the feeding zoo faded and the breeding zoo emerged. Zookeeping changed with the zoo. During the era of union organizing, keepers had little education beyond high school, McKew noted. As the strike wave ebbed and the environmental movement waxed, zookeepers were better trained. "Now, you are besieged with applicants that possess bachelor and master's degrees," McKew noted. And those applicants were now both men and women.[76] The surging demands of conservation transformed zookeeping. Artificial insemination, studbooks, sperm banks, and embryo transplants all became part of the daily work of the animal keeper. Sobriety and physical courage once defined the suitable man for zookeeping. Now, a university degree marked the job's professional aspirations.

McKew's job was in personnel management, not animal breeding. Yet he recognized that the decline in wild animal populations affected animal keeping. The demands on animal keepers with the new emphasis on planned breeding caused keepers to "outdistance their unionized colleagues," and McKew was delighted. Solidarity had once joined keepers with janitors, maintenance workers, and gardeners. Now, he exploited a growing distance between those he classed as "blue collar" and college-educated, white-collar keepers. Unions had come to the zoo to stay, but McKew instructed his counterparts at other zoos to encourage keepers to pursue officerships within their locals. "[I]t certainly would be illegal to seek to place certain employees in union positions," he admitted, but "there is nothing wrong with a zoo administrator who would not hold it against keepers who involved themselves in union offices." A union led by keepers, he suggested, meant fewer strikes.

McKew's thoughts remained secret, internal to zoo managers. McKew was highlighting how zoos could take advantage of animal conservation to prevent the reoccurrence of strikes. His motives might have offended keepers, but the reality he described would not have surprised them. As zookeepers articulated a sense of their work as a "profession," they found a collective home in a new organization, the American Association of Zoo Keepers (AAZK), that coexisted awkwardly with both their managers' AAZPA and their older unions. The AAZK was founded in San Diego just a year after that zoo's major strike ended, and it grew quickly. Despite a shoestring budget, within a year, AAZK members worked in forty-four zoos in twenty-five states.[77] The modern zoo was rapidly changing, Richard Sweeney, the AAZK's executive secretary, told its founding conference: "We're scientific hunting grounds, we engage in the propagation of vanishing species." The zookeeper's responsibilities had grown and, with them, frustration. The zookeeper had relied on "mediocre attempts of professional zoo personnel to provide him with a working knowledge of his profession." The AAZK promised to provide scientific knowledge to those "who shoveled and raked their way to their present positions."[78]

The persistent demands for professional status, and the education and training that accompanied it, led keepers to contrast their work with industrial workers and bury forever the old manly code of the suitable man. "We ARE professionals and we DO intend to demand quality within our profession," Sweeney declared. In the "era of the vanishing species," the AAZK demanded recognition as a profession, part of a team with zoo management in public education and animal conservation. "[Z]oo people are individuals that do not work on an assembly line; they do not sell cars or deliver T.V.'s; or dig ditches," reminded Ed Roberts, a keeper in the Stone Memorial Zoo.[79] In Oklahoma City, zookeepers had long been classified as the equivalent of laborers. "The welfare of delicate and rare living organisms" was in the hands of "the proverbial ditchdigger," complained the AAZK. The stigma of the past labor of zookeepers was shaping the primary care of endangered species. "Shoveling feces" was now only part of the frontline conservationist role of the keeper. Yet a "laborer" cared for "irreplaceable mountain gorillas or a group of rare golden marmosets." By 1973, the Oklahoma City zookeeper had become a "Zoo Animal Technician," classified and paid like "his laboratory counterpart."[80]

When keepers became technicians, they advanced a rapprochement with management at the expense of older ties to zoo workers less concerned with the daily care of animals. In its very constitution, the AAZK promised its "activities shall not conflict with managerial policies." The new responsibilities of conservation demanded "a complete team effort" that joined keepers and management. Sweeney assured managers the AAZK is "NOT a labor organization." Even its by-laws, he noted, declared that the AAZK is not "a bargaining unit for economic reasons."[81] Yet zoo management remained "doubtful of our intentions"—probably rightfully so. Armed with a professional identity, keepers advanced new demands that overlapped uneasily with those of unions. As they asserted their primary role in animal conservation and husbandry, keepers demanded training and education. The claim to professional status linked conservation to daily care, a connection few zoo directors recognized. It also required new training programs for employees once regarded as feces shovelers. The AAZK demanded financial outlays zoos were often unwilling to make and professional pride easily gave way to "professional frustration."

Ed Roberts voiced exactly that frustration. "It's too damn easy for management to holler, 'if he's not able to hack, get rid of him,'" he complained. Yet few zoos would admit they depended on the keepers' "professional capabilities." Instead, zoo management exploited the keepers' natural concern for their animals, demanding overtime or weekend and holiday work. But "is management interested enough in trying to help this man get the Professionalism he is trying to obtain? Do they try and help him by offering courses to develop his knowledge on animals and endangered species entrusted to his care?"[82] Management, on its side, realized that the new era of endangered species erected a barrier of class and identity between keepers and other zoo employees. Professionalism may have drawn keepers into a world they imagined would ally them with management, but those ties were rarely realized.

Management had changed too. Zoos, faced with competition from other, potentially more exciting forms of animal entertainment needed to change. The era of the suitable man had ended. So, too, had the days of the dour Hornaday long past. The director was no longer there to scold employees or fume at littering visitors. If the zoo and its animals were going to survive in competition first with jungle films and, later, television, then the zoo director needed to replace the frown with a welcoming pitch. The zoo official set aside taxonomy to become first an explorer and a field scientist. Later he—and zoo officials developed their own code of masculinity—distanced himself even more from hard science to assume the roles of adventurer and tourist. Just as keepers put down the picket signs to pick up their textbooks and studbooks, their bosses refashioned themselves as the public face of conservation. Keepers and their managers both claimed to love their animals, but they had nothing but suspicion for the people who lived alongside those animals in the world's last wild places.

THE ZOO MAN'S HOLIDAY

Adventuring for the Zoo

IN 1956, George Pournelle, the San Diego Zoo's mammal curator, was finally off to Africa's game lands, from Kenya to the Congo to Mozambique. After World War II ended, Pournelle had a long shopping list: red colobus and owl faced monkeys, dwarf antelopes, and, above all, a pair of reticulated giraffes, "almost a symbol of Africa."[1] Africa's tropics, its vast plains and dense jungles, beckoned to zoo officials like Pournelle. On the expanses of the Serengeti, elephants, impala, and wildebeests searched desperately for water as the dry season began. The jungles of the Belgian Congo were the "true rain forests of Africa," Pournelle wrote, where pygmy tribes hunted elephants with bows and spears in "their age-old manner." In the Congo jungle, Pournelle hoped to trap the elusive okapi.[2]

Like so many adventurers before him, Pournelle snapped a picture to record the journey's beginning. He posed with the relics of the old and emblems of the future. With a pith helmet perched comically on his head, he traced a genealogy back to adventurers who shot animals in European empires. Pournelle pleasured in colonial order, praised their regimes' conservationist efforts, and despaired of locals' relationships with wildlife. The zoo safari was long evolved since its late nineteenth-century origins. Theodore Roosevelt had set off to Africa after posing with his rifle.

He was going to prove his manhood by facing down—and shooting—the biggest game in the world. Pournelle, by contrast, posed at his typewriter, and the rest of the expedition carried cameras, not guns. Roosevelt had promised taxidermy. Pournelle would bring back Kodachrome from a "bloodless hunting expedition with camera."[3] He left on tour with Glenn Pearson, a local car dealer who sponsored the zoo's safari, and Dorothy, his wife. Pournelle was not particularly concerned about testing his manhood in confrontation with animals, nor was he wealthy. Instead, he was a "zoo man"—a common title that for Pournelle and so many before him meant a sense of manly duty to the zoo and its animals.[4]

Pournelle, the zoo man, represented a new species of adventurer on safari: a tourist. From the end of World War I to the years after World War II, the martial manhood best represented by Roosevelt the hunter passed through the entangled evolution of the zoo and its management, wildlife, and empire into the person of the tourist. Like anyone who read enough jungle stories and watched enough animal trapping films, Pournelle carefully scripted the character he was going to play in his dispatches home. He played the common tourist leaving on a dream vacation and the luckiest one of all. The gates of colonial regimes swung open wide for the zoo man. Pournelle enjoyed special access to national parks and animal reserves—not because he was somehow braver but because he had money to spend to restock an American zoo. In the remote regions of Tanganyika, he was proud to be the first "white man" the animals had ever seen. Zoo men adventured for the zoo partly because zoo officials, long dependent on the animal catalogues, stock lists distributed by traders, wanted to travel. They also needed to travel and share their observations in order to compete with the men who supplied them animals. They too wanted to depict themselves as manly adventurers but as scientists, not trappers. Later, like Pournelle, they went as tourists.

By his own admission, Pournelle was not a brave safari adventurer. Once, he got out of his car to read a road sign. Stay in your car, it said, and beware of lions. "Now I'll know better than to do that again!"[5] Still, every holiday had its gray lining. Pournelle feared the changes he witnessed. He admired the dying colonial regimes because their game wardens seemed dedicated and the officials honest. To Pournelle, natives suggested only future disorder, even as they were also touristic emblems of the savage

Africa he—and other zoo visitors—had come to expect. Maasai tribals watching over their flock with a spear only feared work, he said.[6]

After World War II, in the hands of the zoo man, the reborn safari—the zoo-fari—married tourism to conservation and begat "ecotourism." "I wish you could see this," Pournelle wrote home. The sun dappled the slopes of Tanganyika's Ngorongoro Crater, and a jackal yapped in the distance. He enjoyed playing jeep "tag" with a rhino and, even more, relaxing on the porch of the local game warden. In the bosom of colonial comfort, he admired the grazing herds of wildebeests, zebra, and Thompson's gazelles. "Thank God there are places like this left in the world."[7]

The zoo still depends on its zoo-fari, the zoo man's holiday. In Cincinnati and Tampa, the "ZooFari" is a fund-raising party—"the wildest party with a purpose."[8] Philadelphia advertises a camel safari. Experience Africa's wildlife "from the camel's back." In Toledo, the safari is a summer camp, and in Milwaukee, it's a train that winds through the zoo. At the Bronx Zoo, a fiberglass exotic archway advertises a "Zoo-fari," and in St. Louis, a verdant pathway, complete with animal footprints (simulated in concrete), leads to the elephants and hippos. Visitors are encouraged to feel as if they are plunging through a primeval jungle, much like Pournelle before them.[9] Slowly and especially after World War II, zoos began to encourage visitors to imagine themselves as travelers, and the zoo man spun animal stories that helped the everyday zoo visitor become a global tourist, enjoying animals on the verge of extinction. When conservation met visions of manhood and fears about the effects of the end of empire, the zoo became a place to travel—and, along the way, to save the animals.

Zoo Man: Philadelphia, Borneo

The zoo man's typical day in the 1930s: C. Emerson Brown settled into his chair in the Philadelphia Zoo director's office. However mundane, "to feed, watch, and protect the thousands of specimens in our own or any other large zoological garden entails a great responsibility." A deer had died in the night, and a valuable bird seemed ill. While Brown watched keepers disinfect the bird's cage, he hoped the death wasn't an outbreak of disease. Then the lion keeper messengered nervous news that a lion was refusing its feed. Was it ill or just angry? It was his responsibility—and one he refused to shirk—to find out. Back in his office again, an animal dealer arrived to offer specimens from his latest expedition, but Brown could easily comparison-shop in the pile of price lists spread on his desk. From a young Malayan sun bear to a Sumatran tiger to expensive elephants and hippos, he could stock the zoo from the comfort of his desk. There were reports to hear, forms to fill, and bulletins to write, all before the zoo man's day ended.[10]

Yet the tropics have their siren call. Like most Americans, zoo officials rarely had the chance to see wildlife in its native habitat, but they yearned to see the wild tropics. Working at the Bronx Zoo, William Bridges had a bad case of wanderlust, the "urge that sends people out into the far corners of the earth." Like so many zoo men, he wanted to "live hard, work hard, eat poorly, and suffer in various degrees from heat, insects, loneliness, and disappointment."[11] When Bridges finally escaped to the Congo, he turned to the Bronx Zoo's most experienced traveler, William Beebe, for guidance. "You are going to see a great continent," Beebe told him, "in some ways the most wonderful continent." It will be alive in sensuous glory in ways that zoos could never be. "You'll see everything, feel it and smell it and taste it, and each day will be more wonderful than the one before, because you're always seeing new things."[12]

Only a few lucky officials left the office for the jungle, but when they did, their travels were crucial for zoos as they struggled to compete in the business of selling animal adventure. "You know as zoo men, that one of your primary functions, in addition to your zoological collections, is to provide happiness and enjoyment for the people," reminded one official. The zoo man must also sell the zoo to visitors and donors. The zoo man must display a "vital, virile interest" in telling "dramatic stories" about his zoo.[13] The zoo adventurer became the zoo's eyes, ears, nose, fingers, and tongue in the jungle, the luckiest tourist. Beebe reminded himself when the tropics seemed workaday just to close his eyes. "I, Will Beebe, lately of the Bronx Zoo, am in Borneo," he said, "Borneo!" Then he could open his eyes to "see again."[14] Maybe visitors, stranded at home, might also see animals through his eyes, less as scientific specimens then as the stuff of grand adventure.

Jungle Peace: France, British Guiana

William Beebe first came to the jungle to escape the western front. His "soul was sick," diagnosed his friend and mentor Theodore Roosevelt, "to him the jungle seemed peaceful." Beebe initially had been eager to experience the "high adventure of righteous war."[15] When the United States entered World War I, Beebe was a forty-year-old zookeeper recovering from an airplane crash he suffered while training for combat aviation. Only a leave of absence from the zoo and the personal intervention of Roosevelt with General John J. Pershing, the American commander at the western front, helped Beebe get to France. Once there, Beebe flew reconnaissance planes and wrote articles about the trenches from above for *Scribner's Magazine* and the *Atlantic Monthly*.

Shattered by the war even from above, Beebe slipped "quietly and receptively into the life of the jungle." Wildness was a "sanctuary."[16] One of zoo's first patrons, Roosevelt had helped promote the outsized image of the big game hunter. Guided by a strict code of the "fair chase," the elite hunter endangered his own life (and hunting was predominantly a men's game) in ending that of big game. Roosevelt's accounts, and those of other safari hunters, helped Americans measure what it meant to be manly through violent encounters with wild, often tropical animals. The men who supplied animals to zoos, especially Frank Buck, offered a different image of the manly tropical hero, overcoming the resistance of both valuable beasts and native humans, to "bring 'em back alive" for profit.

All Beebe wanted was "jungle peace." He never dreamed of mastering wildlife through the hunter's rifle or the trapper's cage. The hunter knew how to kill the tiger with a single shot, the trapper knew how to catch the elephant in kraal, but the zoo man knew how to watch. Animals, insects, and natives were all "actors and companions" in the jungle drama unfolding around him, Beebe explained.[17] In attack and defense, the struggles and tactics of ants, birds, monkeys, and reptiles reminded Beebe of the characters of modern warfare and politics. Howler monkeys perfected a totalitarian society of male rulers harshly governing their large harem. Leaf-cutting ants formed a workers' democracy of total equality— an insect Bolshevism. Animals developed weapons and tactics analogous to the warring human. Just as the neutral ships flew large flags, the butterfly proclaimed its neutrality in flaunted color. The tarantula built pillboxes; the boa constrictor set an ambush camouflaged in the dappled light of a jungle game trail. In the jungle, Beebe discovered pursuit planes, mines, tanks, and gas.[18]

Like others drawn to zoo work in the years before World War I, Beebe was a childhood collector. Even before he attracted the attention and affection of Roosevelt and William T. Hornaday, his fascination with insects and birds helped him amass a remarkably broad collection of specimens. The collecting urge trained him first to think about nature as a vast hierarchy of genus and species, but when he went to the jungle, he abandoned dry taxonomy to become a participant observer in jungle drama. As early as 1916, the New York Zoological Society founded a Tropical Research Station perched between a British Guiana rubber plantation, government buildings, and a penal colony. Beebe, although eager for the front, was its obvious director. At the war's end, Beebe returned to the jungle and his tropical research station. Battered and shattered, he yearned for "primitive wilderness" and the "comforts of civilization which mean continual health and the ability to use body and brain to the utmost."[19]

He never quite escaped the western front. The war was everywhere in how he observed (and described to domestic audiences) life under the jungle canopy. Beebe knew stories of quiet observation were going to be a tough sell, so memories of the front helped enliven scientific study. In everything from films to pulp novels, ordinary Americans enjoyed images of the tropics as a kind of primeval Eden of abundant foliage and exotic fauna.[20] Buck offered another and even more exciting depiction. In his jungles, animals battled in a daily, savage struggle for survival, and any white man who penetrated the tropics risked poisonous snakes, vengeful predators, primitive tribes, and mortal disease. Beebe once derided that as a fanciful jungle of "poison fang and rending claw."[21]

In his particular corner of the jungle, "[f]or those who think of the tropics as a place of constant danger and disease," Beebe admonished, "mosquitoes and flies, malaria and other fevers were absent." Warm breezes kept his jungle house comfortable, and a truck delivered weekly food from overseas. The white man, Beebe insisted, could live as comfortably in a jungle as "naked" Indian hunters. The zoo man from the "north temperate regions" easily maintained his keenness and health even in the "primeval jungle."[22]

But the tropics were not a tourist paradise—yet. The "daring adventure" of the tropics demanded skills born from years of "jungle experience." The zoo man was a higher breed of man, evolved from the imperial hunter and the animal trader. The "Sportsmen Naturalist" had evolved from the "Shoot 'em and Skin 'em" hunter and from the animal capturer. If "my friend Frank Buck" claimed to "bring 'em back alive," Beebe's slogan was "watch 'em alive," and he defended his own adventure against accusations of "sissiness or any sickly sweet sentimentality."[23] It was the beginning of a competition between Beebe and Buck that would soon reach its climax back in New York.

Buck looked and acted every inch the manly tropical hero, and his bluster matched his muscular frame. Beebe, by contrast, was tall and gawky. "He talks more like a classroom lecturer" than a tropical adventurer.[24] Stillness, quiet, and a keen eye—not muscular force—defined the zoo man. Watching 'em alive introduced a new kind of proud manliness

"Binoculars and Plaster Cast at Portachuelo Pass." William Beebe began his career as a zookeeper but made his name as an adventuring zoo man. Even a broken leg couldn't stop Beebe from "watching 'em alive." William Beebe, *Adventuring with Beebe* (New York: Duell, Sloan and Pearce, 1955).

and just as proud a sense of racial superiority over native peoples. The zoo man depended on acute senses, strained eyes, aching muscles, camouflage, and a fair bit of courage.[25] As Roosevelt himself declared in his praise of Beebe: "Books of this kind can only be produced in a refined, cultivated, civilized society." The savage found joy in killing, but the zoo man's patient observations were the end product of a long evolution. His genealogy, lauded by the great hunter himself, set Beebe's observations apart from those of the "wild fool he meets in the wilderness—black, or yellow, brown or red."[26]

Beebe had little patience for tourists. He was an adventurer and scientist—not a vacationer. Postwar, and personally devastated, he found himself confined on a ship bound for the tropics. Between careful observations of the silver fish that trailed the boat, he regarded his tourist shipmates soaking up the sun. To Beebe they seemed like "effeminate men, childish women and spoiled children," oblivious to the subtle wonders of the tropics.[27]

Even without the drama of staged jungle battles, his lyrical natural histories appealed to readers. "To read William Beebe is to learn of the drama, the romance and the poetry of the Jungle," admired one reviewer in 1925.[28] The armchair explorer could adventure high in the Venezuelan jungle with Beebe and stumble over a rotting stump and into insect drama. Lying at "ant-height" in the grass, Beebe discovered an iridescent blue and gold earwig, just one of many "exotic surprises in a tropic world."[29]

Tropical research was good publicity for the zoo between the two world wars. When Beebe set off for the Galapagos Islands on a cold day in March 1923, the zoo sent along Ernest Schoedsack, the budding jungle filmmaker. Perhaps along the way Schoedsack found inspiration for his films *Chang* and, even more famously, *King Kong*. He certainly met his wife Ruth Rose, an actress who had tossed aside her own theatrical career to join Beebe's tropical adventures as their official historian.[30] Back at home, the zoo served tea as visitors and donors mingled with the rare birds Beebe had collected. Movies of African jungles flickered in the background (and no one seemed to complain about the incongruity of Pacific birds and African jungles). The zoo had come to realize how much was at stake in Beebe's adventures. "We must steadily make friends for the Zoological Society among boys and girls," argued a donor. The old taxonomic display of animals in rows of cages just didn't appeal. Instead, the "younger generation" wanted "daring." When Beebe penetrated the jungle, sailed for the Galapagos, or descended to the bottom of the Bermuda seas in his bathysphere, the public was enthralled.[31] Even the modest Beebe marveled at the "American reception" for "out of the ordinary" zoo adventures.[32]

When the World's Fair opened near the Bronx Zoo in 1939, the zoo realized only Beebe could help it compete with Buck's Jungleland exhibit. Buck had won—and jealously guarded—the monopoly on the display of animals at the fair. Much was at stake in the fair's rivalry between the zoo and Jungleland, Beebe and Buck. After all, if animal adventure could be found only at the movies, why visit the staid zoo? As Buck was completing his latest jungle films, Beebe logged a record dive to the depths of the Bermuda seas in the confining metal hull of the bathysphere. He returned with breathless tales of strange creatures and nervous courage. "Will's dive" demanded the bravery of "anything in aviation or racing," boasted the zoo. Surely, that could compete with an animal trapper. As the zoo scrambled to build its concession, Beebe, his bathysphere, and his tropical research station dominated the exhibit alongside the rare giant panda and the electric eel.[33]

Without most of its animals, at the least the zoo could display the zoo man. Beebe invited visitors to the zoo and the fair, but he reserved jungle adventure for the zoo man, his small expedition, and his wealthy benefactors, like Henry Devereux Whiton. Like Buck before him, Beebe didn't even welcome volunteers. "I am a . . . young man 6 feet tall, weigh 350 and have no ties," pleaded one eager adventurer.[34] He stayed home. Whiton, though, had come from old money and made a great deal more of it as the founder and president of the Union Sulfur Company. Whiton loved both yachting and the Bronx Zoo.[35] He appealed to other wealthy friends, including the explorer and department store magnate Marshall Field and the philanthropist Vincent Astor, to outfit a scientific yacht for Beebe.

By 1925, workers overhauled the *Arcturus* from "stem to stern" into Beebe's floating tropical research station. Long undersea lights penetrated the black depths. Special dredges snagged bottom dwellers, and a long boom dangled researchers far out over undisturbed waves. For the general public, movie cameras and a long-distance radio recorded Beebe's adventures in the Sargasso Sea and the Galapagos Islands. A few elite patrons, like Herbert Satterlee, a zoo patron and one of J. P. Morgan's heirs, dropped in on the yacht as "guests of honor." Later, wealthy yachtsmen like William Vanderbilt retraced Beebe voyages. Tropical adventure remained a game for the very wealthy or the very scientific.[36]

The Forest of Adventure: Guyana

John Morlan was one of those very wealthy. Struck down by the kind of lassitude that only infects the rich, "I haven't the faintest idea of further expanding my business," he said. Rather, he yearned for adventure. Morlan didn't want to go shooting to collect heads and horns for his estate walls.[37] Instead, he had a yacht, a zoo, and a plan. Morlan spent a sizable fortune outfitting a yacht to collect specimens for the city zoo where he was a trustee. Leading an expedition of scientists, he headed for the Demerara River and its thick jungles. The Guyanese jungle was hardly the equivalent of its African or Asian counterparts. The jaguar couldn't match the tiger's ferocity, and howler monkeys posed far less of a danger than mountain gorillas.

Yet this was the jungle Raymond Ditmars knew best, and this was his 1933 novel, a vaguely fictional composite of the kinds of expeditions he had joined. Like his contemporary Beebe, Ditmars developed his fascination with animals as an amateur collector of snakes and other reptiles. Ditmars and his snakes soon joined the Bronx Zoo. As the zoo's new reptile curator, Ditmars was an ideal candidate to transform zoo work into adventure. Snakes were poor business for animal traders. They were simply too dangerous and too perishable to transport profitably. Apart from large pythons that could be purchased from stock lists, Ditmars and other reptile curators realized that if they wanted the rare, poisonous snakes that excited visitors, they had to capture them.

Ditmars's snake hunting turned the necessities of zoo work into stories of jungle adventure. "To the layman," warned Raymond Ditmars, "a tropic jungle is a dream of the garden of Eden."[38] Lush leaves sprouted from towering, vine-draped trees. The flowers were perfumed, and fruit was ripe and juicy. The *Oraloo,* like the *Arcturus,* was a comfortable floating laboratory, but its crew was decidedly not cinematic. In place of the muscular Buck and his trademark pith helmet, the scientists were gawky, and Professor Smith kept rattling the handles of his insect nets. They were not "invading" the jungle, Morlan reassured. That conceit was best left to the big game hunter or the animal trader. They had come to "watch 'em alive."

The ungainly scientist replaced the muscled trader. Still, this was "a thrilling story" perfect for "MEN 10 to 60"—or so the book's ads read.[39] The expedition confronted strange and biting insects, relentless heat, and tropical downpours. While blundering lost in the jungle, Professor Smith, the expedition's entomologist, even faced down a menacing bushmaster snake. His scientific comrades wanted new discoveries, and Morlan sought a giant armadillo for the zoo. The otherwise-gentle armadillo was an untypical protagonist for a jungle story but ideal for a novel about the adventures of tropical science.[40]

Like Beebe before him, Ditmars dismissed wild tales of animal collecting as romantic counterfeit. As the expedition returned triumphant from the bush, Morlan mused about organizing an animal-collecting expedition to collect the jungle's larger animals. Pure fantasy, scoffed his native guide. Animal collecting was nothing more than dull commercial exchange. In fact, while waiting for the returning expedition, the crew engaged in a lively trade with natives. They bought snakes and a young tapir, each captured by natives through dreary happenstance. In contrast to the tense drama of the expedition's close encounters with vampire bats, thief ants, and a bushmaster, the tapir wandered lost into a native's hut. An easy capture, it was an even simpler purchase for the ship's crew.

Buck, it seemed, was either a clumsy explorer or a flimflam man peddling lies on the big screen. Ditmars knew the jungle wasn't rife with man-eating peril. After all, he had once faced down a jaguar. On one of his own expeditions, he played his flashlight across the startled face of a jaguar entangled in mosquito netting. The cat snarled, leaped—and disappeared into the jungle, "like a ghost." The "savage wild beasts" of the tropics would only attack visitors foolish enough to swim with caimans or unlucky enough to tumble over a poisonous snake. Even the bite of the vampire bat, like "almost all tropical terrors," was so survivable its victim could proudly proclaim in pleasant conversation: "The last time I was bitten by a vampire . . ." Mosquitoes and the diseases they carried were the real "dangers" of the tropics. But malaria didn't make for good movies.[41]

The jungle could certainly injure but rarely did. When Beebe led his readers on an imaginary walk through the jungle, a startled agouti scurried into the jungle. A jaguar snarled somewhere in the dusky darkness. The armchair "explorer" eager for stories of man-eating danger would be dissatisfied. Snarls and roars receded before bushwhacking humans. Near the laboratory, red-headed vultures circled, but they were disappointed: "You seem discouragingly healthy."[42] Instead of death, the zoo man offered life through an intense immersion of his audience's senses into the jungle. Ripe mangos hung heavily from a tree shrouded in colorful flowers. The jungle air was perfumed and abuzz with the "continuous hum" of insects and the distant call of a howling monkey. Bothered insects filled the flowery glade with an aroma of "witch hazel."[43] Back at home, zoos could never reproduce the sensual experience of jungle exploration. How could they make its buildings smell like anything but animals and their droppings? The zoo man could, however, depend on the visitors' imagination to help them see not mere groupings of animals but the tropical landscape itself. Ride the subway; visit the tropics.

Africa for Five Cents: New York, Tanganyika

In 1941, Africa was closer than subway riders thought. Reedbucks, bushbucks, and nyalas meandered down the slope to the water hole under the keen gaze of three lions perched on a rocky outcropping. "It might be in Africa," but it was the Bronx. The new African Plains exhibit imported "a Spot of Africa" and announced a new era of zoo design that depended on visitors' imagination and thirst for adventure travel. The "zoological park of the future" opened after the World's Fair and on the eve of American involvement in World War II. African animals herded into a "game-abounding open terrain"—at least as it existed in the traveling mind.

The visitor might "look around quickly for the nearest tall tree" as the lions stalked their prey, promised the zoo's publicity. The game, though, was perfectly safe, separated from the lions by a hidden moat. The explorers' eye—the "observer of African wild life"—recreated a veldt in which "killer and quarry" shared the same water hole.[44] The Bronx Zoo had reimagined the zoo as a safari for the everyman. The zoo planned to expand the water hole exhibit with all its drama of lions and their prey suspended at the breathless moment before the hunt to include rhinos, giraffes, and elephants. Future reconstructions would follow the "green line" of the tropics, visiting the Amazonian jungle, a tiger island, and an Australian habitat.[45]

Visitors delighted in their exploration of the African Plains. Attendance in February 1941 rose three times from the previous January. In all of 1941, attendance topped five million visitors, or, as the zoo pointed out, 4 percent of the nation's population.[46] Lions perched at the edge of their moated island thrilled one "timid visitor" enough that he wondered aloud if they could "jump the ditch." A moat kept that visitor "gazing across Africa" but "safe in the Bronx."[47] "I loved it," wrote a visitor. "I can remember so well when it was a very dull place," she said. "I only realize now how dull it was when you have made it such an exciting place to spend a day."[48]

"Looking across African Plains" (ca. 1941). The animals graze by the drinking hole in the shadow of a simulated primitive African village, all meant to give visitors the chance to imagine a safari to Africa—via the zoo.

In its offer of a safari in the Bronx, the zoo played on the double meaning of "recreation": the pleasurable communing of urban inhabitants with live animals and "re-creation," the representation of tropical lands.[49] The African Plains exhibit, at least virtually, democratized travel to exotic lands. "Tanganyika at our doorstep," the zoo advertised. The 12,000-mile journey to Africa remained a privilege of the elite, but the new exhibit allowed "our millions of visitors" to view the animals of "that unique continent in their own natural setting."[50] The zoo promised Tanganyika in the Bronx; one early visitor, however, looked at the lions and saw Uganda.[51]

Just as reserves in Africa developed complex networks of game wardens that regulated animal populations and interceded in human / animal conflicts, so, too, was the zoo compelled to rearrange its wild population. Soon after the exhibit's opening, the reality of sex and mixed species increasingly gave way to segregation. The zebras, especially the stallions, were the first problem. They stalked the warthogs outside their dugouts. The zebras were the first to disappear, then the warthogs. Their burrowing riddled the plains with caves, and they ended up, far from their continental home, in the kangaroo house. Male bushbucks were soon relocated into a nearby corral after one attacked and killed a young nyala. Among male hoofed animals, only docile nyalas remained amid a dwindling herd of female bushbucks and rheboks. Even the ostriches were sex segregated after a cock ostrich's aggression forced keepers to rescue a young boy who wandered into the enclosure. The lions were also sex segregated. The absence of "disturbing feminine influence" ensured peace among male cats, admitted a keeper.[52]

The zoo also couldn't smell or taste like the jungle, but the visitor could still watch 'em alive, like Beebe and Ditmars deep in their forest of adventure. The obvious limitations of zoo architecture and the wartime context in which the exhibit opened demanded the zoo promote a dramatic vision of nature in which, paradoxically, the African plains could provide peace and refuge. Just as the jungle succored Beebe after the western front, the African Plains provided an entertaining respite for New Yorkers facing the onset of World War II. When the new display offered "an hour of recreation snatched from these troubled days," the lion rested peacefully in full view of the gamboling zebra. At the zoo, the lion was never hungry and never killed.[53] The exhibit was "jungle peace" for the masses. With war, though, an era of popular travel was just beginning.[54]

World War Zoo: Pacific Ocean

The sirens wailed as searchlights crisscrossed the sky. The whine of the first bombs echoed across the zoo's open enclosures. Human houses were spared, but the zoo's enclosures were crushed into rubble of concrete and twisted metal. Panicked animals broke free, adding to the terror of the bombed city. This was the "zoo in wartime" or, at least, the worst nightmare of zoo directors when the fighting began.

What to do with the snakes? In London, the prime minister had barely finished his radio address announcing war before zookeepers destroyed the venomous snakes. They worried the snakes might escape during air raids. Close to the fighting, World War II profoundly affected zoos and their animals. The Shanghai Zoo was bombed as early as 1937, killing rare blue sheep. German zoos were decimated by bombs first and then by orders later. The Hagenbecks' famous Stellingen Zoo was bombed at the beginning of the war and never repaired. Instead, its monkey mountain became machine gun emplacements. Rumors circulated that in 1944 wolves had escaped and zebras and lions roamed Berlin's streets. In Moscow, an incendiary bomb reportedly landed in the tiger's enclosure, but the cat escaped to an adjoining cage. An elephant, in suspiciously imaginative wartime propaganda, used its trunk to put out the fire. In the United States, zoo officials wondered about the fate of their dangerous animals in the event air raids crossed the Atlantic. For a few months in 1942, Bronx Zoo keepers carried rifles and practiced tracking escaping animals.[55]

"What are you going to do with your poisonous snakes?" Roger Conant, the Philadelphia Zoo director, asked William Mann. Mary Bessemer was even more concerned. She lived close to the National Zoo, and the chance of snake escapes during air raids "gives you the creeps." Another neighbor encouraged dropping all the poisonous snakes in American zoos on German cities. Seeking personal profit, a circus owner in Minnesota offered to house the National Zoo's snakes during the war. In the end, the National Zoo decided to trade its vipers to an animal dealer. Both the Philadelphia and the National Zoos created displays of "fixed snakes" with their fangs removed.[56] Floyd Young, the Lincoln Park Zoo director, reassured a concerned public that lions were, in fact, cowards, and even if they escaped during an air raid, they were unlikely to attack people. Nonetheless, he promised that the zoo had collected a substantial arsenal for its keepers.[57] In San Diego, the zoo calmed visitors with promises that only a direct hit could destroy the zoo's strong buildings. Even so, it posted night guards and locked away its animals at night.[58]

In Philadelphia, zoo officials milked the poisonous snakes' venom for the production of serum for the army. Thousands of ordinary Americans in uniform would soon travel to the tropics and confront cobras, mambas, and more. In wartime, ordinary Americans were coming face to face with animals that had once been only exotic zoo specimens. Zoos were ready to reintroduce tropical wildlife to an American public with relatives on the front lines. In 1942, the Philadelphia Zoo exhibited Weapons of the Wild, which compared the natural defenses of animals to human weapons. The hippo stood in for the submarine and rhinos for tanks.[59]

Zoos near large navy or army bases, like San Diego or St. Louis, were popular attractions for the enlisted. In San Diego, they marched to the zoo led by their officers.[60] The zoo was a respite, but it also introduced these training soldiers to the tropical world they would soon invade. When Beebe's research base in Bermuda became a military airfield in 1942, he returned to the Bronx to curate a new exhibit on jungle warfare; 232,000 visitors trooped through Life in the Jungle. Soldiers bound for the tropics "copied down instructions" on the exhibits showing how to find food, water, and directions in the jungle. As he regarded the exhibits' crowds, Fairfield Osborn marveled that the zoo resembled "a gigantic USO annex." Animals from "distant parts of the world" had an entirely new and intensely personal significance. Beebe's bathysphere, meanwhile, helped the navy test depth charges.[61]

The wartime zoo had a military role. Mann was now in uniform as a technical advisor for the Quartermaster Corps, and the Bronx Zoo counseled the army and navy on everything from repelling sharks to the desalinization of seawater.[62] The tropics were now a battlefield for humans, not just animals. The return of the draft in 1940 and the bombing of the American naval base in Pearl Harbor in December 1941 meant millions of Americans suddenly were serving on the tropical front. "Well, you're in the Pacific," wrote Beebe to American soldiers and sailors. "You men of World War II" had plenty of images of the tropics, said Beebe. "Boyhood books" and movies provided "intensely vivid pictures" of jungle life. The problem was: most of it just wasn't accurate. Kangaroos or monkeys lived on Pacific islands only in the books or movies. Now, Beebe told soldiers, "you go ashore and actually walk through the jungles."[63]

"The Pacific is Before You"—and the zoo and Beebe offered to act as the soldiers' guide. As American fighting forces slowly drove the Japanese from their Pacific empire, Beebe was confident he could train soldiers as amateur zoo men. "[T]rack them down, sneak up on them, watch them," but don't, Osborn pleaded, use rare birds and animals for target practice. Osborn gathered the leading scientific men, all under the umbrella of the American Committee for International Wild Life Protection, to prepare a book about the nature of Pacific tropical island battlefields. Zoo men, biologists, botanists, and zoologists wrote *The Pacific World* (1944) for the "men of the armed services." You will encounter lizards that jump from tree trunks, Beebe told them, or mammals that lay eggs. "I have seen all these," Beebe said, and if you "use this book," then "you will too."[64] The committee distributed 100,000 copies to fighting men in the Pacific.[65]

Soldiers did see those exotic animals and many others. The tour of duty had become jungle adventure. They wrote home about the animals they saw, partly as hints to wives and parents about their secret deployments. "If Seaman Joe Doaks sent home a letter describing a caravan of single-humped camels, it was pretty certain his ship" was in Africa. Most often military censors sliced out such giveaway descriptions, but sometimes they slipped. When an animal fighting men described was less known, family members wrote the zoo. At first, the Philadelphia Zoo happily identified animals. Army intelligence eventually cracked the animal code: "Tell them nothing," they warned zoo officials. When Americans went to the tropics as soldiers, the zoo "guarded military secrets."[66] Wait for "final leave" to ask your questions, suggested Beebe. Then come home as veterans *and* as naturalists to share marvels your family never saw in the movies.[67]

In the meantime, when the "Jap has been cleared out," Beebe encouraged soldiers and sailors to train lizards, birds, and small mammals as pets. Record their cries, visit their nests, and discover what "natives know of them."[68] Of course, when enlisted men discovered their new pets didn't make "good companions," a zoo might benefit. One sailor purchased Jocko on the African coast. Name notwithstanding, she was a young baboon, friendly and tame but too big for a sailor's bunk. She ended up in the Philadelphia Zoo. From Okinawa, soldiers trapped and sent poisonous habu pit vipers home to Washington, DC.[69] Even Admiral William Halsey shipped animals home. Grateful for the entertainment the San Diego Zoo provided stateside sailors, he arranged to send a pet wallaby home by airplane.[70]

Meanwhile at the San Diego Zoo, "[o]ur boys" were leaving for the front. Some didn't come home. Keeper Max Brown died in a bombing raid over Sicily, and Thomas Brown Burrell was badly injured by flak over Frankfurt. K. C. Lint cared for the birds until he joined the tank corps, and Ken Stott Jr., the zoo's general curator, served with the "common soldier" in the Philippines. Like Osborn and Beebe, Stott prepared guides to the fauna of the Pacific tropics. He also described for his audience at home a bushwhacking march through the Philippine jungle. The fighting aside, this "typical day's journey" through the wilder parts of the jungle brought "the incomparable thrill of 'exploring' a new land" to the fighting zoo man. The water in the canteen was "evil tasting," laced with chlorine to protect from jungle contamination. The nature, though, was glorious. The white herons flushed pink in the rising sun. Here, the colorful birds of the zoo, like a red-backed sea eagle, were a common sight. As the troops approached a native village, a civet cat careened into the underbrush. The natives seemed primitive, utterly detached from the war, and Japanese soldiers were nowhere in sight.[71]

The Great White Hunter: Nairobi, San Diego

Stott visited the Pacific islands with the army, and in 1950, with the war over, he went to Africa, the "dark continent," as a tourist in search of wild game. His reports, printed in the zoo's magazine, described encounters with dangerous beasts and with natives he considered ill-equipped to shoulder the burdens of civilization. Like his coworker Pournelle, he went to Africa to retire the hunting safari to effete snobbery. Postwar, he traveled as a tourist, not a naturalist. The homebound American, steeped in the lore of African game trails, expected roaring lions and the subservient obedience of tribal peoples. Instead, Stott, the war veteran, confronted a paradoxical mix of modern comforts, surly locals, and hospitable white colonialists.

"Oh, what it takes to be an African 'explorer'!" In his Nairobi hotel, Stott suffered the hardship of two bars, a reading lounge, a hairdresser, a coffee shop, and a dining room the size of an auditorium. "I can see clearly that life in East Africa is going to be rugged at best," he wrote home.[72] Here at the end of the age of empire, Africa had become a "land of macadam and steel and cocktail lounges." The trains were luxurious, the sound of their steam-driven speed broken only by the tinkle of ice in carafes of fresh water.[73] Stott was almost overjoyed to encounter a dilapidated, vintage bus as he crossed Uganda. "I experienced a tremendous, almost jubilant flood of relief," he wrote. This, at last, was the Africa of Theodore Roosevelt and Martin and Osa Johnson.[74]

Accounts of the turn of the twentieth century elite hunting safari had once captivated ordinary Americans. Yet if pre–World War II African travel accented wealth and privilege, then Stott's postwar African tour heralded new popular access to increasingly unfriendly lands. More and more American tourists returned home to regale their friends with moving pictures and souvenirs. "Elephant legs,'" explained one tourist, "make excellent trophies. They can be used as wastebaskets or table supports."[75] The elite safari had become a caricature of its former self, and to Stott, Nairobi with its golf courses, swimming pools, and fancy residences looked like Pasadena. The obvious difference was that the gateway to Kenya's game country teemed with that strange specimen, the "Great White Hunter." The hunters lurked in plush hotel lobbies, conspicuous for their khaki shorts and broad-brimmed hats with leopard-skin bands. They preyed on "tourists" with safari dreams and dollars to spare.[76] The safari, instead of a test of manly courage, marked only a full wallet and a longing for elephant leg trophies.

Leaving Nairobi, Stott wrote his sister: "You may tell Mother and Dad . . . that I will in all likelihood survive the rigors of this perilous country."[77] There was nothing death-defying about the hunt anymore, and the code of the fair chase was long extinct. Stott confided to his zoo audience that Great White Hunters led clients to their quarry, sighted their rifle, and if—as was likely—the inexperienced tourist merely maimed the animal, the "GWH," as he called them, stood ready for the kill. The GWH even provided the long lines of porters that once characterized Roosevelt's safari. Tourists could look back from the comfort of their car to snap pictures of African porters for friends at home. But there was a secret: porters were little more than stage furniture. Land Rovers carried all the needed gear, and porters' baskets were filled with rocks or newspapers—just so the "tourist" could tell tall tales of the "hardships they were forced to endure in this God-forsaken wilderness."[78]

Danger had disappeared, Stott realized as he gazed at big game from automotive comfort. Rhinos charged only when deliberately provoked so Stott could capture his trophy picture. The rhinos, annoyed by Stott's car, snorted and pawed the ground. "I began snapping pictures."[79] Even tribal people who seemed so primitive and threatening to past explorers now appeared just strange. "The women are handsome in a grotesque sort of way," he told zoo readers, but they had lost their savage allure. Stott had a difficult time shaking the image from his mind of Maasai and later Somali tribal women misappropriating telegraph wire. For Stott these women and their strange jewelry were evidence of the continent's uncertain future. "Between the natives' cutting down the lines for bracelets and the giraffe, getting tangled up in 'em . . . the telegraph people have just about given up hope."[80] Local women looked to Stott like "some fantastic type of electrical equipment," not native peoples ready to grapple with the burdens of modernity and independence.[81]

Stott enjoyed the last days of empire with comfortable hotels, the crisp salutes of native policemen, and the infinite hospitality of white colonists, but African independence only filled him with trepidation. Empire, he argued, had produced not the wasteful safari but a high regard for and efficient management of wild game. Even the necessary shooting of a marauding elephant evoked only sadness among British colonists and game wardens, he reminded American audiences. In independent Africa, by contrast, Stott felt besieged. As an American tourist in Egypt, he barely rescued his expensive camera from hostile customs officials. Angry, he withdrew into the isolation of his Nile steamer cabin until the Aswan Dam retreated into the distance. "I ground my teeth. So this was Africa!"[82] As he looked with concern at independent Africa, he relished the fading charms of European colonists who had welcomed him with open doors and still dressed for dinner. Roosevelt's safari account remained thrilling to read. Henry Morton Stanley and his search for David Livingstone were still fascinating. Yet Stott realized that even the Johnsons' more recent books and films pictured an Africa that no longer existed.[83] Game trails had become highways, villages had transformed into cities, and Stott longed for primitive Africa. "It is difficult to wax poetic," he said, "about a Kinguana maid sighing over Frank Sinatra's most recent recording, or a pigmy chuckling over the *New Yorker.*"[84]

Disillusioned and back at home, Stott lectured about the postwar safari, but his audiences were less willing to consign Roosevelt to history. Once a woman touched Stott's arm to reassure him. "I hope I haven't hurt your feelings," she told him, your pictures are charming, but you need a sense of mystery. Armchair tourists and zoo visitors wanted Africa to remain "the stronghold of the blood-thirsty savage and the man-eating beast."[85] When his biggest challenge was winning the window seat on the bus, Stott knew Africa was safe for tourists—as long as independent Africa realized that its economies and its animals needed American sightseers. Zoos, meanwhile, could still depict savagery and man-eaters in their African plains exhibits.

Africa was once the destination of the "extremely rich and the adventuresome." Now it was the "vacationland" of the tourist with an "ordinary pocketbook."[86] The tropics needed its tourists, concluded Stott, and zoos were ready to facilitate travel, real and virtual. "Only by an actual visit can one feel the fascination and see the beauty that is Africa," wrote a tourist during a San Diego Zoo–organized photo safari to East Africa. By the 1960s, in the midst of the era of African independence, affluent Americans like these could follow the "safari route" from Nairobi to the "Seronera Lodge" with its "outstanding native hut construction." This tourist came home not with hides or an elephant leg wastebasket but with sketches of lions on the prowl, a tourist bus snapping pictures of a gazelle, a "beware of elephant" sign, and Maasai tribeswomen suckling their babies.[87] The San Diego Zoo advertised "safari-time" for its members again in 1968. For three weeks, the lucky tour group would see the sights of East Africa and luxuriate at the famous Treetops Hotel and the Mt. Kenya Safari Lodge. "So start dreaming of seeing all the magnificent wildlife of East Africa," advertised the San Diego Zoo. For $1,650, it was an expensive trip, but a life's dream still reachable by many.[88] "Africa comes closer," recalled one tourist from her zoo trip.[89]

That same year, the San Diego Zoo also sent one of its curators to "exotic, strange, and overpowering" India. Local newspapers advertised shikaris, India's hunting answer to the African safari, to tourists. But as he walked around impoverished, crowded cities, he realized India's wildlife, like Africa's, was threatened. In fact, the notion of India's animal abundance was "50 years before the times." This expedition, though, was not to collect animals for the zoo but to visit the game sanctuaries of the newly independent country. He also wanted to inspect "comfortable tourist facilities." Wildlife competed with all the other needs of the postcolonial state, and the zoo was convinced only Western tourists could save the animals. "[W]ildlife," read the curator's report, "can be saved by the foreign tourists just as it is being saved in Kenya."[90]

The San Diego Zoo was especially worried about native poaching. The biggest threat to tropical animals, from large mammals to colorful birds, was the end of "white man's control over the land and the natives."[91] By the 1960s, Fairfield Osborn was convinced tourism might be enough to alter that new reality. In independent Africa, animals might become valuable as something more than a meal. "[W]hile poaching continues to be a very serious problem," Osborn hoped African leaders would realize that "big game is a prime tourist attraction and must be saved." In the waning years of imperial rule, laws that set aside game reserves for the pleasures of European hunters also defined the wildlife management policy for pioneering national parks in the prime game lands of Kenya. In their confines, parks encouraged Western tourism but banned native hunting practices as poaching. Separating native humans from wild animals became the goal of imperial game reserves and then postcolonial national parks. Osborn had good reason to hope the United Nations might "recognize national parks" as the world's assets, only partly owned by independent nations.[92]

Adventure married tourism and begat the postwar zoo. Voicing the goal of conservation, zoo men questioned native peoples' ability to care for their animal neighbors. For everyday visitors, new displays at zoos attempted to replicate the vistas experienced and described by the zoo adventurer. The zoo-fari was a sentimental journey though bygone times to see animals on the verge of extinction.

Rhino! Rhino! Assam, Mysore, Washington

As Americans went soldiering to the tropics, exotic animals stopped arriving in the United States. Two giraffes, the last animal cargo to arrive from overseas and destined for Philadelphia, just barely evaded a U-boat attack.[93] Then Singapore, the hub of the animal trade, fell to the Japanese. "It was the place from which Frank Buck used to bring 'em back alive," lamented one observer after the demise of the port's trade. Rare specimens, like a rhino bound for the National Zoo, were stranded in transit. In 1940, that rhino was trapped in Assam's jungles and traveled overland to Calcutta. Along the way, it struggled against its tight enclosures and suffered in the heat. By the time it arrived at the port, the rhino was in "shockingly bad condition," but by the time the injured prize was well enough to travel, the steamship lines were firmly closed. The rhino spent the war running up lodging bills in the Calcutta Zoo. Only in the waning months of fighting could the army help the zoo by lending a transport to ship the rhino.[94]

The animal trade struggled to recover after the war's end. Henry Trefflich announced he was restocking his New York animal showroom, but optimism that the animal trade would boom again was short-lived. When Noel Rosefelt steamed into San Francisco with a shipment of Malayan animals, the largest "Noah's Ark" since the war's end, zoos could briefly hope that the prewar flush might reappear. One elephant headed to Madison, Wisconsin, to replace an elephant that had died during the war. Another went to Sacramento, paid for partly by the city's children. Even the navy helped make the sixty-four-day journey possible; a flying boat had dropped twelve bales of hay for the elephants and thousands of worms to feed the birds.[95] By the early 1950s, however, Rosefelt confronted the new reality of the animal trade. Collecting animals in Indochina, he and seven baby elephants were briefly held captive by

Communist guerillas. Imperial officials no longer could help smooth the animal trade.[96] His business faded, and soon Rosefelt and his "Oriental Shop" began selling Asian furniture.[97]

Rosefelt went into the furniture business partly because animal populations themselves had dwindled and partly because the imperial system that hoarded animals for Western hunters and businessmen was collapsing. Even "zookeepers [*sic*] joy" when giraffes became available again was "dampened" by news that all shipments of orangutans out of Singapore were banned.[98] There just weren't enough left in the wild. With the war's end, zoos faced interlocking challenges. Of course, they still needed to fill emptied enclosures to attract audiences that now had closer relationships to the tropics than ever before.

Colonial governments that had long eased the flow of animals from jungle to zoo teetered and fell as early as 1947 with Indian independence. Former British India was partitioned violently into India and Pakistan (including East Pakistan, now independent Bangladesh). Other prime game-collecting colonies also struggled toward independence at the same time their game stocks collapsed. Malaya, once the happy hunting grounds of a generation of American traders, was in open revolt by 1948 when British officials and rubber planters tried to reassert their power. British bombers, fresh from the European front, pulverized the Malayan jungles to root out a Communist insurgency during the euphemistically named "Emergency." By 1957, Malaya was independent. Across the Strait of Malacca, Indonesians resisted the return of Dutch rule and won independence by 1949. In Africa, Kenya, torn by rebellion, eventually achieved statehood in 1963 as part of the independence wave sweeping across Africa.

The old ways of collecting animals changed as animal populations also shrank. The era of the trader's ark packed with animals was passing, like the hunting safari, into history. Zoos still sought publicity from bringing back a rare specimen or, even better, a breeding pair. In 1948, as India transitioned from raj to republic, Ralph Graham, the Brookfield Zoo's assistant director, arrived to capture two rhinos. The capture and transport of an independent Indian rhino proved very different than that of a Raj rhino. Before the war, when the St. Louis, Bronx, and Philadelphia Zoos sought Indian rhinos, they commissioned Buck who, in turn, negotiated with colonial officials and relied on the knowledge and muscles of locals like Ali to capture, then transport individual animals. Now Graham and his zoo begged the fledgling Indian government for permission to remove a breeding pair from Kaziranga, the rhino reserve in the remote reaches of Assam.

Graham was off on a "once in a lifetime adventure" and sent home a tourist memoir about animal encounters during the chaos of independence.[99] Graham had little to tell his audience in Chicago about the rhinos' actual capture. In fact, he wasn't even present when the second rhino tumbled into a muddy pit trap. Rather than animal danger, the biggest perils he encountered were the bloated bureaucracy of independent India, the laziness and superstition of native workers, and the venality of merchants. He narrated a comedy of errors in which Graham, the intrepid tourist, negotiated monsoons and airplanes that detoured miles out of the way to avoid flying over East Pakistan. India emerged from empire, in Graham's estimation, as a nation comically condemned by its infernal weather, primitive technology, and corrupt peoples. Local craftsmen haphazardly completed the traveling crate Graham carefully designed. Then they rebuilt it, this time under Graham's constant supervision.[100] His truck driver (Graham called him "Fatso" in his letters back to the zoo) mysteriously misappropriated gasoline rations. Graham even blamed corrupt India for his failure to return home with any other animals for display.[101] A Calcutta dealer had promised him specimens including two baby tigers but sold them to the London Zoo instead, evidence not of clever bargaining or maybe Graham's low offered price but of Indian lawlessness.

Graham navigated India but mastered its nature without the help of loyal assistants or natives equipped with intimate knowledge of animals and their ways. If Buck depended on the lithe and dedicated Ali and Lal, Graham hired the rotund Kashi Ram. (He would later name the male rhino after him, admitting he knew only two Indian names: Ram and Mahatma Gandhi.) Ali knew the Malayan jungle by instinct; Ram, like the new Indian forest officers, couldn't even pinpoint Kaziranga on a map.[102]

The female rhino—Kamala Rani, he called her—was in a precarious state. She was heavily pregnant. Maybe she should have been released to give birth back in the wild. Or she could have been stabled in Assam until she gave birth. However, Graham wanted to escape the coming monsoon and sent her on the rough road and river journey toward Calcutta and then home to Chicago.[103] Along the way, her calf was stillborn, and Graham was furious. He had ordered round-the-clock supervision, but when she went into labor, her attendants were fast asleep. "I was utterly sick, and almost to the point of uncontrollable fury."[104]

Soon after Graham crated his suffering rhinos for shipment to Chicago, independent India began preparing another animal transport. In 1950 the Indian Prime Minister Jawaharlal Nehru offered two baby elephants as gifts to the "children of the United States." Ashok and Shanti were captured in Mysore and sent to Bombay to await a long airplane flight to Washington and the National Zoo. Shanti, though, had hurt her eye, and by the time she recovered, the elephants had grown too much to fly. Like thousands of animals before them, they traveled to America by boat.[105]

Yet the crowds that welcomed the young elephants to Washington knew these elephants were different than other animal arrivals. "[S]eldom has a gift from abroad caused so much comment or been more truly welcome," said the State Department's press release. Shanti and Ashok were diplomatic currency and the property of India. As the local Indian embassy reminded the zoo, the male elephant, Ashok, named after an emperor, recalled India's glorious independent past. The female, Shanti, was the Sanskrit word for "PEACE."[106] Through his political gesture, Nehru laid claims to India's animals. The rhinos for Chicago and the elephants for Washington (and American children) were both important symbols of the new nation. When the Indian government offered them as state gifts, the animals ceased to be imperial commodities.

And Nehru expected animal gifts in return, though they would take a dozen years to arrive. "Americans," read the press release announcing a return gift, were "so used to exotic animals in our Zoos." American animals in the age of independence could also be exotic. In Washington, Theodore Reed assembled a collection of American animals, including a bison pair for the Delhi Zoo, a fitting counterpart to Nehru's elephants. In 1962, they took off for Delhi scheduled to arrive just before the visit of American First Lady Jacqueline Kennedy. At first, the diplomatic corps hoped she would present the animals, but she was reluctant for reasons that were never recorded. Instead, the press was informed the animals would arrive "by coincidence" just before her visit to India. Perhaps, in the careful choreography of diplomacy, the American government wanted to temper its own messages of friendship.[107]

So, instead of Kennedy posing with American animals, the newspapers published pictures of the "zoo man" sent along to care for the animals. With his broad cowboy hat and long Western slicker, W. T. Roth leaned easily against a puma's crate. A New Delhi newspaper remarked that this "zoo man" looked strikingly like Lenin with his goatee and high forehead, a less-than-subtle jibe at the current zoo curator and former trapper's all-American outfit.[108] The gift of animals—first Graham's rhino, then Nehru's two elephants, and finally the bison and puma—signaled a shift in the animal game from big colonial business to postcolonial diplomacy.

Yet the contrast between Roth and Babujan, the mahout who cared for Ashok and Shanti, still rooted the exchanges in the colonial past. Both men had occupations that recalled the older age of empire. Roth might modestly joke that animal trapping was safer than "rush-hour traffic in New York," but he had enjoyed his share of jungle adventure before he went to work at the National Zoo. Once a spitting cobra had blinded him for a week. He had also narrowly escaped a mauling from a "man-eating tiger." Babujan, meanwhile, joined a long line of native men sent along to care for their animals. In Washington, no one had made any "official arrangements" to pay him. Instead, William Mann gave him "a little spending money," bought him some clothes, and offered a "good watch" as a gift. "He seemed quite happy when he left," Mann reassured the Indian Embassy.[109]

Animal diplomacy was designed to secure ties through the exchange of symbolically important animals. What could better represent India than rhinos and elephants or the United States than bison? When George C. McGhee, the assistant secretary of state for Near Eastern, South Asian, and African Affairs, officially welcomed Ashok and Shanti to Washington, he hoped the two elephants would "learn to like their ... adopted country" but never forget "their native land of India." From India, but on display in America, the two elephants linked the "independent democratic peoples of the world." But Graham's frustrations in India, republished as a publicity booklet by the Brookfield Zoo, told a different story, less about strategic relationships than about the new reality of the animal game after the end of the empire. Graham knew his Assam rhinos were now endangered wild Indian animals, India's to give or to protect.

The rhinos must have felt stressed by the locals who crowded around their pens. Graham was convinced that superstitious natives believed touching their horns would get them into heaven. Graham estimated that "2500 people" pushed and shoved each other around the rhino crates, all "seeking a pass to Hindu Heaven."[110] Of course, visitors back at home would also crowd around those rare rhinos, tossing peanuts rather than touching the horns. But, Graham insisted, this was an entirely different kind of relationship. The zoo visitor was learning to watch them alive. Like the zoo man or the ecotourist, the visitor learned to love the animals. Graham insisted that when zoo visitors enjoyed animals, they helped save them. War gave way to anxiety about the future of animals as well as to mass tourism. Zoos eased that transition. They toured before anyone else and recast the zoo man as the bold tourist rather than a scientist. After World War II, zoos turned to television, the perfect foil to animal traders who had monopolized the big screen.

My Wild Kingdom: St. Louis, Chicago, South Africa

The zoo had to change, and Marlin Perkins knew it. In 1942, he was looking for a new job, this time at the Lincoln Park Zoo. Ever since his first year as a laborer at the St. Louis Zoo, Perkins knew he was "first and foremost a career Zoo man."[111] He had advanced through the ranks in St. Louis, evolving from laborer to keeper to curator of reptiles. With his promotion, he spent half a year learning about reptiles from Ditmars, the man who would soon help save his life. As he practiced the mundane tasks of caring for live exhibits, he realized that zoo work could be tedious and the animals somnolent. One of those tedious tasks was using tweezers to remove the parasitic mites that plagued captive snakes. There was also always danger in zoo work, like "hammering nails," Perkins explained. "One day you bring the hammer down on your finger instead of the nail." In 1929 for Perkins, that hammer was a Gaboon viper.

The Gaboon viper is a "quiet enough snake"—except when it freed itself from the keeper's grasp and lunged for Perkins's finger. Then the Gaboon viper could be deadly. One of the largest and most poisonous pit viper species, the St. Louis Zoo's specimen had reached four feet long. Its brown, mottled body was twice as heavy as a diamond-back rattlesnake's with fangs more than an inch long. The bite felt like a bee sting, "magnified a hundred times." As the pain shot up his arm and into his chest, Perkins knew he would probably die. As Perkins and his assistant sliced into his finger and desperately tried to suck out some of the poison, George Vierheller reached for the telephone.

He called Ditmars long distance in New York. As Perkins rushed to the hospital, Ditmars counseled on the right type of antivenom. The next few days were touch and go, and Vierheller kept calling New York for emergency advice. Perkins's pulse dropped, his breath was labored, and his pupils were dilated. The doctors guessed he was dying, but after three weeks in the hospital, Perkins was back in the zoo's new reptile house, a little thinner and paler and a rising star at the zoo.

Perkins was now the "Snake Man."[112] At the point of the viper's fangs, Perkins had learned a lesson that would guide him for the rest of his career as he advanced from a well-known reptile curator to a television star: visitors were fascinated by the zoo man when he recounted his animal encounters on the other side of the enclosure. The whole city seemed to hang on Perkins's fight with one of the most dangerous poisons in the world, and finally, when Perkins breathed easier again, so did St. Louis. Perkins "engaged in a struggle for life as dramatic as that waged by a soldier in battle," marveled the local newspapers.[113]

The zoo's reptiles suddenly seemed exciting all over again. Perkins, the Snake Man, had a special power to fascinate the public about mundane zoo work. As he recovered from his viper bite and as the zoo's chimpanzee show became more elaborate and popular, Perkins came up with an idea about how to turn the typical struggle to force-feed a large python into a popular event. At virtually all zoos of the era, pythons fed only when food was forced down their throats. Zoos did this disagreeable force-feeding in private. Perkins, though, turned it into theater. Soon, just like the chimpanzee shows, newspapers published previews of the big show, and in 1937 *Life* magazine came to photograph 2,000 spectators watching the Maharanee of Wangpoo's lunch.[114]

Perkins realized the zoo man must also be a showman. The public didn't want a herpetologist. They wanted "to be entertained, to be amused, and perhaps to learn a little." They especially wanted to know what it was like to handle the snakes and survive a viper bite. Soon, Perkins was a mainstay on St. Louis radio, broadcasting live from the zoo. Once KMOX radio endured minutes of silence; Perkins and an electric eel had blown the station's radio tubes.[115]

From St. Louis, he became director at the Buffalo Zoo as it was completing its last WPA projects. Looking for a new job again a few years later (this time at the Lincoln Park Zoo), he recognized the postwar zoo needed to change. The "old idea" of the zoo was a collection of rare animals where a few people might find some "small pleasure" in viewing them. The new postwar zoo needed to be "scientific, progressive, exploratory," showcasing the animals and the people who worked with them through children's clubs, lectures, media, movies, and—soon—television. He offered an exciting vision, and by 1944, Perkins became the Lincoln Park Zoo's director.[116]

As he walked around the Lincoln Park Zoo, hemmed in by Chicago's skyscrapers and the lake, Perkins realized that the "lifeblood of a zoo is publicity and promotion."[117] This smaller urban zoo depended more than anything else on Bushman, its famed, if surly and aging, gorilla. The zoo's concrete enclosures were small, and its real estate limited. It could never build the vast vistas of the Brookfield Zoo, its nearby competitor. Perkins wanted to transform the zoo visit not through reconstructing the old cages but through television. Chicago had only 300 television sets when Perkins began broadcasting in 1945. At first he brought the animals to the station, like the bullfrog he dumped out onto the presenter's desk. In 1949, the televised "chalk talk" became *Zoo Parade,* a friendly introduction to the way the "zoo man" enjoyed his zoo. The following year, NBC picked up *Zoo Parade,* and a local zoo show had become a national hit.[118]

It wasn't the first time zoos used popular media to refashion how visitors experienced zoo animals. Roger Conant in Philadelphia offered his own long-running radio program.[119] Ditmars made an early film of a cobra's attack. By the 1950s, a survey of the nation's largest twenty-seven zoos found that almost half promoted their own television or radio program. Such programs highlighted the zoo as a tourist destination.[120] "Visit the animals" that had come "from the dark forests of Africa" and the "steaming jungles of Burma and Brazil," the San Diego Zoo advertised.

New animals had arrived in San Diego, and *Zoorama* invited TV viewers to view them "in surroundings that closely parallel their natural habitats."[121]

Zoo Parade couldn't—yet—offer a realistic TV safari. The Lincoln Park Zoo was just too small, and television technology was just too limited. Instead, Perkins could help visitors experience the zoo through the eyes and hands of the zoo man. Perkins knew visitors wanted a close relationship to animals. Certainly, they could already toss peanuts or marshmallows. They could ride a few selected animals, and in new children's zoos, they could pet animals and their babies. On television, Perkins, though, helped visitors cross to the other side of the enclosure. "Did you ever hold hands with a chimpanzee?" he asked during a show on animal health, as he entwined the ape's long fingers in his own. In the monkey house, a rhesus monkey clambered onto Perkins's shoulders. The animals on live television accepted Perkins's caresses because at the zoo, he countered the drabness of the cages by encouraging people to touch the animals. *Zoo Parade* was a "training program" for keepers to handle animals, for animals to tolerate the human touch, and for visitors to explore the zoo all over again.[122]

With the affable Jim Hurlbut, the program's cohost, playing the role of the curious zoo visitor, Perkins learned how to "lure" visitors to the zoo. Television proved key as zoos reimagined the typical visitor less as a science student and more as a tourist, eager for naturalist adventure. On television, *Zoo Parade* was "potent publicity." By 1951, attendance at the Lincoln Park Zoo had jumped by 500,000. Across the country in cities where *Zoo Parade* played on NBC, zoos also enjoyed noticeable leaps in attendance.[123]

A generation before, Beebe had jealously guarded the zoo man's expertise. Now, Perkins shared the "special things a zoo man must know." Watch *Zoo Parade* instead of Jack Benny (competing in the same time slot), and learn how to purchase rare animals, like a West African brush-tailed porcupine, or how to keep rare prizes comfortable. Perkins realized the "secret desire" of children and their parents to be zoo men. He played on their fantasies of intimacy with animals as he let the gorilla baby suck on his fingers and the python enwind around his arm. Viewers were jealous, repulsed, excited, and a little bit educated—especially when Perkins was bitten again.

Perkins had broken his cardinal rule: when handling dangerous animals, take your time, even if it was a hectic day at the zoo and the TV show was about to begin. The show planned to demonstrate how keepers milk snakes' venom. Like the Gaboon viper years earlier, the snake wriggled free and, again, Perkins was bitten. Once more, the public was fascinated. This time, however, many people insisted they had actually seen the dangerous mistake. For years after, Perkins recalled, visitors recounted watching the bite live on television. The truth was that the accident happened offscreen before the cameras rolled.[124] Visitors and viewers had convinced themselves they had witnessed the ultimate in zoo dangers, and here was the key to *Zoo Parade*'s enduring popularity: the zoo man had opened the cages and helped the public watch the animals while he touched them alive. Somehow, it made Perkins both approachable and brave.

On-screen, Perkins appeared gentle and folksy with none of Buck's bluster but also none of Beebe's bookishness. He was a "greying impresario," not a scientist or a hunter. In fact, he had never even shot a rifle. Still, viewers and visitors thought he was the "bravest man in the world." For the moment, long microphone cords that occasionally tripped the presenters tethered the show to the zoo. Incrementally, though, *Zoo Parade* evolved into the zoo safari, illustrating how animals passed from jungle to zoo. Perkins and his cohosts would explain how zoos and tourists could help preserve vanishing species in the world's newly independent tropics. The show's advertisements, trading cards, comic books, and, of course, the Zoo Parade board game linked the urban zoo visit to tropical adventures.[125]

Perkins's hair was just as gray in the comic books, but his features were just a little more chiseled. After introducing the show in front of the lion cage, he left for the veldt, and along the way, Perkins explained to his young readers how the lion and lioness hunted. As evening in Africa closed, Perkins and his readers lay in wait near the water hole. The buffalo and antelope had gathered. They were nervous and sniffed the air, but they "need their evening drink." In the tall grass nearby, a lion roared and the hoofed animals panicked, turning back directly into the ambush. "Just what the lions hoped for." The zoo director, like Perkins, "must be a man of many talents." A television episode set in the lion house explained how the zoo man cared for the lions while the comic book explained how lions

"Zoo Parade: The Board Game" (1954). The gray-haired Marlin Perkins invited ordinary zoo visitors into the exciting adventure of bringing back animals alive to the zoo.

hunted around the water hole. The board game, meanwhile, explained how big cats ended up in the zoo. First in comic books and in board games, later on television, he brought along his public. "[E]very hunt is a new invitation to adventure."[126]

Finally, with color cameras and a bigger budget, Perkins could make comic books and board games come alive. Hurlbut and Perkins were off to Africa again. In South Africa's Kruger National Park, Hurlbut played the excited tourist, full of questions for the local park ranger. Perkins was the educated tourist, full of textbook knowledge but just as thrilled to be in Africa at last. They visited herds of animals "saved from extinction," jumping from impalas to warthogs to sable antelope like a visitor rushing past zoo enclosures. Then they watched tourists caught in a traffic jam of sorts. Their cars moved slowly as a pride of lions weaved through the traffic and they enjoyed another "spine-tingling dramatic experience" as lions kill an impala just alongside the road. The next stop was a water hole—like in the comics—where baboons, zebras, and wildebeests gathered warily, afraid of prowling lions. Back in their camp, Perkins and Hurlbut realized that, in modern Africa, "game and civilization" were just not compatible. In Africa, the solution seemed obvious: game reserves like Kruger attracted both tourists and animals. As the gates of their comfortable tourist camp closed to keep out the predators, Marlin and Jim headed for tea.

By the 1960s, viewership for *Zoo Parade* declined. With the leap to color television, Perkins's wild safari to Kruger helped plant the seed of an idea for a new show. Perkins headed from Lincoln Park back to the St. Louis Zoo as its new director. In *Zoo Parade*, Perkins played the zoo director, sharing secrets with the public; everyone could be a zoo man. Premiering in 1963, *Wild Kingdom* featured Perkins, the zoo man and a tourist, enjoying the new conservationist safari. The nation's TV viewers watched Perkins wrestle anacondas, herd hippos with a road leveler, and lasso giraffes.[127]

Perkins had turned his board game into a television spectacular as viewers could watch a constellation of conservation efforts. In Africa and Asia, he chatted amiably with dying breeds: white park rangers, wild animal ranchers, and imperial game wardens. Black or Asian workers stood silently nearby, quiet hired hands in the conservation game. Perkins knew he was the luckiest tourist in America and a pioneer of ecotourism. Once, while walking along the beach following Komodo dragon tracks, he said quietly to himself: "You have recorded these wonders all over the world. Marlin, you are a very fortunate fellow."[128]

Wild Kingdom turned Perkins, the zoo man, into the educated ecotourist. His television safaris became family holidays. His wife Carol joined him on his adventures. "It was always much more fun when she was along." Soon, Carol began organizing her own safaris for St. Louis tourists. She guided them through Kenya, Uganda, and Tanzania, and they "had a wonderful time." Marlin, meanwhile, introduced *Wild Kingdom* from a homey living room set with the help of the St. Louis Zoo's famous chimps. Before he became uncontrollably resistant, Moke was his first chimp cohost, forever tethering the zoo-fari to all those other forms of zoo entertainment.[129]

Chai, Bwana? Kenya

Around the same time Perkins began filming *Wild Kingdom*, the retired Stott returned to Africa. Trappers since 1907, Hugh and Jane Stanton had hosted Stott on his first tourist safari. In 1952, Hugh had trapped Sally and sent the black rhino to the San Diego Zoo. Now nine years later, he had retired to a new wild game business. Hugh's Kenyan trapping camp became one of the continent's first ecotourist resorts. The lowland thorn forest was virtually leafless in the dry season. The Athi River slipped lazily in its muddy banks to the Indian Ocean, 200 miles to the east. The tourist endured a bone-shaking sixteen-mile jeep ride just to glimpse Mt. Kenya or Mt. Kilimanjaro. But for Stott, on a family holiday, Bushwhackers Guest Camp was still "easily the most wonderful spot on earth." Bushwhackers' cottages, kitchens, bathrooms, and attentive servants offered "the essence of unspoiled Africa at modest cost."

"Chai, bwana?" asked the servants early in the morning. The tourist wanted to stay in bed, but a verandah and its animal views beckoned. The tea steamed while a bushbuck drank at the river until startled by a distant lion roar. Gray dawn turned to bright morning, and a zoo's birdhouse worth of calls echoed. A velvet monkey was oblivious to the humans, not ten feet away. "He seems to know," Stott mused, "that Bushwhackers belonged to his kind long before it did to yours." Somewhere in the ownership transaction from primate to American tourist, unspoiled Africa never seemed to belong to the African human. Yet native humans were still there in Bushwhackers as poachers who killed Cora, a rhino the Stantons had trapped, domesticated, and rented for film shoots and whose daughter Hugh sold to the San Diego Zoo. They were also the servants bearing morning tea and lavish snacks. Dancing drummers called tourists to dinner. "Kuenda sabali?" they called, "where are you going on safari?"[130]

Belle Benchley, Stott's old boss, never did get to go on safari. Then, at the 1953 AAZPA convention, the other zoo directors chipped in and bought her "a beautiful white leather train case" to celebrate her retirement as the nation's first female zoo director. Luggage—and she planned to use it. She had spent a career looking out of her office window at the world's animals. She knew more than most people in California about places and people she had not yet seen in person, and she had sent her subordinates like Stott off to Africa. Others, like the male keepers who had worked under her leadership, had served on tropical battlefields and jungle fronts. They had marveled at birds soaring through the kind of jungle canopy she had seen only in cages.

After a zoo career, Benchley went touring. She—and many other women and many other zoos—had helped raise the animals at zoos that, more than ever before, fashioned themselves as nurseries for vanishing species. She had enjoyed her own "very rare experience" but had never plunged into the dark jungle or searing desert. Rather, she was a woman in what was still the zoo man's world. Fittingly enough, it was Lee Crandall, author of *A Zoo Man's Notebook,* who presented the professional organization's official gift. Benchley knew her own zoo adventure was "coming in to work with a group of men—being the only woman that was doing it and the first one doing it." She was the first in a movement that would change the profession and the zoos they ran.[131]

MY ANIMAL BABIES

Caring for Endangered Species

D ID YOU EVER SEE a baby warthog?" As the executive secretary of the San Diego Zoo from 1927 to 1953, Belle Benchley fondled a range of creatures from infancy to old age. She disparaged the adult warthog as an ugly, ornery beast. "How could anything so hideous as that creature ever have found favor with the Creator," she wondered. Its infants, by contrast, were cute. At the zoo, the warthogs lived "in style" in a house that opened onto a terraced hillside where they enjoyed a pool and a mud wallow. One early morning, despite her disdain for the ugly adults, Benchley rushed to the hogs' building to watch the warthog "in the accepted fashion of mother pigs," nursing her two newborn, squirming babes. As they climbed over the mother, pushing their snouts into her side, Benchley felt the new mother's pleasure. She seemed to be "thoroughly enjoying her new young maternity."[1]

When Benchley opened the door to the hog's grotto, the mother led her two babies into the Southern California sun. A few hours old, they already had a warthog's characteristic tough hide. Their hoofs were tiny, and their blunt heads seemed too large for their bodies. In this spring morning before the arrival of visitors, Benchley's delight gave way to dismay. One of the babies was a runt, and although he struggled mightily to get his share of mother's milk, his mother did little to push aside the larger baby. Perhaps it was her pride in her healthier baby. Perhaps, Benchley insisted, it was simply "her gentle indifference of character." Benchley realized that if the poor runt was going to survive, she must intervene. The runt needed a surrogate mother. Benchley reached into the enclosure and, gently, carefully, lifted the squealing, hungry runt from its animal family and into the human world. Benchley and the other keepers fed the tiny hog on milk and cod-liver oil; he lived only a month.

Spunky, as Benchley named the remaining baby, by contrast, grew quickly. By the fall, Spunky clambered and wallowed alone in the grotto. When he tired of his romps, he nuzzled against his mother's fat body. These bonds of motherhood proved too tight when the mother warthog gave birth again and ignored her new infant in favor of her firstborn. Like the runt before him, the new baby seemed on the verge of starvation when keepers removed him to the care of Emily Burlingame, one of the zoo's dieticians. Burlingame had often "mothered" zoo babies, and she cud-dled and cared for the pink, squealing hog—Mickey, she called him—like a "human baby."[2] She carried the devoted baby home with her at night, and during the day the hog, grunting softly, followed her around the zoo.

Yet Mickey grew, and families of different species don't last. By the time Mickey was four months old, the vigorous warthog was too active to stay with Burlingame in her home, and keepers returned him to the warthog grotto. Spunky, at first, seemed delighted to have a new playmate, but his mother was angry. Benchley wondered if the hog mother sensed "some slight towards herself and her own motherhood." The mother warthog charged Mickey, breaking his hind leg. Burlingame removed the quivering hog to his human home where he continued to grow, losing his pink infant playfulness for the strength and warty coarseness of older warthogs. Too old for his human babyhood and rejected in the warthog haven, he was sold to another zoo far from his mothers: the human who raised him and the warthog who rejected him. "Oh, but I did love my little wart-hog baby," complained Burlingame, "I just hugged him tight in my arms."[3] Maybe Mickey even liked the human love.

The trials, successes, and frequent failures of pregnancy, birth, and parenting revealed the challenges of imposing the domestic "civilization" of the zoo on wild animals. Almost since their inception, when the stork arrived, zoos celebrated notable animal births and newspapers delighted in stories of animal romance. Even so, animal births and breeding took on new significance in the years just before and following World War II when zoos could no longer depend on colonial Africa and Asia to replenish aging and dying stock. The first world war transformed the global animal trade; the second paralyzed it. Americans who had once thrilled to stories about animal traders in darkest Africa or jungly Asia now scorned the overt racism of such stories. Horrified by Nazism, many Americans, who sought postwar peace through a day out at the zoo, also believed in a shared humanity that united races in a "family of man." In the postwar era, the emergence of children's zoos and the fascination with animals being raised by humans suggested animals could live in that family of man.[4]

As zoos moved increasingly toward breeding the animals they displayed, they offered images of animal infancy that stressed similarities to human babies. Everyone knew the lion cub tearing up the furniture or the gorilla wearing diapers was eventually going back home to the zoo, but Americans still delighted in funny images of animal babes reared in the human home. Americans loved it when animals seemed almost human—especially if those animals were cuddly babies or contemporaneous chimpanzee shows. The comedy of animals placed into the tender arms of white women in their homes evoked a new intimacy between Americans and their animals. That comedy also obscured the tragedy of animal extinction; vanishing species became human wards. Even as zoos encouraged the creation of nature reserves in newly decolonized countries, they also admitted animals were going to have find ways of living alongside humans and sometimes within their families.

The idea of the universality of childhood across species helped fit zoos into the larger fabric of postwar and Cold War American society. Soldiers returned home after months and years away from their families. The GI Bill provided low-cost mortgages that helped, in particular, white veterans purchase homes in suburbs that often restricted black ownership. In the new insecurity of the Cold War, Americans retreated home to raise families. They married younger and, to use the terminology of zookeepers, they bred early. Babies were certainly on American minds, and as it turned out, they cared about human and animal babies at the same time—and remarkably often in the same places and in much the same ways. The home and the nuclear family became a symbol of national stability. Americans seemed all the more bound to hearth and home when they succored animals no longer capable of taking care of themselves.

Conservation was reared in the postwar home. There, Americans learned to become the stewards of animals, the last defense against their extinction. Conservation had and still has contradictory meanings and goals. Sometimes the aim is to keep animals flourishing in their traditional habitats. Sometimes conservation keeps species alive in an entirely new habitat: the zoo. Postwar Americans believed the future of the world's wildlife was now in their hands and demanded a woman's touch in animal care. Women found their way into zoos and eventually into paid zoo-keeping when conservation became domesticated. From bring 'em back alive to breed 'em alive, the zoo was changing again. The humor in the descriptions of animals living like members of human families represented a serious reform of the postwar zoo. In the giggle of the ridiculous, the sigh of selfless tenderness, and the intrusion into precarious life, modern conservation emerged.

My Animal Queendom: Chicago, Singapore

Two baby orangutans snarled traffic. Ten thousand people an hour streamed into the monkey house on this early spring 1940 day. Human children were pushed out of the way by adults, eager for a glimpse of the resting apes. Young boys crawled underneath taller legs to get closer to the bars. For those who couldn't face the busy Chicago traffic to get to the Lincoln Park Zoo, WGN Radio described every move the babies made. Unfortunately, the two exhausted young apes did little but sleep. Occasionally, the female—"that little beauty"—reached out from her improvised nest of burlap sacking to grab a grape.[5] The two young orang-utans were, at once, exotic and familiar. In their sleep and games, they reminded at least one visitor of the playfulness and gentleness of the human child. "I wish I could take her home forever," she sighed. A policeman tried to control the crowd and soon gave up. "It's an impossi-bility," he shrugged.

The Lincoln Park Zoo needed those crowds. Just when those orang-utans captured Chicago's affection, the world was again at war, and the established routes that brought such animals from Africa and Asia were again disrupted. Nine years and a world war later, a shipload of 950 animals traveled from Singapore to New York. Along the way, the usual trouble began when a python escaped its container and attacked his handlers. Off the African coast, a dozen monkeys escaped and climbed into the rigging. As the ship moved from the equatorial tropics into the north Atlantic, the larger primates suffered. Some fell ill; others refused to eat. In all, 15 percent of the shipload perished at sea. On the New York docks, a caged leopard swiped at a Brooklyn teenage spectator, the final casualty of the long voyage. In all, the trip was "anything but fun" but typical in its dangers and death. However, "Jungle Jenny," the trader who had assembled this menagerie, was out of the ordinary.[6]

Jungle Jenny—Genevieve Cuprys—was the rarest animal on board: a twenty-one-year-old animal trader and a woman. She had come from good animal trading stock as the foster daughter of Arthur Foehl, an animal trader working out of Singapore. One of the first to get back into animal trading in Singapore after the end of the war, Foehl unexpectedly died while organizing a large shipment of animals.[7] Foehl had been trying to find his successor in the business. He had tried out his brother, Edward, and considered his young nephew. When Foehl died, though, the business fell to Jenny. Business, for a while, thrived, not least because of the novelty of a female animal trader. Henry Trefflich hired her as his agent in Singapore and emblazoned his stock lists with colorful accounts of her dealings in the Far East. In 1948, his firm boasted that Jenny was bringing back alive orangutans.[8]

Jungle Jenny, a postwar salute to a bygone era, offered a feminine contrast to Frank Buck. Gone were the pith helmet and the "movie star" masculinity of the prewar trader, and in their place she promised postwar femininity. She never forgot she was "a woman and an American." Except when she went "upcountry," she said, she wore a dress and sandals. In Siam, she caught six elephants. In Malaya, she collected orangutans and cobras.

Newspapers were fascinated with Jungle Jenny, but only briefly. They delighted in describing the "small girl" and her perilous encounter with a thirty-two-foot-long python. Soon, though, with the twilight of the animal trade, Cuprys traded jungle for home and family. Her romance began when an angry black leopard escaped from his cage. The cat pounced, and she shoved her elbow into its mouth in an effort to save her throat. "I knew I was going to die," she recalled to reporters, "so I relaxed." Miraculously, the cat relaxed as well, and she rushed to a hospital. Surreptitiously, she crept from her bed to negotiate the purchase of a replacement big cat. Back in the states to recover, she met her fiancé in the office of a wild animal veterinarian. "I've always like the animal business," she told Trefflich. But "I won't go back into it—unless my husband . . . Well, you know men."[9]

Bring 'em back married; domesticity, love, and family reshaped the animal business. Marté Latham sought to escape the drudgery of her middle-class marriage through adventure in the wild animal trade, bringing back animals from South America to American zoos. The press was eager to describe a married and female animal trader. They begged her to agree to a publicity stunt that could combine both motherhood and adventure. They wanted her to raise a chimpanzee baby in her home. The odd family—"my animal queendom," she called it—soured quickly.[10] Latham herself recalled flying into rage and beating an animal that just couldn't adapt to the suburban ideal.

On the other side of the gender divide, Peter Ryhiner yearned for the days of Buck when Americans enjoyed a free run of the markets and jungles of the colonized world. He also missed his wife. Buck's divorce had been amicable, easy, and a mere paragraph in his autobiography. For Ryhiner, the rise and fall of his marriage—fleeting bliss in the jungle—mirrored the rise of his career and the fall of the business. Ryhiner was on television enough to be recognized occasionally, but he was never a star. And he was broke.

After the war, the colonies were no longer the animal free-for-all Buck described. Malaya, where Buck had made his fortune, was in revolt—the British called it the "Emergency"—when Ryhiner arrived. He hoped to collect apes, tigers, and elephants. Openly displayed portraits of Stalin mocked a British regime that was disappearing in a hail of bullets and crowds of protests. The natives were more likely to stab the trader in the ribs than to call a "white man" *tuan*, Ryhiner noted bitterly.

As European empires fell and, with them, the animal trade, so, too, did his marriage. Marcia was beautiful in Ryhiner's eyes, but she was mixed race and felt ill at ease in the decaying colonies. America was no better. Domestic reporters found in her mixed-race beauty the perfect foil for exotic animals. Yet she hated the inevitable demands of photographers to pose with Ryhiner's captured snakes. Married to a man who depended on publicity at home and access to animals abroad, Marcia was caught between Asia's imperial past and its independent future. When Ryhiner dragged her back to Asia in search of the big game, Marcia asked for a divorce, and Ryhiner realized that animal trade could not mix with either marriage or the end of empire. "Dead, dead," Marcia told him as they sat down at one last marriage meal. "I tell you that the animal collecting business is dead." Leaving behind his ex-wife, he returned to India. One last hope, one last expedition, and everyone would know his name "like Frank Buck." Yet this was no longer a business of huge cargos but of smuggled animals: the "hot okapi" smuggled from the Congo to Uganda or the rhino spirited out of Sumatra. Divorced, penniless, and a lesser light advising nature documentaries, Ryhiner looked to the future of the animal game and realized it was in zoos with breeding programs. He dreamed that someday he could buy enough land to build a game park. The days of $100,000 cargoes had ended, empires had fallen, and the era of breeding had begun. He hoped it would be "far more inspiring than any Frank Buck picture ever made."[11]

Babies of the Dark Jungle: St. Louis, San Diego

Despite a wartime pinch on animal shipments, in 1940 the St. Louis Zoo enjoyed the pleasant problem of crowding. Births had increased enough that the zoo could sell off some of its babies to struggling animal dealers. The zoo earned a healthy $2,500 profit by selling two antelopes, a Malayan buffalo, and two Siberian ibexes. The zoo kept the babies of animals no longer available on dealers' stock lists.[12] Director George Vierheller enjoyed his well-earned reputation as a showman. Relentlessly promoting its famed chimpanzee show, Vierheller had masterminded the zoo into a national tourist destination.

He had an idea for the best chimpanzee act of all. In 1944, a baby chimp had been born at the zoo, but for fears it might die, Vierheller kept the news secret until it was about two months old. Then Vierheller introduced the little star to the city in a news conference that included the mayor. The celebrated birth became a stage skit. The chimpanzee mother, Alice, a veteran of the show, brought her baby onstage to the delight of the audience. Jitter, the proud father, handed out cigars.[13]

The act was good theatre and also typical of the way zoos increasingly celebrated animal maternal tenderness and proud paternity. Whether in arena seats or crammed around an enclosure, visitors could watch parents and children and imagine humanlike emotions. Zoo directors, in turn, looked happily at the gathering crowds. Motherhood, they declared, crossed the boundaries of species. In St. Louis, Maggie, an orangutan, was "like any modern mother" with her baby George. As photographers snapped the first pictures of the rare baby, Maggie posed, cradling her newborn in her arms. Photographers and keepers alike described her expression as "pride and self-conscious joy." Maggie was also protective and angry at the exploding flashbulbs. Maggie seemed such a good—that is, human—mother. The keepers reported that Maggie gave George a daily sunbath and treated him to small nibbles of solid food. They even joked that she did daily exercises to "get back her figure."[14]

When zookeepers joined visitors in looking across moats and bars to discover animal families that seemed comfortingly human, there was a real, often fatal cost. Zoos typically housed animals in nuclear families composed of babies with doting fathers and mothers, often denying the multiplicity of ways animals, in fact, cared for their young. Especially in the early years of deliberate zoo breeding programs, zoo directors enforced monogamous relations and after births encouraged nuclear family units. Apes, especially, were match-made in couples to encourage a zoo "romance." Once animals were born, zoos and reporters shared stories of indulgent mothers and doting fathers, like Jitter handing out cigars. The animal enclosure had come to look a lot like the idealized American family.[15]

All those color stories of zoo babies encouraged zoo officials to breed animals actively, but they also made zookeepers' jobs harder. As zoos realized that animal babies could draw crowds and turned away from wild cargo, they confronted the challenges of breeding 'em alive. Efforts to breed animals in nuclear families also stressed the animals, adults and babies. When the orangutan Bimbo's baby cried, her father howled and

shook his cage. Reporters insisted he was simply voicing his pleasure—
and the crowds still gathered. In reality, such fathers often represented a
threat to newborns, especially if they were kept too near mothers and
babies. Sometimes fathers attacked the youngsters; sometimes, their mere
presence provoked mothers into killing their babies.[16]

Zoo directors and keepers recognized that while they may have
removed animals from the natural struggle for survival, they forced them
to breed in surroundings that created new tensions. But sometimes it was
too late. Prying visitors eager to see newborns could panic a mother into
a helpless effort to protect her baby. Crowds gathered to see a young
hippo gamboling on the muddy banks of its water hole. Its mother rushed
to provide it cover underwater and, in her hurry, ruptured its liver. Only
then did the San Diego Zoo realize hippos needed time to be acclimatized
to eager humans.[17]

Zoos celebrated animal births, but behind the stage shows and colorful
newspaper reports, zookeepers worried. Would a mother reject her
youngsters? Would she be able to nurse? Sometimes, warned zoos, animals are not the best mothers. When the "stork flew low" at the
Philadelphia Zoo in spring 1945, it kept keepers busy. The Barbary sheep
were typically the first to give birth. Elands and camels arrived next, then
a black leopard. Even a domestic cat meant to hunt mice gave birth in the
enclosure of a deer named Pepita. Pepita displayed the mother urge,
joining the cat in caring for the kittens. She licked the kittens clean under
the mother's watchful eye. The black leopard mother, however, abandoned her kittens. Some animals "do not make good parents," admitted
the zoo.[18]

The idolized nuclear family in combination with stories of animal
mothers rejecting their babes helped cement images of zoo animals and
"vanishing species" as wards best cared for within the confines of
American domesticity. Zookeepers and directors sometimes sounded like
social workers of sorts. They scolded captive animals for behaving like
poor, even dysfunctional mothers. One Sumatran tiger in St. Louis gave
birth to triplets. She was, however, an "irresponsible" mom who "gadded
about ignoring the babies," read the reports. The babies went home to a
keeper's wife.[19] A jaguar in San Diego successfully raised her first litter.
With her second litter, however, she became, according to Benchley,

"shiftless," refusing to clean her babies. Burlingame, as with the baby boar, became the surrogate mother.[20] Maybe the human family home and women's nurturing could save zoo babies? In an era when Americans idealized home fires, the domestic ties that bound exotic animals to foster women delighted a public beginning to tire of the tall tales of animal capture.

Ma Doesn't Want Her: Philadelphia, St. Louis

"Any one seen Charlie Campbell?" shouted Pat Menichini, a lion keeper at the Philadelphia Zoo. In 1940, the black leopard had given birth, and only Campbell, the zoo's head keeper, could save the kittens from their mother. This was the leopard's third litter, and the zoo hoped the twins would survive this time. The first litter simply disappeared without a trace. The mother "neglected" the second kittens when they lived, the zoo reported, and when they died she chewed on one. With this third litter on its way, Menichini, Campbell, and the other keepers were on the alert with a secret plan; they were going to take the kittens and have a surrogate raise them. The zoo was already closed when the kittens' cries rang plaintively through the lion house and Campbell came running. The mother, however incapable, was unwilling to give up her young without a struggle. As Campbell and the other keepers grabbed a nearby metal scraper, used normally for cleaning cages, she hissed and growled. The keepers found the kittens near death, hanging by their heads near the roof of the cage. "No wonder they were crying!"[21]

That night Fred Patton, a zoo employee, took the kittens home, and his wife became a mother to the young leopards. She and the zoo knew little about how to feed leopard cubs, but they seemed hungry most of the time and took a milk formula designed originally for orphaned puppies from a nipple meant for premature humans. Within a week, the leopards were growing and Mrs. Patton was "rather like a hen with ducklings." She endured sleepless nights and endless loads of laundry and for her efforts suffered scratches and bites. The zoo announced the leopard births with a photograph of their proud mother: Mrs. Patton, bottle in hand, looking adoringly at the leopard she cradled in her arms. Soon after, the kittens succumbed to strange but brief illnesses. Newspapers reported the Pattons were "heartbroken."[22]

Mrs. Patton, like so many other the wives of zookeepers and officials, provided maternal care for animals rejected by their natural mothers. Zoo animals taken from their natural mothers were often reared at home but in keeping with strict gender norms that still guarded zoo work as above all a man's job. Zoo wives did the night feedings and placated crying apes, cats, and other animals with improvised bottles—all under the glare of relentless publicity. Reporters laughed at the playful paradox of tropical animals living in American homes, and zoos eagerly promoted the seeming domestic bliss of animals and people living in close quarters. A stream of newspaper and magazine reports turned typically staged family snaps of adoring mothers into examples of cross-species motherhood.

For a lion cub in St. Louis in 1944, surrogacy ended more successfully. "Miss Cubby Lion" was going home, but not to her mother. "Ma doesn't want her." The cub was only three days old when her mother rejected her and the cub went home to the wife of Moody Lentz, one of the zoo's long-time keepers. This was a familiar task for her. Moody had come home before with orangutans, polar bears, and tigers. For the young lion, it must have been a strange disruption, punctuated by a few days of sickness. According to the local newspapers, the lion fit well into the Lentz home. Animal foster care could be a loving story of shared domesticity. The little lioness was a playmate (if a bit rough) for their five-year-old son, Jerry. "Cubby" seemed to love human company, especially anyone wearing "silk hose." She enjoyed shredding stockings. As she grew, Cubby would be returned to the zoo, from one domestic setting to another. She would soon have her "own little bar-covered cottage in Lion Lane."[23]

There was a reason the St. Louis Zoo recast the lion enclosure as a cottage. As the war neared an end and then as soldiers returned home, family and domesticity took on an outsized importance. This focus on the home was magnified once more in the uncertain context of the Cold War. The home, if a symbol of stability, also became a political marker of the "free world." The Soviet Union may have had its factories, but the United States had the nuclear family in its own four walls of the suburban home. The family also became a dominant metaphor for new understandings of racial difference among peoples at a time when strict conceptions of racial hierarchy were tainted with Nazism. Nazis, after all, and their crimes had depended on a strict grading of races. When the popular photographer

Edward Steichen displayed the photographs of his 1955 "Family of Man" exhibition to rapt audiences, he described racial difference instead as an example of exciting biodiversity within the human race.[24]

From the Family of Man, it was an easy jump to the Family of Animals. Both stressed sentimental bonds of affection and care like those that linked Mrs. Moody Lentz to yet another animal. "Baby ape thrives on tender care of zoo employe's [*sic*] wife," proclaimed one 1947 report, admiring Mrs. Lentz's devotion to an undernourished orangutan. "Rusty," she named him, "is almost like a human baby." Rusty grudgingly accepted the grasp of the professional keeper, Moody, but he whined pathetically if his "'Mom'" entered the room. The new family posed for pictures. With only her blond hair visible, Mrs. Lentz lowered her face toward the ape. Nestled in his blanket, the "affectionate baby" grasped his human "mother."[25] The purity of a mother's love and the child's touch, all typical of family snaps of a newborn child. Only this time the child was an ape and the photographs appeared in newspaper pages.

In a postwar world as racial relations merged with familial relations, interracial and international adoptions became a way Americans saw themselves as "rescuing" the less fortunate, as the historian Laura Briggs notes.[26] Animal adoption seemed a natural, if amusing, way of caring for zoo specimens. For zoos, animal adoption was also good publicity. Color stories accented by animals cuddled in mothering arms helped zoos fit into the domestic spirit of the era. During the Cold War at the twilight of empires, zoo animals were no longer colonial subjects fiercely resisting their capture. Rather, they were babes like Oofy, a chimpanzee bound to humans by affection and care, a victim of a dysfunctional mother, and a ward dependent on a zookeeper's wife.

The son of stars from the St. Louis Zoo's famous chimpanzee show, Oofy was being raised "just like a baby" by Mrs. Frank Florsek. In late 1945, while the trade in wild chimpanzees remained at a standstill, the zoo was desperate to save a valuable animal rejected by his mother. Oofy might star in the ape show someday, but now in his infancy Oofy helped redefine the zoo. As soldiers returned home to long-promised domestic rewards, the care of a baby chimp in a house with four human children earned photo spreads in all the St. Louis papers. The "simian addition"

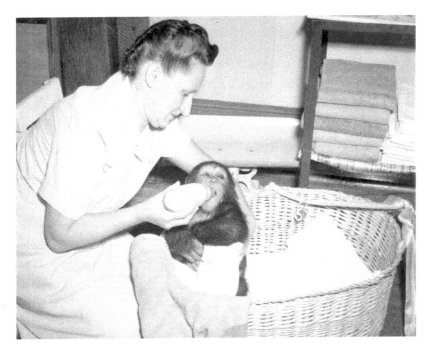

"Feeding Oofy." In 1945, Oofy, a baby chimpanzee, was raised like a human baby to the delight of zoo visitors and St. Louis newspaper readers.

to the already crowded Florsek household enjoyed the same care and diet as the "Florsek's four children got at his age."[27]

At night, Oofy slept beside Mrs. Florsek's bed in a baby basket handed down from the other children. He played with their toys. In fact, unlike his new siblings, he could "manipulate [a] big rubber ball with all four 'hands.'" Unlike the other Florsek children, Oofy lived a public childhood as reporters rang the doorbell, cameras in hand. They wanted familiar family snaps of postwar childhood, only with an almost-human baby. In his newspaper photo album, Oofy napped "with a contented smile lingering on his face" while a beaming sibling, Betty Florsek, brushed his fur. Above, in the iconic pose of motherhood, Mrs. Florsek, her hair tied back into a sensible bun, cradled Oofy while he sucked greedily on his bottle. From a photo of his diaper change—"Oofy's hand-like feet make things a bit confusing sometimes"—to an oversize snap of his nighttime slumber, curled up in a bassinet, clutching his blanket, the modern zoo family was proudly on display.[28]

Our Appreciation to the Ladies: Washington

Another young woman clasped the infant to her breast. The malnourished baby chimp couldn't have been more than a few weeks old when it was offered a human breast and someone snapped a picture. This photo was meant to disgust and enrage. Specifically, it condemned the destructive native trade in wild apes through familiar ideas about racial similarity linking Africans and their ape neighbors. The argument was easy to follow: African humans were killing or selling their cousins, not wild animals. After all, the babe seemed to suckle so easily and naturally.[29] Unlike the nameless African woman who for profit placed a chimp to her breast, Lucile Mann's mothering was meant to demonstrate the role of the zoo as the guardian of the world's helpless species.

"Breast-fed Chimpanzee." The image of the breast-fed baby chimpanzee contrasted dramatically with those of animals raised by surrogate American mothers. Anna Mae Noell, *The History of Noell's Ark Gorilla Show* (Tarpon Springs, FL: Noell's Ark, 1979).

"I wonder," remembered Lucile, "if anyone started housekeeping with as many and as novel interruptions as I did." Her "domestic routine" combined the tasks of the housewife—"made strawberry jam"—and the exotic—"Bill brought my galago home." Recently married, Lucile accompanied William, the National Zoo's director, to an animal dealer's warehouse in Manhattan. On the top floor of the shop was the "nursery" housing a young gorilla. The ape was the most expensive prize for sale, but Lucile saw only a "solemn child." Twin orangutans also excited her maternal feelings as she watched them sleeping entwined under a blanket. She held them in her arms and "that brief moment gave birth to my own desire to have a pet monkey." Her maternal urges were lucky for the National Zoo as it transitioned from William's prewar expeditions to its postwar dependence on the animals it could raise from infancy.

Monkeys and other animals passed through the Manns' small Washington apartment. Zanzi, as Lucile nicknamed the galago, was a type of lemur William wanted to tame enough to tolerate petting from zoo visitors. As a youngster, Zanzi was a "funny furry little thing" who enjoyed his fruit and did not seem to mind when Lucile tied him to a chair. In the confines of their apartment, he never intended to hurt, though his favorite game was chomping on Lucile's hand. As he grew older, playful chewing gave way to vicious bites. After he drew blood, the Manns returned Zanzi to the zoo he had left as a baby. He could hardly have remembered the cage and proved ill-suited to captivity. Within a few months, he died.[30]

Susan, a young lion cub, took Zanzi's place after her mother refused to nurse. Every three hours even at night, Lucile warmed powdered milk formula. By the end of ten weeks, Susan had grown into a playful cub with teeth that could draw blood and claws that tore stockings and ripped rugs. The demands of raising a lion kept Lucile chained to the home. "My time may or may not be valuable, but she had every minute of it," Lucile complained. Both Susan and Lucile likely felt a mixture of confusion, discomfort, and genuine affection. Susan followed Lucile around the apartment, whining if she was left alone. For her part, Lucile bemoaned the destruction wrought by a growing big cat. Susan's food bill came to just two dollars a week; repairs to rugs, furniture, and clothing cost much more.

Still, Lucile was proud of her accomplishment. She raised her little lioness "as carefully and methodically as a human baby." Like the human

child, Susan's life was scheduled around doses of cod-liver oil and feedings from sterilized bottles. Yet "[a]ll the domestication in the world
could not cover up the little savage that she really was," wrote Lucile.
With her first bone, Susan pounced and "killed." Soon, Lucile was happy
to play with her lioness with "iron bars between us."[31] William's expeditions to Africa and Asia had linked the zoo to the adventures of the animal
trade. The tender care Lucile offered wild animals in their small apartment helped make zoos seem more like comfy homes. She helped domesticate the zoo. Yet it was stressful and constant work. The first few ten-
uous weeks of a captive animal's life, especially if its natural mother
rejected it, depended on the kindness of wives and the scant few female
workers actually employed by zoos.

Zoos had "outsourced" the medical care of newborns to zoo wives. For
Theodore Reed, who succeeded William Mann as director, this was cause
for celebration. At the National Zoo, Lucile and keepers' wives gave med-
icine, kept watch (often all night), and fed babies for weeks at a time.
Esther Walker raised monkeys, bats, and flying squirrels. Louise Gallagher,
wife of Bernard Gallagher, an ape keeper, specialized in chimpanzees and
gorillas. Nettie Stroman reared twin gibbons, and Margaret Grimmer a
hyena and snow leopard. Apes, Reed noted, were just as "difficult as
raising a child." Occasionally, zoos paid such women for their labors. In
St. Louis, Mrs. Frank Frosek and Mrs. Moody Lentz each received a pit-
tance, a few dollars for expenses. In Washington, Louise Gallagher was
paid a small amount for her work with gorillas. Instead of a salary, in 1965
the National Zoo presented surrogate mothers a certificate with "our
appreciation to the ladies."[32]

The labor of zookeepers' wives in their own homes marked the postwar
zoo as the breeding grounds for animal conservation. These wives cared
for animals neglected not only by their natural mothers but also by native
peoples abroad. The disturbing image of the breast-feeding native woman
intent on selling her charge for profit contrasted with the domesticity
afforded by the modern zoo—or so zoo officials claimed. Mambo, wrote
William Bridges in 1953, the Bronx Zoo's public relations officer, was a
lucky little gorilla growing up in the zoo. As Mambo swung and played in
the Bronx, Bridges could hardly imagine he was "once a sad and lonely
baby in a faraway village in Africa, with not enough to eat and nobody to

play with." Mambo was born in a jungle but captured by hunters who realized that "people in America would pay a great deal of money" for a baby gorilla. By the time "a man from the zoo" actually arrived, Mambo was badly malnourished. Angry, the zoo man paid for the baby, "though not nearly as much as they wanted." That the zoo man had secured a bargain for a valuable specimen mattered little.[33]

Clinging to life, Mambo remained under the care of "the woman who was in charge of the Nursery." Inside the zoo's hospital, she recreated a tender home both for Mambo and for cameras. Beaming for the camera in front of a background that provided no hint of sterile, medical surroundings, the unnamed nurse cradled the young gorilla, nursed him on a bottle, and brushed his fur. He was a lucky baby to have found "such a good home in the zoo."[34]

The Family Life of the Cage: Washington, San Diego

For some women like the nurse mothering Mambo, World War II offered an opportunity to move to the other side of the enclosure, but change was achingly slow. During the war, a host of new occupations, including in munitions factories, suddenly were open to women. More women than ever before applied for zoo work during the war. "I realize that most of the staff in a zoo are men," wrote Jacqueline Sabin, "but due to the present war conditions, I thought perhaps some openings might be made for women." She asked little—"a living wage" and the chance "just to be with" animals at the National Zoo. She boasted impressive qualifications, including undergraduate degrees in botany and zoology. William replied with the form letter of rejection he sent to all female applicants: "[W]e have never employed women as Keepers in the National Zoological Park."[35]

The working world of the zoo before World War II was, to the regret of such eager and well-qualified women, a man's world. "The work is really quite hard manual labor, and I really do not think that you would enjoy such a position," William wrote in 1944.[36] Wartime labor shortages helped women push open the door to malodorous work, but zoo directors steered them toward jobs raising orphaned and rejected animals. Zoos also hired a few women to nurture child visitors. "I do enjoy contacts with

children," assured Grace Wood. Other women applicants like Beatrix
Moore insisted they were equal to the physical and sensual demands of
zoo work. "I like snakes. I do not mind odors," she promised in 1944.[37]
William was still discouraging. His zoo was willing to depend on informal
female labor, including that of his wife, in caring for animals, but it would
only hire women as secretaries.

In fact, it was through the nurturing and secretarial roles that Belle
Benchley ascended to become the first (and still one of the very few)
female zoo chief executives. Benchley started at the San Diego Zoo in
1925 as a substitute bookkeeper. Meanwhile, the zoo's founder Harry
Wegeworth had begun a long and difficult search for his successor. The
zoo turned first to Frank Buck, hoping his fame as an animal collector
would benefit the zoo. It was a failed effort, and the zoo acrimoniously
terminated his contract after only one year.[38] He was better suited to
celebrity than administration.

Turning from the animal trader to the zoo mother, Benchley was "the
solution to our executive problem," realized Wegeworth. Benchley
explained her promotion slightly differently. She found the zoo so "poorly
organized" that she assumed new responsibilities daily until she became the
zoo's head, but a new kind. William Mann enjoyed jungle adventures,
whereas Benchley sought "intimate knowledge" about the zoo's specimens,
especially its babies. Mickey, a South American tapir, "never had a doubt
that I am her real mother," she boasted. Mickey arrived at the zoo thin but
active. Soon, she began to decline. As sickness became mortal crisis,
Benchley assumed the role of surrogate mother. Arriving early at the zoo,
she cooked Mickey's breakfast of salty oats and patted her gently as she ate.
Benchley was an executive in charge of men, "themselves a new order of
being whom I came finally to understand and respect." Yet she described
herself as an "affectionate foster mother."[39] It took the zoo a while to finalize
her promotion to director; for years, she remained its executive secretary.

An interloper in the world of the "zoo man," Benchley focused on the
"family life of the cage." The zoo, she said, must be "dedicated to the
children" on both sides of the enclosure. As much as she delighted in her
quiet hours before or after closing time when she could observe her
animal charges, she also enjoyed the "laughter of the little folks who run
freely" and feed their animal "friends."[40] With Benchley as the pioneering

zoo mother, her zoo acquired a reputation for its animal babies. "When zoo directors chide me," she recalled, "for mothering my animals and claim that is why we have so many babies in San Diego, I plead guilty." Through her eyes, the zoo featured romances, tenderness, and tragedies that brought visitors, especially children, ever closer to "the story of preservation of the species and the story of reproduction." One pigtail monkey suffered a tragedy that easily resonated with a human audience. The monkey had already raised a "delightful" son and was expecting a second birth. The father seemed particularly affectionate toward his children. At first, the mother was tenderly devoted to the new baby. Then a keeper arrived in Benchley's office with the baby, battered, bruised, and dead. The mother had killed it, as she did another baby two years later. Even experienced keepers missed what the motherly Benchley noticed: the monkey mother was not producing any milk. She could not bear the whimpering of her starving babies and had killed them when she could not feed them. "These were," said Benchley, "mercy deaths in her mind."[41]

Benchley hired other women to help mother the animals, including her invaluable assistant Emily Burlingame and Dr. Joan Morton Kelly. Kelly had established herself as a psychologist while working at Penn State University before Benchley begged her to care for and study the zoo's three new infant gorillas. When she entered zoo work in 1953, Kelly became both a "mother emeritus" and a "teacher" (rather than a keeper, a title still assigned only to men). Like chimpanzees in the zoo show or human babies in the suburban home, her gorillas ate at the table with a cup and spoon. And they knew enough not to pick their teeth, reassured Kelly.[42] Women workers, highly qualified like Kelly, came to the zoo as mothers caring for the animals zoos increasingly wanted to raise. In the ties of care and motherhood, women workers, animal babies, and human children transformed the nature of American zoos, and even William Mann recognized the shift. Now William suggested to female applicants that they apply to the Bronx Zoo, which was constructing a children's zoo.

Shirley Visits the Zoo: San Diego, Washington

Shirley's dad worked for the phone company, but on weekends he loved snapping pictures at the San Diego Zoo. On some mornings Shirley got up very early and went hand in hand with her father. Sometimes she asked her dad to buy her a snack, and her "animal friends" helped her eat the peanuts. A regular visitor, her dad must have been well known to the keepers because occasionally they lifted Shirley over the fence to enjoy a rare intimacy with the animals. Once she rode on top of the hard shell of the giant tortoise, and another time a keeper boosted her onto the Asian elephant whose bristly hairs stuck through her clothes.

"Good bye, little friends," she called out to the animals as she and her dad made their way to the zoo gate. Shirley was, in all her politeness and cuteness, fictional. Benchley published *Shirley Visits the Zoo* in 1951 to reintroduce the zoo to its most desired visitors—children. Five years later, the San Diego Zoo joined most other large and mid-sized American zoos in constructing a children's zoo. Pittsburgh's Highland Park Zoo boasted a children's zoo as early as 1951. The Brookfield and Philadelphia Zoos opened their children's zoos in 1953.[43]

Brookfield's children's zoo was one of Chicago's most popular attractions, drawing 611,000 children in 1957.[44] In the children's zoo, kids no longer had to perch precariously on the bars and fences to see the animals. When animals remained in enclosures, they were built to children's scale and adults had to stoop. Other animals roamed free, at least within the confines of the crowded children's enclosure. The young crowds enjoyed displays plucked from nursery rhymes or bible stories. They patted domestic animals in farm scenes. More rare animals, typically babies, were also on display. San Diego's children's zoo boasted lion cubs for children to pet. Baby monkeys and chimpanzees joined children's zoo displays, and in Milwaukee a dad's club formed to buy exotic animals, including a hyena, for the proposed children's zoo. In the Bronx Zoo, a "play ring" provided children the "valuable experience of touching animals."[45]

Children—adults too—loved to touch the animals, but there could be a tragic alchemy when the human desire for physical intimacy met animals' natural traits. In 1958, Julie Ann Vogt, a two-and-a-half-year-old toddler, was enjoying a day out at the National Zoo with her grandfather. Perhaps she begged for peanuts to feed elephants or monkeys. Vogt was standing too close to the lions, and a big cat reached out its paw, swiped, and decapitated the girl. Her grandfather was distraught, and the public was

"Shirley Visits the Zoo" (1946). Anticipating the fun of the children's zoo, this children's book featured the adorable Shirley playing with her animal friends. Belle J. Benchley and G. E. Kirkpatrick, *Shirley Visits the Zoo* (Boston: Little, Brown and Co., 1946).

furious. As Reed defended the lion against a public demanding retribution, he knew he faced a "crisis at the zoo." "You are stupid," an angry
third grader wrote Reed. "You should put electric bars or kill the lion that
killed the girl. I am never going to your unsafe zoo."[46]

Reed and the zoo's keepers hated being called "murderers," but they
didn't want to lose a lion just for the crime of being wild. Something in
that third grader's letter touched another raw nerve; Reed knew the zoo
couldn't afford to lose young visitors. Underfunded for years, the free
National Zoo was in sad decline. As the Vogt family lawsuit unfolded in
court, the zoo reexamined one unsafe exhibit after another. The hippos
stayed inside just in case children might tumble into their pool. Other
animals had wire mesh strung around their barred cages.[47]

Some letter writers insisted the lions were still wild creatures, even if
they were captive in substandard cages with little resemblance to their
natural habitat. Killing was "inherent in the nature of a wild beast—even
one which has lived in captivity for many years!"[48] Others demanded
punishment. "Regardless the purchase price," wrote one irate visitor,
"they should be destroyed."[49] Visitors didn't really want dangerous wildness. Increasingly, they expected friendship with animals, especially for
their kids.

Vogt and her grandfather were not much different from other visitors.
Visitors tossed peanuts and other food to elephants, sea lions, hippos,
and lions. Zoos, of course, abetted (and profited from) the careless feeding
of animals. Despite the "don't feed the animals" signs, kids expected to
share their peanuts. In 1956, Bobby and Nancy were bouncing off the
walls when their father Jerry (just as fictional as Shirley and her father)
agreed to take them to the San Diego Zoo on his day off. Jerry, Bobby,
and Nancy were heroes of a zoo story about how kids and their harassed
parents should enjoy children's zoos. For weeks, the kids had begged
for a trip to the new children's zoo where kids "can pet the animals."
Sure enough, after they slid down the slides, they came to the "Direct
Contact Area" where "gentle" animals roamed "under the watchful eyes
of young, girl attendants." The kids listened to a few facts about guanacos
while feeding them peanuts, popcorn, and other treats. "Daddy, animals
are wonderful," the children exclaimed when, tired and happy, they
headed home.[50]

If visitors in the grown-up zoo disobeyed signs forbidding the feeding of animals, they also ignored warnings of the dangers posed by zoo animals. Reed complained that Vogt's grandfather, like so many other parents and grandparents, likely lifted his child over the railing near the lion cage. In the children's zoo, the railings were low enough and the animals hopefully gentle enough that the kids could touch, feed, and "otherwise make friends with them." "This intimacy makes it much more interesting to children" than the grown-up zoo with its heavy bars and deep moats—and dangerous animals.[51]

Zoos encouraged intimacy, the sensual closeness of pats and caresses, regardless of animals' desires. "Your youngster," the San Diego Zoo told parents in 1956, two years before Vogt's tragic death, "can learn much through intimate contact with animal life."[52] At the Brookfield Zoo, the children's zoo was "like stepping into a fairyland" with its small buildings, tiny moats, and animals mingling with children. "You can pet a lion," the zoo advertised in 1957—the year before Vogt lost her life. The animals herded around the refreshment stand where children could buy candy, peanuts, and popcorn to share with the animals. Nearby, children gathered to pet Diane and Daphne, lion cubs who "squall like angry human babies." Even Rosie the rhinoceros was available for petting. That intimacy, exciting to the children and required of the animals, helped testify to the universality of infancy. Julie Ann, Shirley, Daphne, or Diane, "[t]hey are all babies."[53]

The Best Man in the Building: Chicago

Inevitably, when women like Virginia Havemeyer found paid work caring for animals, it was in the children's zoo. "Like the old woman who lived in the shoe, Virginia Havemeyer has so many children she doesn't know what to do," the Brookfield Zoo boasted. Havemeyer, however, knew exactly what she had to do: raise some of the zoo's most fragile animals in the midst of the zoo's most popular attraction. A mother herself, the zoo reassured, Havemeyer was also "foster mother to 300 baby animals." Baby animals, Havemeyer said, needed the same loving care and strict discipline as humans. When the zoo published a picture of Havemeyer pushing twin orangutans in a baby carriage, it linked caring zoo work to mothering.[54]

At the Brookfield Zoo, sixteen women, including volunteers, heated formula, sponge-bathed the animals, and sterilized bottles for newborn big cats. "Women seem to respond to these little things. . . . They have fun doing it," said the zoo. The zoo's male keepers still wore military-style uniforms, but at the children's zoo, women workers dressed in blouses and skirts.[55] Terry Blueman came to children's zoo part-time in 1965 and found full-time work two years later. Eventually, she moved from the children's zoo to the aquatic bird house and, later, the reptile house. Gail Schneider followed a similar path toward full-time work. She started working during summers in the children's zoo before succeeding Havemeyer as its superintendent. Women who worked at the Brookfield Zoo were remarkably qualified. In fact, they often boasted more formal education than their male counterparts who had followed an older system of apprenticeship and advancement from the ranks of helpers or groundskeepers to animal keepers. Ann Owens relied on a degree in zoology to work in the Small Antelope House and the Wolf Woods. Mary Jo Fleming had a degree in biology, and Gerry Radaszewski eventually found a job in the reptile house with her zoology degree and a specialization in wildlife management. She recalled, however, that she felt "discouraged in college because that field 'was not open to women.'"[56]

Even at the Brookfield Zoo whose hiring practice as late as the 1970s was still the exception in the world of the "zoo man," female keepers couldn't escape the mothering role. They could leave behind the children's zoo for the mammal or reptile houses, but motherhood was harder to evade. "The women keepers at Brookfield are certainly liberated," wrote Gay Kester, one of the zoo's female keepers in 1973, "in the sense that they have no job restrictions." She was particularly proud that they had disproved all those reasons William had cited in refusing jobs to female keepers. They were not in the slightest "squeamish." They pushed the same wheelbarrows of manure and tended colonies of rats (to be fed to reptiles and birds of prey). Even so, zoo publicity lauded the domesticity women imported into the previously male world of zookeeping. They did a more thorough job of cleaning, noted one male keeper, adding the "personal touch" to feeding and cleaning while "looking quite feminine." Still, male keepers struggled to define some work as men's alone, particularly care for larger ungulates, carnivores, and primates. There "is a place in the zoo for women," agreed one male keeper, just not in the pachyderm or primate houses. Any woman who could work with elephants, apes, or the big cats "would have to be very husky, and if she is that big, she belongs on a wrestling team, not in a zoo."[57]

Across town at the Lincoln Park Zoo, Kathy Silhan had a degree in anthropology but found few paid opportunities. In fact, she volunteered for two years because the zoo refused to hire women keepers. Finally, in 1973, the zoo took its "step out of the past by allowing girls to become Keepers." Before then, they could only work in the children's zoo. Soon keepers like Pat Sass, Caryn Schrenzel, and Andrea Poveda moved from the children's zoo to other houses. Sometimes men called them "the best man in the building."[58]

"[T]he best man"—this little pun about marriage suggests how women's care work, first in private homes and then in zoo nurseries and children's zoos, had changed the zoo. The zoo was reborn as a family raising animal babies to the delight of human children. Yet at the exact moment women found zookeeping work, typically under the banner of motherhood, the focus of care shifted from nurturing animals to breeding them. As zoos increasingly seized the mantle of protecting the world's vanishing species, a major question loomed: where did the conservation of animal life begin? Did it begin at birth when a baby needed mothering? Or did it begin long before conception? Just as women became paid keepers in the grown-up zoo, zoo officials shifted their focus from the romance of child rearing that encouraged fantasies of crossspecies love to fertilization and planned breeding. From motherhood, zoos prioritized the biological goal of animal husbandry (the breeding of animals) over mothering (the nurture of zoo babies). Already in 1967, Theodore Reed claimed: "[E]xcellence is measured in terms of animal husbandry."[59]

Noah's Ark: Washington, French Equatorial Africa, Colorado Springs

Early in the morning on a fall day in 1961, Melvin Kilby, the keeper at the National Zoo's gorilla house, was growing alarmed. The gorilla Moka refused to cradle Tomoko, her new baby son. When she "appeared to lose interest," Theodore Reed faced a difficult decision. "[T]he baby should be taken from its parents," he concluded. On his orders, keepers lured Moka and her mate Nikumba outside with treats while others removed, cleaned, and handed Tomoko to Louise Gallagher. Tomoko had taken another step in a journey away from the African jungle, one his parents had begun in their infancy.[60]

Moka and Nikumba came to the National Zoo from French Equatorial Africa. A French mining expedition captured the infant Moka in 1954, and natives trapped Nikumba when he was just ten months old. Later, a wealthy zoo patron purchased the young pair during an expedition to colonial French Africa. Tomoko took yet another step away from gorillas' jungle habitat when he left the zoo for a crib near Louise's bed. For the first seven months of his life, Tomoko lived like a human baby. He wore diapers and baby clothes. By the time he was ten weeks old, he cooperated with his dressing and changing, obligingly lifting his legs and holding out his arms. He played with rattles and enjoyed a game of peek-a-boo. "On the ninth day, to use a human term, he 'smiled' at his foster mother," and by his fourth month, the bond had grown even stronger. If Louise put on her coat to go outside, her gorilla baby cried. Soon he was eating cookies, crackers, and candy and responding to his name. He had "what might be called in human terms, his first temper tantrum," the zoo reported.[61]

When zoos promoted breeding programs for the world's vanishing species, directors and keepers implicitly admitted that for many animals their habitat was truly the zoo, if not the human home. As the zoo described Tomoko's challenging childhood, the obvious remained unsaid; Tomoko was never going "home" to the jungle that had once been native to his ancestors. Even his transition back to the zoo was difficult. Even with Louise in sight but on the other side of the bars, Tomoko buried his face in a favorite blanket. He had been raised as human, then forcibly removed from the only habitat he had ever really known. The jungle was a long way off for Tomoko, and he fought his keepers when they tried to remove the clothes and diapers he had worn since he was a few hours old. He cried for his surrogate mother. He wanted out of his cage, but the tears and screams that a few days before got him lifted from his crib were ignored.

Tomoko's birth was a triumph for the National Zoo; he was only the second gorilla born in captivity in the United States. In the aftermath of Julie Ann Vogt's death, Tomoko, crouched under his blanket, reaffirmed the prominence of the National Zoo in international campaigns to save the world's vanishing species. Gorillas were expensive, popular, and, soon after World War II, virtually impossible to capture in the wild. As zoos turned in earnest to captive breeding programs, gorillas, and the other large apes, were obvious priorities. It is easy to imagine Reed and Kilby's nervousness when Tomoko was born and their excitement when he survived. Within the careers of keepers like Bernard Gallagher and their wives like Louise Gallagher, gorillas, once hard even to keep alive, could now be bred and raised from infancy. When the American Association of Zoological Parks and Aquariums (AAZPA) offered a prize in 1956 for the most unusual birth, Colo, a gorilla born at the Columbus Zoo, was an obvious choice. Colo immediately went home to the zoo superintendent's wife, who dressed the endangered animal in a bonnet and dainty dress for the newsreel cameras.[62]

Fairfield Osborn was hopeful zoos could become a permanent "reservoir" of vanishing species. The point of the zoo, Osborn said, was to breed more animals; at least, they would be alive in captivity. At first, directors like Don Davis of the Cheyenne Mountain Zoo in Colorado Springs was confident that "every pair of animals in the zoological garden has a reproductive potential." Under his direction, the zoo worked energetically to achieve a "much higher birth rate," but by 1960 only ten of the zoo's thirty-one species of primates were reproducing. Davis was disappointed. He blamed inadequate or fattening food and, as in the case of chimpanzees, masturbation. The zoo's orangutans, he argued, simply did not have the knowledge of "technique" they might have learned in more natural family groups.

The problem of the birthrate had superseded the challenge of raising newborns. Husbandry had trumped mothering. Davis was willing to sacrifice the illusion of nuclear families if it would help his breeding programs. He separated males from females to encourage sexual interest and sometimes even removed breeding pairs from public view. Davis looked to a future, in fact, that might not even depend on natural attraction between mates. Hormones, even artificial insemination, he wondered, could one day increase the birthrate.[63]

We can now raise "breeding stocks of rare and endangered species," boasted Osborn prematurely. Breeding meant a permanent shift in the balance of power between dealers and zoos. By the late 1950s, Osborn needed to be so optimistic because he realized the animal trade could no longer supply the nation's zoos with "rare and exotic animals of unusual interest," like rhinos, gorillas, even komodo dragons, that drew visitors but were approaching extinction in the wild.[64] Animal dealers, once almost equal partners in the AAZPA, were now regarded with suspicion of preying upon the last of "vanishing species."

Domestic politicians were also growing wary of the animal trade at a time when visitors seemed more interested in domestic babies than in wild cargo. In 1958, Congress debated a bill that would have prohibited the import and sale of exotic ruminants. Zoo officials were alarmed: "Every Zoo in America is faced with an *EMERGENCY*." The act would mean the end of elephants and giraffes in our enclosures, the AAZPA testified to Congress. The AAZPA was mounting a rearguard defense of the trade, and zoo directors knew it.[65] Abroad, a rising tide of conservationist concern also focused on the global trade in fauna and flora. Just five years after Congress considered its restrictive bill, the International Union for the Conservation of Nature (IUCN) launched deliberations that eventually produced the Convention on the International Trade in Endangered Species of Wild Fauna and Flora (CITES).

Formed in 1948, with only tepid support from American zoos, the IUCN sought the conservation of species and decried the encroachment on habitats, especially in newly independent tropical countries.[66] The IUCN drew from the legacy of colonial-era efforts to conserve tropical nature. Indeed, leaders of the IUCN had come especially from the Society for the Preservation of the Wild Fauna of Empire, which had succeeded in the creation of game reserves and the restriction on native hunting in British East Africa. The two organizations not surprisingly shared similar goals, including the regulation of big game hunting and the creation of national parks, and they also shared a suspicion of native hunters. The IUCN faced entwined challenges. There really were not enough big game animals left to sustain European and American recreational big game hunting. Moreover, the game parks the society had helped found were now located within the borders of independent nations. The IUCN sought to argue that the problems facing wild game, especially human intrusion and poaching, were the world's challenges. The IUCN, bolstered by growing conservationist concerns, turned to focus on the animal trade, no longer a legitimate business facilitated by colonial game laws and trapping permits.

Though CITES would not come into effect until 1975, zoo directors realized they faced whole new levels of oversight over the import of wild animals. The IUCN codified the loose idea of vanishing species into the Red Data List of "endangered species." The Red Data List, first published in 1964, was bound and distributed in a three-ring binder. Optimists noted that species could have their sheet removed from the binder; realists knew other species would more likely be added.[67] In 1962 the AAZPA adopted its first resolution condemning the commercial trade in endangered species and urged greater cooperation with the IUCN. In fact, when the association strengthened the resolution in 1967, zoos and dealers had to agree to its terms if they wanted to remain as AAZPA members. Certain animals, like orangutans and Sumatran rhinos, were permanently off the list of tradable wild animals.

The resolution had an element of self-defense. Millions of exotic animals were imported into the United States in 1967 (with live fish representing the overwhelming majority). Of the 74,304 mammals, almost all were primates. The number of endangered animals imported to zoos, the association insisted, was minimal compared with the number of these primates imported as pets or laboratory animals.[68] In the wild, many other animals were killed for meat, hides, and fur, the organization complained. Zoos weren't the largest consumer of animals. In the case of certain animals (rare birds, for example, harvested for their feathers), zoos' protestations made sense, but nevertheless zoos knew they needed to husband their own stock. If zoos were indeed the "Noah's Ark of Future Generations," then animals needed to breed on board. And everyone knew that the boat could never again land on the shores of native habitats.[69]

Studbooks and Cocktail Tigers: Washington, Seattle, Toledo

Two years before CITES came into effect, Theodore Reed was worried again but this time about the "whole bloody zoo profession." He told his

old correspondent Marvin Jones: "I think we are going to be a thing of
the past if we do not get off of our collective rear ends." Zoos, Jones had
long argued, needed not just animal babies but careful plans for animal
husbandry.[70] Jones wasn't optimistic. Privately, Jones condemned Reed
and so many of his zoo colleagues. "[O]ff the record," he said, "neither
Ted Reed or most US zoo men are real died [*sic*] in the wool conservationists, unless it works for their benefit."[71] Certainly, animal breeding at
zoos could blunt bad press or offset the effects of a dying animal trade, but
could it really save whole species? Without the kinds of planned breeding
Jones wanted, populations would still dwindle. It was one thing to
encourage two wild-caught animals to mate, but what about their progeny?
Would they understand the very technique of breeding, let alone be able
to raise their own children? Just as bad, what kind of animals would zoos
breed or, worse, inbreed? As small populations inbred, could subsequent
generations remain healthy?[72]

Jones was the nation's most dedicated zoo visitor. He had learned to
love zoos as a child and began writing to zoo directors asking for advice
and offering suggestions. By 1944, just sixteen years old, he was volun-
teering at the Bronx Zoo and had begun a lifelong quest to trace the
ancestry of zoo animals. He continued building animal family trees even
after he was drafted into the army in 1951 and during his time serving in
Vietnam. Jones was obsessive, and zoo directors were occasionally
annoyed. Sometimes they thought of him as a "snoop," but he understood what real records would reveal.[73] When he began tracing family
trees, he unearthed the dirty roots of an animal business traders wanted to
leave buried.

In the animal trade, paperwork and permits could obscure animals'
pasts and genetics. Sometimes animals advertised as wild caught were
actually zoo born, and surplus zoo animals could end up in pet shops
specializing in exotic animals. Too often zoos accepted an illegal specimen or passed off a hybridized surplus animal as purebred. Exchanges of
animals even between zoos often concealed an animal's mixed heritage.
Were zoos, Jones wondered, breeding a strange hybrid of Siberian,
Sumatran, and Bengal tigers just because the subspecies would mate and
just because zoos wanted to display tigers? It was one thing to raise baby

animals. It was quite another campaign to breed them scientifically. Was the "cocktail" tiger the freak of conservation?[74]

Soon the U.S. Department of the Interior began collaborating with the AAZPA on regulating the import of endangered species. In 1972 the Woodland Park Zoo in Seattle, for example, submitted an application to import a breeding pair of lesser mouse lemurs to the department's Fish and Wildlife Service as well as to the AAZPA's Wildlife Conservation Committee. The committee balanced the potential for successful breeding and the impact on wild populations with zoo's natural desire to exhibit rare species. The zoo's desired lemur import slowly worked its way through a new system of conservation regulation. Maybe the "planned husbandry" promised by the zoo could actually help the species survive. However, the committee wasn't entirely convinced of the dealer's claims that the lemurs were, in fact, captive born. The zoo had produced a great deal of paperwork, but there were still missing documents.[75]

Zoo directors were often reluctant to ask questions about desired animals' origins, even their legality. "All this is going on now," complained Jones in 1973, even as zoos were "'talking' conservation."[76] Of the 291 mammal species listed as endangered in the Red Data List by 1972, 162 lived in zoos, but only 73 had reproduced, and only a small handful of those had secure reproduction rates. Zoos were still "consumers rather than producers of wildlife," admitted a National Zoo curator.[77]

Like a growing number of zoo officials, Jones was interested in family trees, not animal families, and breeding, not caring. Jones organized his records not into family albums but into "studbooks," matchmaking guides for husbandry based on the family trees of endangered species captive in zoos. By the late 1960s, zoos across North America and Western Europe began compiling studbooks, albeit incomplete, that traced the lineage of specific endangered species captive in zoos in order to encourage husbandry that maintained the integrity of the species and the size of the gene pool. This was zoos' "greatest contribution to zoological science," Jones argued. Now zoos could understand how to maintain a purebred animal "race."[78] Specific zoos and individual keepers or officials (as late as 1972, these were all men) accepted responsibility for maintaining the studbook for a specific species from tigers to gorillas to anoas. Jones maintained four of the thirty-four official studbooks, evidence of his own tenacity.[79]

Yet how did zoos choose to create studbooks of certain species but not others? Partly, the decision favored species like the Przewalski's horse that bred relatively easily in zoos.[80] Others, like gorillas, were charismatic species. Visitors liked to see gorillas and their humanlike babies. Zoo directors assumed an awesome new power as arbiters of birth that coexisted awkwardly with their need to keep the zoo turnstiles turning. Breeding programs, once left to chance and keeper's wives, were now managed at the very top of the zoo administration, those select few officials who organized husbandry across zoos.[81]

A single birth was good for the headlines, but a species could only be "secure" if birthrates remained strong into the second and third generations, according to Jones. And here was the problem. When zoos conducted a census of endangered species in their collections, they realized that individual zoos simply didn't have the populations to maintain sustainable breeding with the exception of less popular herding species. Zoos faced tough and potentially unpopular choices. Jones had found that American zoos by the 1970s displayed 885 different species, but zoos just weren't big enough. However they did the math, they simply couldn't breed the animals they needed. Zoos needed smaller collections with bigger groups of animals. If they wanted to breed their own collections, Jones estimated the zoos couldn't maintain more than 320 species.[82]

Suddenly, the demands of visitors to see certain charismatic animals, like elephants or lions, conflicted with conservationist imperatives. Zoos didn't like to think of themselves as consumers, but what would happen if they also became producers? Who would buy the surplus animals? Some popular animals like zebras, not comparatively rare in the wild, could always be sold, but others like Père David deer that lived almost exclusively in zoos would languish on surplus lists. The old urban zoos simply didn't have the empty cages to breed the world's endangered animals in order to preserve the species, let alone fill the exhibits. "[T]he best hope for significant gains in captive breeding," concluded John Perry, the National Zoo's assistant director for conservation, lay outside of cities in safari parks.[83]

Visitors had permanently entered into a kind of intimate ecosystem in which the cuteness of animal babies and overall touristic appeals became factors of survival. Which species should zoos display? Questions of appeal to visitors became at least as important as the depth of the threats they faced in the wild. Zoos by necessity maintained and bred populations of species like elephants or bears that did not face exceptional human pressures but were expected by zoo visitors. "A few popular species may sustain themselves, simply because of demand," admitted Perry. Unless zoos made difficult decisions, "[a]ll these resources combined . . . will not, however, add up to species survival."[84]

Zoos developed a division of labor between zoo conservation led by administrators, charged with the exchanges of animals that managed planned breeding, and zookeeping, focused on caring for animals during the breeding process. As one keeper described the new reality of zoo work in this new era, "[t]he preservation of any endangered species in any zoological garden is initiated by the zoo director," but the zookeeper was "the one person who gives them all his attention, his experience, his love and knowledge." Increasingly, "his attention" was actually "her attention."[85]

As husbandry moved the challenges of species survival to the administration building, keepers seized the mantle of care work to suggest less that they were mothering animal babies than that they were caring for endangered species, "those helpless living things." Through the American Association of Zoo Keepers (AAZK) keepers voiced demands for the recognition of their work as "professional animal care."[86] Conservation, they insisted, demanded women's and men's care work when it was up to keepers to feed, shovel manure, learn when the rare animal was ill, and educate the visitor who might have wanted little more than to toss a peanut. Mothering, they admitted, had little to do with conservation strategies so dependent on husbandry. The hard work of maintaining the studbook often fell to keepers as well. "I have no greater consumer of my time," noted the keeper of the clouded leopard studbook. But at least he was helping "the survival of an endangered species."[87]

So, by the 1970s, Glenous M. Liebherr had a task female zookeepers a generation before could never have imagined. She came to the Toledo

Zoo with a degree in animal science, the kind of training now expected of zookeepers. Yet she was still a novelty: a "woman zoo keeper." Animals "enjoy attention," she told local newspapers, as she defended zoos against critics. They were happier in the zoo than in the wild where "there's not enough open land left." Even in the wild, some animals "just aren't good mothers and desert their young." At the Toledo Zoo, Liebherr saw herself as a conservation worker at the front lines of animal husbandry. She announced a zoo achievement to her fellow keepers across the country: the Toledo Zoo had artificially inseminated a chimpanzee. Coco was the most popular animal in Toledo, so the zoo wanted to save his "blood line," but he had never tried to mate. However, Coco was, Liebherr noted, "an expert at the art of masturbation," and Liebherr helped design the artificial vagina that collected his sperm. The zoo had saved the genetics of its popular primate, and the keeper had pioneered a new method to "preserve endangered species." The chimpanzee mother gave birth to a healthy baby girl but just didn't seem to know how to care for her baby. The keepers removed the baby for "hand-rearing," but nobody bothered anymore to record who did *that* work.[88]

A.D.O.P.T.S.: Toledo, Detroit

Around the time Coco masturbated into the artificial vagina, zoos had begun to admit that raising animal babies in children's zoos wasn't a particularly good idea. Keepers and directors alike knew the children's zoo was the "one place where people can satisfy their natural desire for close contact with animals." They could stroke the lions or play with the chimps and other "orphaned young." Once it seemed "logical" to turn children's zoos into nurseries. Parents knew as well that children's zoos were popular. "There are no bored children in the animal contact areas," admitted one assistant curator. "This is what visitors want, enjoy, and remember most."

But would human contact help endangered species breed? Visitors, young and old, loved seeing the "baby monkey" suckled by a nurse in a rocking chair. The monkey could never become a person, but its infancy in the children's zoo, tickled and hugged by humans, had removed it from the monkey species in all but genetics. It would struggle to readapt to life in a monkey colony. "It is then doomed to a life as a solitary cage occupant."[89]

Animal mothering had returned to the enclosure or to nurseries behind closed doors, but we still adopt animals. The Toledo Zoo, Coco's old home, "shares with you the joys of 'animal parenthood.'" For just $45, you can join the "zoo family" by adopting an animal. Virtually every zoo invites visitors to become zoo "parents." Zoos are still families, more than a half century since Oofy and others went home to keepers' wives, but adoption is now in the hands of education and fund-raising departments.

Fund-raising, publicity, zoo animal care, and global conservation still meet around animal adoption. Conservation, according to the Cleveland Zoological Society, "is about *connecting people with wildlife*." Today's zoo parents get more recognition than those unpaid wives. Parents get an invitation to a "Zoo Parents Picnic" or an adorable plush toy. At the St. Louis Zoo, a $100 adoption earns a Zoo Parents frame to hold the family snap of their animal child. At the Detroit Zoo adoption is the "opportunity" to save endangered species around the world and captures how officials came to defend their zoos as the last best hope for endangered species on a crowded planet. "A.D.O.P.T.S." means "Animals Depend on People to Survive."[90]

Animal adoption lives on along the safari as living history at the zoo. Even at the end of the era of the animal trade, alongside the construction of children's zoos and the creation of breeding programs, Americans still yearned to see wild animals. Delight in the cuteness of zoo babies persisted alongside the thrill of seeing charging wild animals. The twilight of the animal trade in the 1950s also marked the era of decolonization. Ordinary Americans and zoo officials alike worried that the end of empire ushered in an era of extinction. Zoo officials were worried. At a time when domestic critics of zoos spoke with louder voices, adoption was one way of reshaping the American zoo. Wild adventure was the other.

---------------------------- 9 ----------------------------

DANGEROUS SAFARI

Conservation at the End of Empire

I N 1956, on the East African veldt, a rhino confronted a truck, a man, and a camera. The rhino's charge smashed the jeep's hard metal door. Stunned for a moment, he dashed a short way into the bush and paused. His massive sides heaved. Perhaps he believed that he had driven off the harassing vehicle, but the jeep roared closer again. Even worse, the man armed with a movie camera inched ever nearer. With poor eyesight but sensitive smell and hearing, the rhino must have felt assaulted by the acrid exhaust, the grind of the jeep motor, and the crunch of its wheels as it crushed the savanna's hard crust. He also sensed the strange aromas of the cameraman sweating in the beating African sun. Might the clicking of the camera be the firing of a gun like those of legal white hunters and illegal black poachers?

The rhino was stressed, confused, and angry. He turned his bulky frame with remarkable agility to meet each new threat and, resorting to the resistance he knew best, he charged again. The cameraman just reached the relative safety of the jeep ahead of the rhino's long horn. Again, the car roared its approach, and again the rhino charged. This time he hooked his horn into the damaged door, and as if he were tossing a mere branch, he flipped the jeep onto its side. As he vented his anger and frustration on his tormentor, the cameras rolled. Months later when the film was cut and edited for an American audience, the charging rhino was the essence of wildness and the appeal of a changing continent.

For Lewis Cotlow, the man with the camera, a damaged jeep was a small price to pay for the signature shot of his 1956 film *Zanzabuku*. "Perhaps it was foolish" to provoke a rhino, Cotlow admitted. "But a moving picture of a rhino is nothing unless he's charging," and this charging rhino graced the movie's posters.[1] Chrysler, the film's sponsors, must also have been delighted with the proof of the resiliency of a jeep against the ravages of an angry, charging rhino. Americans, Cotlow realized, still craved the violence of the African bush, the reactions of animals under stress. He had discovered the secret of the successful animal film: grazing zebras or wildebeests were less exciting than a vast herd in panicked stampede. Any good wild adventure film needed shots of rhinos, recalled another filmmaker while on safari, "preferably charging." The first rhino he encountered was "a sissy." He charged admirably—but then stopped and swerved twenty feet from the truck. That footage ended up on the cutting room floor. Finally, he found a "lady rhino" who charged to protect her calves.[2]

While Cotlow's rhino charged, the world was changing. Even the rhino might have recognized the changes when empire gave way to independence. More hunters, tourists, and filmmakers than ever before crowded the veldt, eager for a last colonial safari experience before independence. There were also fewer rhinos than ever before. *Zanzabuku* at once excited and comforted American audiences still insistent that Africa should be a primitive wildlife paradise. The "maddened rhino," said one reviewer, captured the "mystery and splendor of Africa."[3]

Animals were Cotlow's unwilling and frightened actors, goaded into charges or terrorized into flight. His camera lay still when animals slept, nuzzled, or grazed. Africans, too, appeared on-screen only to play the parts he assigned them. If animals were meant to be wild and violent, African humans appeared set in tribal ways. First the film featured a spear-wielding leopard hunt, then—inevitably—a stomping, drumming dance. When Africans departed from Cotlow's script, for example, when they became revolutionary soldiers, he turned the camera off and picked up a gun. White humans completed Cotlow's cast of characters. When Cotlow did the narration, they profited from the land in brave yet respectful accord with its animals and primitive peoples.

Cotlow hurried to Africa, an eager tourist worried that the violence of the independence struggle would hurt the animals and end the picturesque African way of life. In his cinematic eye, Africa appeared a land of "savage splendor," manipulated into a kind of a game park in which animals, savages, and white ranchers all represented stops on a tourist trail. Come to the game park, Cotlow said, before its animals and primitive humans disappear. Like so many others, from filmmakers to animal traders to zoo directors, Cotlow believed the extinction of animals was inevitably linked to the continent's independence.

Times were changing in the United States as well. Postwar prosperity that sent visitors flocking to zoos had stagnated into uncertainty by the 1960s and '70s. Cities were hit especially hard, devastated economically and divided racially. Often, as in Philadelphia, zoos were located in the center of cities in the heart of increasingly poor and segregated areas. For zoo officials and critics alike, the city itself seemed to damage the animals. The same racist language of slums and ghettos that condemned urban neighborhoods also came to describe shabby zoos. Phase out the zoo— critics urged zoos to clear the animal slums.[4]

The rhino, traders, natives, zoo directors, and visitors were all caught in a moment of transition and the racialized conflict that engulfed American urban centers and African savannas. None could grasp the full tumult of change that linked the end of empire beginning after World War II, the collapse of animal populations, the impoverishment of American cities, and the slow decline of American zoos. Stocks of those large animals zoos wanted (from orangutans to rhinos) had plummeted. Animals that still trickled into the United States were too often illegal, their export licenses forged, altered, or otherwise nonexistent. Zoos were, by no means, the only consumer of wild stock, but they spoke loudest and with scientific authority in response to new governments that slammed the door on animal exports. Always fraught, never equal, the long-distance relationship between zoos and native peoples faraway had irrevocably changed.[5] Zoos regarded natives less as invaluable suppliers of animals, wise to the ways of the jungle, and more as ill-suited guardians of rare natural treasures.

So the last expeditions set out at the twilight of empire, worried about the future and nostalgic for a fading past. Cotlow went in search of a rhino, and the National Zoo hunted for bongos. Other zoos and traders sought as many animals as they could collect before the end of empire and the animal trade it fostered. Zoos needed those animals desperately to revive zoos, themselves facing uncertain futures. Exhibits and enclosures, modern in the Great Depression, now seemed depressing. Even worse, the public was growing uneasy, wondering if animals suffered in their spare surroundings. The rhino paced or stared mournfully out of his concrete enclosure. He would never charge again.

What had happened to wildness behind bars? Zoos directors had their own visions of brighter futures—outside of the cities in safari parks set in bucolic suburbs. The bygone era of the colonial safari could live again when car doors replaced enclosure walls. The zoo in a new incarnation as a wild, accessible safari park, absent the guns or rebellious natives, represented one seductive answer to the crisis of the urban zoo. It salvaged savage Africa (and the tropical world, in general) from the agony of liberation and rebuilt it, without its cities or its modernity, in American sprawl. In the heart of the California rose the biggest safari park of them all, a wild and primitive tropics, all for the perceived protection of endangered species, far from the city and from the reality of African independence. At the end of the age of empire and in the midst of the American urban crisis, the safari park—the model for a new kind of zoo—resurrected savage splendor as the natural state of both animals and people.

The Point of Meat: Congo, Kenya

Lewis Cotlow was a missing link, connecting the postwar tourist back to the early twentieth-century elite hunter. He was an "everyman" with a passion for the primitive. Unlike his explorer ancestors, Cotlow lived at the fringe of New York's moneyed aristocracy. In fact, he sold them insurance. His interest in travel was sparked by a job as an inspector with the U.S. Shipping Board.[6] He saved enough in the following years to afford lengthy travels to the tropics. An insurance salesman with wanderlust, he imagined himself as the last of a dying breed of explorers. In the years spanning World War II, Cotlow searched the world for primitive humans and wildlife, both of which he believed were rapidly disappearing. In 1937, he retraced the footsteps of his friends Martin and Osa

Johnson to the Congo is search of gorillas and pygmies. After the war, he returned to Africa with Armand Denis (Frank Buck's former cameraman) to make his first wildlife film. In 1977, the aging Cotlow won the Explorers Medal from the New York Explorers Club, its highest honor. In celebration of a bygone era, the club feted Cotlow with roast lion and marinated boar.[7]

Cotlow, as he tucked into the tastes of the veldt in a Waldorf-Astoria ballroom, remembered dinners past. Back in 1938, he had hosted a banquet of the Circumnavigators Club with "dishes invented in jungles" and prepared by New York's finest chefs rather than by African porters in gorilla country. They baked lions in clay, and waiters paraded frozen wild boars across the ballroom. Now, many years later, black-tied explorers traded stories of "explorations in the tropics" in front of a chaffing dish of "red deer."[8]

Cotlow played the role of explorer, but in his workaday background he was more a tourist. Abandoning the elephant gun for the camera and the elite hunting safari for a tourist adventure, Cotlow pinpointed the places on the African map that could be visited by any contemporary tourist with some means: the Ituri rainforest for gorillas and pygmies, the savanna for the lion, and the animal trapper's ranch. Cotlow's tourist trail stretched 15,000 miles around Africa, oblivious to its emerging national borders. His tourist itinerary visited key animals, famous primitive tribes, and disappearing oases of colonial comfort. Like carrion on the veldt, the smell of extinction lingered over Africa at the moment of its independence. As Cotlow and a game warden tracked his rhino, they stumbled across poachers' camps. Meat still hung over the smoldering fires. "African hunters," the warden explained, "came into the valley after the animals. They went after every edible creature they could catch." Cotlow was convinced colonial game regulations were fast becoming mere paper laws.[9]

Zoo officials and animal traders like Roland Lindeman, director of the Catskill Game Farm, also feared African independence. In midwinter 1950 Lindeman set out for Kenya on an once-in-a-lifetime safari tour with his wife. He knew this was one of last expeditions to "obtain a substantial number of rare and rapidly vanishing animals." "[E]xtermination of certain species," he advised zoo directors back at home, "is more likely and, in certain territories, taken for granted." He arrived on the tense eve of the

"Zanzabuku" (1956). This poster for Lewis Cotlow's feature film *Zanzabuku* high-lighted its signature scene of the charging rhino.

Mau Mau rebellion and realized Kenyan independence was inevitable, "certainly not more than a year or two." He was worried. "Game in the wild state could soon be decimated, if not exterminated," he said, because the "average African considers game first from the point of meat." Already, natives were "poaching" valuable rhinos with "poisoned arrows." With independence, Africans would be able to buy the guns to kill for meat and profit—at least this is what he "gathered from many natives."[10]

Cotlow and Lindeman had both come to Africa to see animals and native human tribes on the verge of extinction. When prewar travelers in the bosom of empire had described natives as primitive, they cast them as backward, lesser races necessarily bound for obliteration in the march of civilization. World War II and its slaughter, however, transformed the meanings of the primitive by turning colonial racism inside out. Now primitives had become tourist attractions, dying, colorful races unfortunately bound for extinction. "We envoys from civilization look at all the different tribes and decide that one is better than another," Cotlow wrote, but he had little patience for the biological racial hierarchies of the imperial age, even if his new appreciation for native peoples featured its own racism. He liked some tribes more than others. Some had the "most magnificent physiques." Others had those "firm breasts, and lithe bodies" that made for excellent film shots. The "sad truth," however, "is that we are witnessing the end of primitive man."[11]

The wild animals that lived alongside tribal peoples seemed on the same road to extinction. World War II and its crimes propagated under justifications of racial superiority had turned familiar colonial tropes into odious racism. Still, independence and the growing consciousness of the pressures upon animal stocks gave new license and new vocabularies to criticize modern natives, especially Africans. Zoo officials and adventurers like Cotlow no longer viewed Africans as loyal assistants or primitives imbued with an innate understanding of animal nature but as independent Africans, at best, indifferent to wildlife or, at worst, poachers.

The situation wasn't much better in Asia, mourned a San Diego Zoo associate curator. "[T]hose native people involved with game management in Southeastern Asia are of little assistance." The game wardens were uneducated, but the animals were still worth something, if only as hides, horn, and meat. "The populace requires capital simply for the maintenance of life," he reported. It was "almost amusing to speak of conserving the rhinoceros in Malaysia." As he gazed upon crowded Indian cities or Malaysian revolutionaries, he was convinced that outside of the zoo "wildlife conservation" was an excellent idea but "a nonfunctional dream in certain parts of the world."[12]

Like Cotlow's cornered animal, rhinos, both African and Asian, were caught in the contradiction between colonial independence and tourist development. In the prime African game lands, American tourists like Cotlow, white ranchers and animal traders, and black Africans struggled over who would manage the continent's animal stocks. The demands of the animal trade, sport hunting, subsistence hunting, farming, and ranching each envisioned different fates for the dwindling herds.[13] Laws central to the colonial state especially in eastern Africa had long favored the white hunter and, in the interest of regulation, concentrated the animal trade in the hands of a few white ranchers like the brothers Carr and Lionel Hartley.

Lionel had first introduced Cotlow to the thrills of rhino encounters during Cotlow's 1946 safari. Lionel understood exactly what the American tourist wanted: the perfect picture of a charging rhino. Lionel carried the big .475 caliber gun; Cotlow shouldered the camera, but he never did get the shot. Instead, the cornered rhino kept on grazing. Finally, the "belly full" rhino turned and trundled lazily into the bush. A decade later, Cotlow still craved that signature shot for *Zanzabuku*. And Carr Hartley was going to help him.[14]

Carr was a "big, sun-bronzed" man. He was a last relic of the British Empire: white, toughened by his struggles against nature and natives, and obstinate in the face of rebellion. During his 1956 visit promoting his animals to American zoos, Washington, DC, high society treated him to a whirlwind of parties. Boasting to reporters, he held up his left hand, scarred from Africa's wild beasts and missing a finger chewed off by a cheetah he called "darling." He showed the ugly scar on his head where he had been bulldozed by a rhino. "I've never run from an animal in my life," he boasted, but human rebels were a new and different threat.[15]

Carr needed to go home, as much as he liked American adoration and craved zoo business. His 22,000-acre Kenyan ranch where his wife, four kids, and 500 prize game animals lived had become a war zone. Between approximately 1952 and 1960, the most fertile areas of the Kenya colony (home to both white settlers' sprawling ranches and to Kikuyus, one of the colony's largest ethnic communities) were in open rebellion. The Mau Mau insurgency flourished among the almost landless Kikuyus, and the British response was merciless. By 1953 the British constructed camps to detain Mau Mau sympathizers or, just as commonly, ordinary Kenyans. White settlers were drafted into the counterinsurgency campaign, and animal trappers like Carr and John Seago were particularly valuable. After all, they knew how to track and shoot. Seago had long tracked animals for American zoos across the Kenyan veldt, but now it was too dangerous for white trappers to venture far from European ranches. "Also a great deal of time will be taken up by Patrols," Seago admitted in 1953.[16]

Carr's ranch household had become a frontline in the rebellion. One day, a trusted family servant was revealed to be a Mau Mau sympathizer. Carr's sister-in-law (widowed after Lionel's death in a plane crash) fled to Carr's ranch after rebels ambushed and killed her mother and step-father in their ranch house. Still, even in the midst of a rebellion, Cotlow wanted to film on Carr's ranch. He had saved his own hard-earned money from selling insurance and cajoled Chrysler into providing a truck to transport him on a grand tour of Africa's nature and primitive humans. Wildlife pictures were more popular than ever before, and Cotlow had thought up a theme for his new film: a young white boy returning home from school to his father's animal ranch. The boy would help tame the animals and dance with friendly—and subservient—natives. Carr's son Mike was the ideal choice. He even had the requisite blond hair.[17] The only problem was that in 1956 neither Carr nor the British colonial government wanted Cotlow anywhere near the game ranch on Kenya's high plateau.

"It is just too dangerous a situation," protested the local police chief. "Mau Mau bands infest these hills and swamps," but the "slim thread" of Cotlow's film depended on Hartley's farm, his animals, and his son's "winning smile."[18] Cotlow knew he would have to spend more than a month "right in the heart of the Mau Mau country." When Cotlow had visited Carr's ranch about a decade earlier, it had been an oasis of colonial comforts, but by 1956 it had become an armed camp. The animals still roamed in spacious enclosures, but barbed wire tangled around the ranch's cottages. For once, Cotlow felt like a captive. Cotlow once gave a series of lectures to American audiences titled "Through Africa Unarmed." Now he carried a revolver "as protection against—not wild animals, but men—the Mau Mau terrorists," and he felt besieged.[19] During Cotlow's stay, the ranch's inhabitants spent a nervous, wakeful night after receiving a tip about an impending attack. Then police rounded up twenty of the native ranch hands; some confessed to being Mau Mau followers. The head animal trainer was drafted into the militia—just when Cotlow needed him to handle a cheetah for a key shot.

On-screen, Carr's ranch, a colonial fortress, transformed back into a happy outpost of the animal trade. Native Africans, no longer potential rebels, leapt on command from rickety trucks to corral giraffes and even rhinos. They also played roles as colorful primitives who surrounded young Mike in a frenetic dance. When a captured and orphaned lion cub seemed hungry but unwilling to accept a bottle, a native woman suckled it on her breast alongside her own human baby. On-screen, at least, colonial Africa and its animal trade lived again. Mike, after all, was learning the trade. With the Mau Mau safely offscreen, the old order was reestablished, and traders like Carr had ample animals to sell including a giraffe pair snared in a dramatic chase.

Cotlow's truck raised dust clouds as it raced across the dry bush. Impossibly fragile, the rattling machine struggled to catch the graceful giraffe or the lumbering rhino. Still, leaning out of the passenger seat of the "catching car," the animal trapper calmly settled the lasso over the head of the terrified animal. Armed with a tool borrowed from the Wild West, the trapper in his truck no longer needed indigenous knowledge of the manners of wild beasts. The native assistants, instead, clung tightly to the truck's roof and leapt off only when the rhino was safely secured. Cotlow's camera lingered on brief, fleeting moments of intimacy, such as when the trapper stroked the muzzle of the giraffe or the rhino's nose as workers maneuvered the animals into crates. For native workers, Cotlow suggested, it was just another rough job. For the trapper, it was a labor of love and conservation. In that gentle touch, Cotlow showcased the caring nature of the animal trapper devoted to the wild game he captured and sold. The trapper, Cotlow insisted, should be the guardian of the continent's animals—even if he shipped them abroad.

The white rancher, not natives, introduced Africa's animals to American moviegoers. Carr's reality, however, was bleak. Perhaps he agreed to host the American filmmaker, despite the very real danger, because, simply, he needed the money. The returns from the animal trade were declining, and costs for trapping and shipping animals were rising in an era of restive labor and rare animals. Even worse, zoos had less money to spend, so Carr needed filmmakers' dollars. His animals became Hollywood veterans, appearing in *King Solomon's Mines* (1950), *Mogambo* (1953), and *Hatari* (1962), among many others.[20] By the 1950s, in fact, animal traders spent much of their time as experts working for American television and film productions while the capture, domestication, and transport of animals became almost a side business.

The animal trade was under siege, and zoos knew it. With independence inevitable, zoo men rushed to Africa on their last expeditions to collect animals before their extinction. "[W]e should get as much as we can possibly house in the next few years—if not before," advised George Speidel, the Milwaukee Zoo's director, as he set out for Africa in 1962. Caught in the headwinds of change, animal traders were already suffering from the "Africanization" of game management, he argued.[21] The AAZPA agreed. If endangered species needed protection, they must be removed from a wild that had become a battleground in liberation struggles. In this new era of decolonization and conservation, the organization advocated the "judicious collection and placement of specimens in zoos . . . capable of propagating these species in captivity." Dying at the hands of poachers, endangered species needed removal to the zoo sanctuary.[22]

Traveling with the animal trapper John Seago and his white assistants, Speidel enjoyed vast vistas of colonial wildlife. A herd of twenty-five giraffes grazed on acacia trees while zebra and impala roamed nearby. Then the expedition encountered Maasai natives on the road, and nature tourism confronted rebellion. One raised a finger and yelled "freedom." "He was laughing in a very funny way. Not in a hand shaking mood." That Maasai man's call for independence interrupted Speidel's wildlife reverie, turning his thoughts from plenty to extinction: "It seems a shame that all of this might be in danger of complete extinction." So, the worried Speidel set out in search of a rhino. "It's the most exciting thing that I have ever witnessed," and he took "a lot of pictures on the rhino chase."[23]

The Great Bongo Hunt: Kenya

In Kenya, John Seago could also see the beginning of the end not only of
the African rhino, hunted and harassed to the edge of extinction, but also
of the cozy arrangement between professional game trappers and the
British colonial regime. Seago had moved to Kenya after World War II to
restore his broken health and joined Carr Hartley as part of a small, close-
knit group of trappers and game ranchers. By all accounts, Seago was a
dependable trapper, carefully (and at substantial cost) domesticating wild
animals before sending them on the stressful voyage to the United States.
His animals might cost a bit more, but Seago had become a favorite of
American zoo directors, who trusted him to deliver safe and healthy rare
animals. He supplied rhinos to Washington and giraffes to Milwaukee,
and in return he enjoyed close friendships with his American customers.

Yet Seago never trusted or particularly liked the native Africans who
lived around him and occasionally worked for him. From one camp, he
gazed downhill upon a native village of "mud huts." They provided him
"labor and lots of incidents."[24] His antipathy toward both Africans and
their independence sprang from his concerns about animal conservation.
"African politicians," he wrote his American friends in 1958, "were still in
the thug stage."[25] As Kenya struggled toward independence, Carr Hartley
no longer could afford to domesticate wild specimens before sending
them to America. Seago, as well, wondered if he would have to relocate to
(still-colonized) Tanganyika.[26] But Tanganyika soon became independent in 1961, and Seago struggled to maintain his business in free Kenya.
"We have met the changing conditions as best we can," he reassured his
American customers, but he still worried "agitators will work people up
so conditions become impossible."[27] When Seago recuperated from injuries and illness back in England, he had a chance to reflect on Kenya's
"race relations" and the future of its animal trade; "multi racial societies,"
he concluded, were just an "academic dream."[28]

A trapping expedition, by contrast, seemed to Seago to be a microcosm
of a colony—and a stable one at that. On the game trail, "black and white"
could work together with the white trapper in charge. "We make no
suggestion that we like them or they like us," but he was convinced his
black African workers looked to him "to make all the plans and to be

responsible." In this miniature empire, leadership "gives us the right to expect obedience and loyalty," Seago insisted. When his animals arrived alive in America, Seago believed that he proved racial hierarchy "works itself out how somehow." Empire with its regulations and wardens had saved wild game. Independence, he worried, would kill the animal trade and slaughter the animals.[29]

Seago, Hartley, and other traders had a stake in maintaining a colonial system that had cozily regulated the trade in animals. Seago's work with animal conservation societies, film consulting, and vigilante fighting Mau Mau rebels punctuated his animal collecting.[30] Even when colonial governments battered by criticism of the animal trade limited exports, traders like Seago, in quiet collusion with American zoo directors, found an easy loophole. Seago simply applied for licenses for the list of animals he thought a zoo might want and set about capturing the specimens. Those the zoo didn't want could be sold elsewhere. After all, they could hardly be returned semi-domesticated to the wild.[31]

Seago, like Cotlow, recognized the inevitability of independence, but he still hoped it might apply only to humans. Through his friendships with American zoo officials and flush filmmakers, Seago hoped he could revive the older colonial order, at least as it applied to wild animals. Seago imagined traders and zoos working in concert with each other and with filmmakers to maintain the animal trade. Humans might be free, but animals and their jungles and savannas, Seago hoped, would remain a resource open only to Western consumers. In this new order, natives were the obvious enemy. Though he had harsh words for unscrupulous American and European traders who shipped animals too frightened and too ill to withstand transnational travel, Seago saved his ire (and bullets) for those native hunters he dismissed as poachers.

At the National Zoo, William Mann and his successor Theodore Reed had long admired Seago and echoed his fears of independent Africa. Seago was Reed's natural choice in 1968, five years after Kenyan independence, when he launched the zoo's expedition this time to capture bongos—gentle ungulates, endangered and elusive, that lived in the same cool Kenyan highlands that once provided shelter to Mau Mau rebels. In fact, seven years earlier, Seago had postponed a trip to collect bongos for the Milwaukee Zoo fearing confrontations with Mau Mau.[32] Bongos were

rare animals, too small to present much danger to native hunters who
prized their meat but rare enough to represent a significant capture for the
National Zoo. The bongo was an obvious target for the zoo. The animal
was vanishing in the wild and rare in zoos; only eight bongos had ever
been on display in the United States. Years before, Reed, still a young
zoo official, had visited a bongo on solitary display in the Bronx. It lived
only a short while, but the visit must have lodged somewhere in Reed's
memory.[33]

Reed turned to Seago, who was struggling to maintain his business
in postcolonial Kenya, with a plan for capture and conservation. Seago
would capture and domesticate a breeding pair of bongos. Reed promised donors and visitors wild bongos, but in truth, the zoo needed animals
habituated to humans and to a traveling crate. It was unlikely, Seago
warned, that all the captured bongos would survive the expedition to save
their species. Reed sought a grant from the National Geographic Society,
promising hard science and animal conservation. Seago would, he
assured, conduct an intensive survey of bongos in the wild. After their
capture, Reed would, as a trained veterinarian, collect blood samples,
observe their mating, and monitor their labor and delivery. In the privacy
of his friendly letters to Seago, Reed was more honest: all this detail about
breeding was simply the "glop" a foundation with scientific pretensions
required.[34]

The National Geographic Society had funded zoo expeditions before,
but the bongo hunt was different. Hardly a haphazard collection of wild
animals for short-term display, the great bongo hunt promised that
Seago's captures would preserve the species. The expedition would not
only bring back alive a breeding pair but also remove poachers from the
field with enticements of work, wages, and meat. The men cutting the
cages and tracking the animals, Seago and Reed agreed, were merely
poachers in waiting. The great bongo hunt—or so Seago and Reed
argued—offered endangered animals an escape by turning African
poachers into assistants in the service of the white trader.

The excitement of the hunt's beginning descended into tedium. Seago
and his convey hacked their way past white-owned ranches in the fertile
Kenyan plains toward the clouded hills. Here the undergrowth became
lush and the trail harder to cut. The forest was dense and the air thin.

Seago and his workers gasped for oxygen as they climbed into the bongo's realm where they laid traps and waited. In Washington, Reed feverishly demanded updates, but the news was slow. Elephants chased hunters into trees. The wrong animals stumbled into the traps: first a bushbuck, then two buffalo. Occasionally, bongos sniffed around the traps; one even went into the cage, but not far enough to trip the door.[35] For Seago, the wait was bittersweet: the longer he stalked bongos, the more he paid out in wages. Yet he also believed that for each day his laborers stayed waiting to bring back animals alive, they were restrained from poaching.

Seago and Reed had reasons to be impatient. Washington was not the only zoo seeking to preserve the bongo, and Seago was not the only trapper on the mountain. A local Kenyan rancher, Alan Root, was also hunting bongos, hoping to sell his specimens to Milwaukee. Each zoo raced to be the first to trap and conserve the rare animal. Finally, in June 1969, the telegraph Reed had been waiting for; a bongo had tumbled into the trap. Another animal followed a few weeks later.[36]

Then the hard work of turning wild animals into zoo specimens began. The two animals had to be stripped of their wildness before they could be shipped. Slowly, they became accustomed to the humans who now fed them, but just as the bongos were settling into their new captive reality, crisis gripped Kenya. In Washington, Reed read the news of the assassination of Tom Mboye, a key figure in the Kenyan independence movement and now a minister in the new Kenyan government. Reed was worried, but primarily for African wildlife and his white trapper friends. "My first concern is the safety of you all," he told Seago, "the second is my concern for the bongos; and the third is wondering what will happen in relation to all the national parks and the wildlife in Kenya." Only then did Reed "get around to being concerned with the plight of the country and the inhabitants thereof."[37]

Finally away from a changing continent, the bongos arrived alive in Washington (along with a shipment of other animals from Carr Hartley) where these shy animals became the popular symbol of American zoos' concern for African wildlife. A curious crowd gaped at these rare prizes, and Reed sent a delighted telegram to Seago, back again on the game trails of Africa. By late 1979, the two bongos had mated several times. Far from the Kenyan mountains, the bongos' future was now tied to the health of the Washington zoo.[38]

But that future, like that of Kenyan animals swept into the disruption of independence, seemed far from certain. Reed wanted to be in Africa. Pacing his dull office in Washington, Reed longed to be alongside Seago enduring the trials of searing temperatures, the agony of near misses, and the thrills of the ultimate capture. He wanted to be on safari because his zoo was in crisis.

The Naked Cage: Washington, Philadelphia

Reed needed those bongos for a zoo still reeling from Julie Ann Vogt's mauling death. In fact, a few years before, a slight reduction in the fund set aside as compensation to her family helped Reed put a down payment on the endangered Grevy's zebras Seago wanted to capture in a Kenyan "trouble spot" and to preserve in zoos.[39] Even financially, African independence, the decline of the animal trade, conservation, and the crisis of the zoo at home were linked.

The lion's attack on Vogt scarred the National Zoo. During the Great Depression, WPA labor had constructed modern, open displays. During the 1960s urban crisis, Reed covered them up with chain link fences. The evidence of the zoo's decline was everywhere in the weedy paths and crumbling buildings, and Reed confessed to Seago his anxiety that he was presiding over the extinction of one of the nation's finest zoos. The bongos' arrival and their mating felt like a relief after the bad publicity of the past decade.

In Philadelphia, another big cat became the symbol of another zoo in crisis. Tasha, a female tiger, slipped into the moat of her enclosure, precisely the kind of concrete pen zoo people had come to regret. "[M]edieval concrete pits," one keeper called them.[40] At the bottom of her concrete prison, Tasha appeared in shock or stupor. She ignored baits of meat and just sat, a dismal mascot for a crumbling zoo.[41] Above her, first keepers, then visitors, and finally the city's newspapers wondered what to do about the stranded tiger. "Mr. Zoo Keeper," wrote one visitor, "she looks like she is about starved." The letters poured in, each suggesting some kind of a solution.[42]

"How would you like to be in her place?" demanded another visitor.
"Have pity on her." Some saw in Tasha's stupor the decline of the zoo
itself.[43] Keepers, the zoo veterinarian, its director, and its PR executive
struggled to present a "favorable image of the Zoo and its animal management," but the zoo couldn't deny its problems.[44] More than ever before,
the zoo was dependent on visitor entrance fees, but the crowds dwindled.
By 1971, with attendance falling, the zoo temporarily closed its doors on
Mondays and Tuesdays. Animals, like Tasha, grew old, and as they aged,
their infirm slowness seemed to mirror the zoo's spiral toward irrelevancy.[45]

Visitors and keepers admitted there was nothing wild about Tasha anymore. Instead, an ever-louder chorus of critics claimed she and other animals suffered mental decay. When the journalist Paul Wiener visited the
Philadelphia Zoo, he discovered only neurotic animals. It was a quiet day
at the zoo with only a few elementary school groups wandering the
grounds, including a group of the "mentally-deficient." As he watched the
group wander the zoo, Wiener saw an analogy of captive, mentally starved
animals. In this "circus-prison," captive animals and disabled humans
showed similar "antics." It was a brutal comparison. Both animals and the
"mentally-deficient," he believed, had forgotten the "education their parents had tried to force on them." Monkeys and apes vomited on the floor
and then licked it up. Other animals masturbated or picked at their food
scraps. A large cat screamed when a keeper used a power hose to clean its
cage. As Wiener fled, his mind drifted back to his childhood zoo visits. "I
often wish," he wrote, "I could stop thinking and enjoy the sickness."[46]

Finally, after nineteen days of public debate, the zoo made a decision.
Keepers darted, drugged, and dragged Tasha out of the moat back into
the enclosure where she would live for three more years before succumbing to lung cancer, killed ultimately by urban pollution, the zoo
said.[47] Her sad journey into the moat not only contradicted the naturalistic pretensions of the zoo's display but also provided a sad, distorted
image of the tiger's wildness. Cotlow's charging rhino in his savage
splendor represented a disappearing Africa. Just as poignantly, a neurotic
tiger with lung disease embodied urban zoos corroding into decay.

Near Tasha's enclosure, a chimpanzee sat in a cage on a concrete floor,
its hair thinning to reveal its disconcertingly pink similarity to the human.
Absent anything that might have recalled its native habitats, the chimpanzee

smeared feces on the wall and used its opposable thumbs to draw abstract pictures. A gorilla regurgitated and swallowed its scientifically calibrated food, devoid of anything a gorilla might seek out and enjoy in the wild. From concrete enclosures to caked food, it was obvious animals at the zoo weren't really wild. No wild gorilla ever chewed its cud.[48]

When zoo visitors stopped imagining zoo animals as wild beasts and instead as mental patients, zoos knew they were in trouble. The tragic trail from wildness to urban neurosis began with the animal trade, the industry that once had enthralled Americans. Now it seemed only to threaten the very animals "from which man derives such benefit and delight," argued Desmond Morris, the popular author of *The Naked Ape* (1967), a best-selling book criticizing civilization through the natural history of the human ape. Before then, Morris had served as the curator of mammals at the London Zoo, and he witnessed firsthand the effects of the animal trade. Each animal seized from the wild meant one fewer animal to breed. *Zanzabuku* unwittingly pictured the hidden tragedies of the animal trade. As Cotlow again raced across the dusty veldt, this time to capture, not simply to provoke, a rhino, he and his trapper guide selected the youngest, fittest specimen, in this case, the female of a breeding pair. Cotlow and the trapper drove her mate, confused and angry, back into the bush, and she went crated to a zoo. In a cinematic flourish, the cycle of mating and birth had been irrevocably disrupted.

If this rhino lived in a threatened continent, the chimp, gorilla, and tiger in Philadelphia lived in the "naked cage." Between the open veldt and the urban zoo, wildness became madness. Morris, the former zoo-keeper turned apostate, shifted his critical gaze back to the zoo. In 1968 in *Life,* Morris described "the naked cage" where captive, decrepit, and caged animals paced the concrete. Zoo animals, he agreed, might live longer, but at what cost? They had become shells of their wild selves, looking like a rhino, gorilla, or chimp but acting like urban prisoners: pacing, masturbating, smearing shit. Morris provided the testimonial of a zoo man; *Life* supplied the photographic evidence. As it lay near its crumbling wall, a chimpanzee stared, with anthropoid sadness, at the camera.[49] Morris had written a manifesto for a growing movement critical of zoos. Zoo officials, leafing through the magazine's dramatic photos, were worried.

"The Naked Cage" (1968). A shocking image that accompanied Desmond Morris's critique of American zoos. Desmond Morris, "Shame on the Naked Cage," *Life* 65 (November 8, 1968).

In a parody of zoos' efforts to breed animals, sexual dysfunction offered evidence of animals' mental decay. A solitary hyena mated endlessly with its water dish while a lion flung itself against its concrete walls to masturbate. Other animals housed in breeding pairs mated incessantly. A lion secluded with a single lioness far removed from a larger pride or even stimulating surroundings became "supersexual." What else could he do to occupy his time? Just as poignantly, Morris asked, what was the point of breeding animals when the zoo produced only empty vessels, genetic authenticity but with extinguished wildness?

In the era of expanding empires, zoos had provided cities with metropolitan polish. In the age of independence, did declining American cities need their zoos? Writing for a new organization, Friends of the Animals and the Committee for Human Legislation, Bernard Fensterwald, a prominent Washington lawyer, asked, "Isn't It Time To Phase Out Zoos?" Maybe he wondered about Tasha trapped in her moat when he described the maddening life of the concrete cage. In their overcrowded, depressing urban cages, zoo animals were "unhappy, psychotic examples of their species." Any zoo visitor—like those irate over Tasha—could easily diagnose their psychosis. They watched the "species in the wild" captured on film or television. The wildness of Cotlow's rhino, however provoked, measured the mental decline of its imprisoned cousins.[50]

From Morris to Fensterwald, there was a common theme. Animals caught in the naked cage had degenerated from wildness into madness. They were still animals but hardly examples of wild life. Concerns about crumbling urban cores merged with worry about the behavior of aging animals in elderly zoos. In 1971, the Humane Society joined the critical chorus. Over three months, the society covertly visited seventy-one city and private zoos across twenty-three states, and its findings were predictably alarming. Its investigator, Sue Pressman, posed as a tourist. Another apostate, a zoo worker turned critic, she snapped photos of a bear avoiding his dung-covered floor and a tiger blinded by cataracts. In Pittsburgh, three Siberian tigers crammed into a filthy cage, and elephants lined up "wall-to-wall" in their enclosure. The society estimated that fully a quarter of large and small American zoos needed reform and rebuilding.[51]

Zoo officials like Reed read the exposés. Animal breeding helped justify the modern zoo, but buildings constructed in the 1930s, modern in youth, seemed prisons in old age. "Oh yes," worried one zoo director to his fellow officials in 1973, "you make a great deal of noise about barless moated exhibits and then put the poor animal right back in that concrete bathroom."[52] In Philadelphia the zoo's plight recalled the desperation of the Depression. As in the 1930s, attendance slipped and the zoo's animals aged into surly seniority. Massa, the famed gorilla, glared out from behind bars and later through an extra chain-link fence added as extra precaution after the Washington tragedy. Like the Washington lion, Massa remained dangerous because of his constant anger. In his confinement, the wildness of nature twisted into dysfunctional resentment toward visitors and his keepers.[53]

By the time Wiener visited the Philadelphia Zoo, even the zoo's official defender had begun to question publicly its very existence. Jack Chevalier worked as the zoo's spokesperson before resigning in a maelstrom of scandal and recrimination.[54] Joining the ranks of zoo apostates, he raked the muck of crumbling zoo enclosures to question whether the zoo could still care for its charges. As one of the first zoo public relations specialists, Chevalier received "hundreds of letters" complaining about zoo displays and, more specifically, about Tasha's crisis. He typed out his letter of resignation on zoo stationary and headed for the local newspaper offices still carrying the zoo's insignia on his satchel. Trash littered the zoo's walkways, and birds died from poisoned food. Cages, he accused, only helped the zoo lab "reap a harvest" of neurotic casualties for its studies of "hypertension and frustrations among animals." Massa was the most obviously damaged. His notoriously ill temper, a mere decade ago the subject of newspaper jests, now seemed neurosis. Alone in his "same naked cage"— Morris's phrase had caught on—his pathetic condition deserved euthanasia, according to Chevalier.[55]

Publicly, the zoo hinted darkly that maleficence must have been behind Chevalier's attack. Behind the scenes, though, zoo officials were nervous. After all, Chevalier's criticism not only echoed other attacks but also appeared to be gaining traction. "I was on the run for days as I defended the zoo on many television and radio stations," recalled director Roger Conant.[56] Conant admitted his zoo was in trouble, but Americans still wanted zoos and endangered species needed them, he claimed. According to Conant and so many other officials, the problem with zoos was not animal captivity but American cities. Urban ailments had infected the zoo.

The 1960s and 1970s were difficult times for American cities like Philadelphia. The construction of highways in the 1950s had only accelerated the migration of jobs and white populations to sprawling suburbs. In urban cores, infrastructure crumbled, as African American frustration with economic deprivation and segregation increased. Abandoned experiments in public housing only seemed to further shred impoverished communities. By the late 1960s, riots in some of the nation's largest cities, including Philadelphia, highlighted the painful effects of racialized inequality. Once, zoos had brought cosmopolitan sophistication to growing American cities. Now whites who had fled cities (and seemed less and less willing to venture downtown to zoos) decried urban living in close quarters and in squalid conditions. Neighborhoods, they argued, had become slums and ghettos, segregated places characterized by broken families, crime, and ill health. Even the sympathetic researcher Kenneth B. Clark described the "ghetto" in terms that evoked a broken-down zoo with its "walls" and its inhabitants as "subject peoples." As critics cast crowded cities as dystopias, zoo animals also seemed to be victims of urban problems: psychological breakdown, pollution, even crime.[57]

Zoo directors and critics alike applied the same language that disparaged American cities as decaying slums and dangerous ghettos to explain the fate of city zoos. Visual studies scholar Lisa Uddin notes that zoo directors shifted blame for zoo conditions to the city and its growing African American population. The zoo might be an "animal slum," zoo officials admitted, but that was because they were set within the "urban jungle."[58] Officials twisted ideas of animal neurosis into an urban language of racial dysfunction. The ills of the city had slithered through the turnstiles into the zoo. In a much-quoted report, the Philadelphia Zoo described its animals as troubled residents who suffered the same health problems as urban dwellers.[59] According to the zoo, the lungs of animals like Tasha proved that the problems of the zoo were part of the torn fabric of the city and less a fundamental contradiction in the enclosure of animals. Animals could thrive in captivity, just not in cramped urban surroundings. Morris agreed. Human city-dwellers, in this age of urban riots, showed all the same effects of mental decay as zoo animals, he argued. They "attack their offspring, develop stomach ulcers . . . or commit murders"—just like zoo inmates. As well, the zoo animal, Morris said, "exhibits all these abnormalities that we know so well from our human companions."[60]

The city, Morris concluded, was a "human zoo," and the zoo was a slum. "Cramped, slum-like conditions produce slow, torpid animals," argued one critic, in cities as much as in zoos. "Why do people en masse behave like pigs and barbarians?" asked Roger Conant, angry with visitors who fought and stole. The Philadelphia Zoo, a declining zoo in a troubled city, even conducted a laboratory study proving that crowding mice into cages, like people into densely populated cities, increased aggression and neurosis. The mice experiment and the crime wave that seemed to engulf the urban zoo all showed the "horrifying episodes" that happened when "people are crowded together."[61] Human violence even targeted zoo animals. Conant remembered a "hoodlum-vandal element" that hurled coins at the alligators and soda bottles at the hippopotamus. Any animal in an open cage had become a target. "[E]ndless trouble," the director complained.[62]

Like Morris, Conant, the "professional biologist," learned lessons about city dwellers from watching animals packed into zoos. Animals or humans "crowded or en masse, tended to behave like uncontrollable savages."[63] "Zoos," concluded the Philadelphia Zoo in 1967, "constitute experiments in urbanization." And that was their biggest challenge.[64] The city produced its own neurotic, decayed peoples, and animals suffered. Conant fumed about the local teenagers who crowded the zoo, ignored the guardrails, chased the birds in their enclosures, and poked whatever animals they could reach. Sometimes they wielded knives. Things were bad enough that on busy Easter Sundays, Conant requested police to park a patrol van outside the zoo gates ready to drive to the nearby jail. The zoo also purchased two new animals: ferocious guard dogs.[65]

Conant's choice of the word *savages* was deliberate. In all its racial connotations, *savages* evoked Africans spoiling their wildlife just as much as African Americans crowded into urban slums and fouling the zoo. Critics often pinpointed blame for the descent of the zoo into slum on African Americans. They also applied the same arguments about the pathological effects of the city on African Americans to zoo animals. Morris's critique of African Americans echoed that of Daniel Moynihan in his landmark report on "The Negro Family," published just three years before Morris turned his ire on the zoo.[66] Each detailed the breakdown of the family instinct and the emergence of violent pathologies, linked, inevitably, to urban life. Morris found an analogy of black urban violence, even the Black Power movement, in the pitched battles of monkeys in a cramped cage or decaying monkey island.[67]

In 1965 one "white citizen" railed in a letter to his newspaper: "Just what (else) is it that the Negro wants?" Everywhere one looks around the zoo, he wrote, there were African Americans—as if they were not supposed to be there in the first place.[68] The implication was clear: African Americans had come to the zoo not for an educational day out but to infect the zoo with urban crime. Chevalier, similarly, criticized the zoo as a crime-ridden slum. "Purse snatching, thefts, fights and vandalism are rampant." The zoo was as unsafe for its animals as for its child visitors—a "mugger's paradise," according to a news report. Once a young crook scooped up one of the zoo's turtles and, in perhaps the strangest example of zoo crime, used its hard shell to hit a cashier on his head. Dropping the turtle, the thief cleared the cash drawer and fled.[69]

The "children are terrified," reported a teacher after her class was robbed by "young Negro thugs." "Roving gangs" were targetting school visitors, and the zoo was helpless. The zoo needed its own "slum clearance." The phrase wasn't just a simile. In Philadelphia the city council debated zoo improvements in the same meetings when it discussed problems in the city's poorest and predominantly African American neighborhoods.[70] Soon riots that roiled Philadelphia and other cities in the 1960s engulfed the zoo, which was located in a poor, largely African American neighborhood. Conant received tips that "bands of hoodlums planned to start riots." As the zoo struggled to meet the "menace," keepers and officials rotated on night duty to protect a zoo that felt under seige. Conant patrolled with his guard dogs—frightening the enclosed animals. "What might we expect in the mental climate of that period?" Conant asked.[71]

Like Cotlow's rhino, zoos were panicked. On the defensive, zoo directors resorted to a primal instinct of fight or flight. In fact, they did a great deal of both. By 1976 zoos had banded together to form the Zoo Action Committee, a new organization, to defend the role of the zoo as the breeding grounds for the world's wildlife in the face of global threats. In 1975, 112 million Americans visited their zoos, yet twenty-five organizations, from the Friends of Animals to the Society for Animals Rights, had become "anti-zoo," Zoo Action complained. Unfortunately, like the animals it was trying to save, "our zoos, are literally threatened to the point of extinction."[72] Zoos also fled, joining white flight from cities to suburbs.

Wildness Conservation: Milwaukee, San Diego

"Welcome to wonderland," announced the new Milwaukee Zoo. Away from its old urban confines, the zoo moved to the suburbs in 1958. Speidel's vision, nurtured on safari with Seago, had finally been realized. Highways paved over the old downtown zoo, and a new zoo was born in the suburbs. Pass from your car into wildness. Of course, "[i]f you had the money and the time," you could jet off on safari. "[H]ere on the edge of Milwaukee," though, predator and prey roamed in apparent freedom. Like an African veldt, "one of the wildest in all the world," a pride of lions rested on towering red rocks while below antelope, ostriches, and zebras grazed warily. *"But you don't have to travel to Africa."* Here were Speidel's rhinos, collected with Seago in Africa and grazing alongside a pair of elephants.[73]

Some zoos had stayed in urban cores. Others had fled. Flanked by expansive parking lots, the new suburban zoo appealed to those who could afford a car and the time needed to go on a zoo-fari. The new zoo visit, rather than being interwoven into the park system of the urban core, demanded a full day out. The old zoo visit could be a short stroll accompanying a visit to an urban park. If the county zoo's relocation escaped Milwaukee's center city, its architecture promoted escape from the continent, an illusory experience of travel.

Who needed a cramped city zoo when films like *Zanzabuku* pictured the charging rhino? Zoos once brought back from "'darkest'" Africa the "exotic" prizes of "intrepid adventurers." Now, Morris agreed, the African adventure had become an affordable "holiday safari." Even more cheaply, the movie's wide screen or the TV's small screen brought wildness into the home.[74] *Zanzabuku* provoked animals into an exciting and seemingly natural wildness. The rhino's rough charges and terrorized rage appeared vital next to Massa's surly irritation or the lion's murderous swipe. American audiences, fresh from the movie hall, had rising expectations for their zoos. They expected zoos to breed their animals, put cute babies on display, and save vanishing herds. They wanted to see their animals in wide-open spaces. For many of those who had abandoned cities during decades of white flight, visitors wanted to drive to zoos far away from the city.

In San Diego the zoo had begun dreaming of a safari park in 1959 in the midst of the era of African independence. Finally, on the mesas above San Diego, it found its haven. Fourteen hundred feet above sea level, its arid expanses cooled by sea winds, the planned safari park perfectly replicated East Africa—except without Mau Mau rebels or urban rioters. Just as important, a highway wound its way through the 1,800-acre site of the new Wild Animal Park. Even before the park opened, visitors leaned against their cars at a highway turnoff to gaze at hundreds of animals, including a herd of white rhinos.[75] "Just like South Africa—just like Kruger," marveled some visitors as they got into their cars for the drive downhill to San Diego and its suburbs.[76]

Meanwhile in Africa, trappers began to collect the park's animals. The zoo tried, when possible, to collect those animals overpopulating Africa's game reserves. It wasn't that Africa's elephant populations were soaring. Rather, they were concentrated in smaller and smaller realms, hemmed in by humans and often enclosed by fences. "Good game management" in Africa meant some animals headed for California.[77] The rhinos were collected in South Africa from herds overpopulating the Umfolozi Game Reserve. The zoo's Operation Rhino hoped that on their ninety-acre terrain, the rhinos would breed.[78] Other safari parks also springing up alongside America's highways were less hopeful about their breeding programs. Some of the new game parks promised conservation but were voracious consumers of wild-caught animals. One Kenyan game warden, for example, grew suspicious of the new African Safariland opening in suburban Florida. The park required "literally herds of a large number of species." Private and profitable, such parks typically refused to share details on their births and deaths.[79]

In San Diego, Africa and Asia returned to savage splendor. Away from inner-city America, the tropics were wild again. The San Diego Zoo devoted much of its budget, attention, and publicity toward the creation of its safari park, far from the city and due to open in 1972. Land movers chewed the ground into new vistas, and helicopters dropped spools of wire fencing; it was truly remote. Water holes where herds of animals would soon gather burbled up from dry ground, fed by underground pipes. In 1968, ask a group of children what came to their minds when they heard the word *Africa,* and most would scream out "lions!" The others would say "elephants." Two Ghanaian brothers, one at university in southern California and the other in Philadelphia, learned an important lesson about what Americans really thought of independent Africa. Africa, they protested, had its modern cities, but Americans only pictured "jungles, lions, elephants and other wild animals." Americans wanted Africa, free or not, to remain the land of Tarzan, spears, safaris, and tom-toms. In fact, one of the Ghanaian students admitted that the first lion he ever saw was at an American zoo.[80]

When San Diego built its safari park, it argued that the best home for tropical wildlife was in suburbs outside of American urban cores. The new safari park's architecture also denied the reality of the postcolonial modernity of African and Asian cities. In the safari park, zebras raised clouds of dust as they galloped in a panicked retreat. In the zoo's publicity, they looked a great deal like those herds featured on film and television. Rhinos charged with the same impressive power, and lions prowled in natural cover so visitors could imagine the impending ambush, but not so hidden in the underbrush that they were obscured from view.

In the safari park, "Nairobi" looked nothing like its African namesake, even if it was still the gateway to safari country. Gone was the hurly-burly of a growing independent capital swelled by migrant populations. Just off the parking lot, Nairobi was a village of primitive architecture and gift shops. Its Simba Station was the ideal place to begin the "grand tour." Notice the colonial 1920s design as the monorail departed for "your five-mile-trip through the continents of the world"—at least with large animals and interesting natives.[81] Visit the Indian elephants first and then their African cousins. Hatari, named, of course, for the film, was the one splashing in the pool. In the brush, the shy oryxes try to hide from the

"Wild World of Animals" (1973). In an image reminiscent of Cotlow's safari film, this guide to the San Diego Zoo's Wild Animal Park invited visitors on a virtual safari through wild Africa and Asia. Zoological Society of San Diego, *Wild World of Animals* (San Diego: Zoological Society of San Diego, 1973).

train as it pulled into the Congo fishing village where visitors could imagine natives pulling fish from the water with hand-woven traps. If you are truly lucky, the rhino will grind the packed ground into dust as it charges or squares off with a waterbuck. Nairobi again, you have just enough time for a visit to the gorillas. Handicrafts from Asian and African artisans are for sale at the International Shopping Bazaar. Remember, the park was growing, reproducing "many of the endangered species." "Our children in the future"—like my own daughter—can enjoy these wonderful animals in the mesas above San Diego, this pastoral, primeval Africa.[82]

"Somewhere under the overburden of modern twentieth century concerns," wild animals needed a place to live, the Wild Animal Park told its first visitors. Perhaps that "somewhere" was in the brushy outskirts of San Diego. Maybe it was in African game reserves. Somewhere in an age of *uhuru*—the Kenyan call for freedom and independence—animals and people needed "only the freedom to live."[83] But could they live together?

When Africa came to the hills above San Diego, the Wild Animal Park helped mark the long transformation of American zoos from eager consumers of tropical wildlife to institutions that claim to protect and produce. For much of the twentieth century, zoos celebrated the arrival of ships crammed with wild specimens. As the reporters crowded around and confused leopards swiped at anyone who ventured too close, these wild animals seemed to have brought with them the allure of tropical wildness. As the century came to a close, tropical nature was alive and breeding a car trip away from San Diego.

Of course, not all American zoos could or did relocate from urban cores to suburbs or from cities to safari parks. Many, such as the zoo near my childhood home in St. Louis, remained, but even in urban zoos the end of the animal trade was obvious. St. Louis, Philadelphia, and other zoos rebuilt their exhibits beginning in the 1970s' crisis (and construction continues today) to accentuate the reality that animals are managed there but visitors can virtually travel abroad. Visit gorilla forests. Gaze across African plains. Experience the Amazonian rainforest. Hike the Himalayan highlands. Ride a railroad across African rivers and Southeast Asian deltas. The animal arrivals the zoos now celebrated were births, not purchases. Animals still come and go at the zoo. In short news releases, zoos might note that individuals have gone to other zoos as part of species survival plans: animal husbandry at its best.

Animals could be separated irrevocably from their original habitats and, later, traded to other zoos, but adventure was harder to disentangle as zoos moved from the consumption of animals to their conservation. The St. Louis Zoo stayed in the urban Forest Park, but its zoo map today offers an adventurer's world tour. Visit the River's Edge, Big Cat Country, Sun Bear Forest, and Andean Bear Range. Dine at Café Kudu or the Safari Grill, and shop at the Safari Shop or Tropical Traders. Read, as well, its mission statement: "The mission of the Saint Louis Zoo is to conserve animals and their habitats to animal management, research, recreation and educational programs that encourage the support and enrich the experience of the public." Once America's lust for adventure engaged the animal trade. From bring 'em back alive to breed 'em alive, conservation remains entangled with tropical adventure and our lust for a certain kind of wildness. Visitors still seek that charging rhino.

Even after the end of animal trade, Frank Buck lived again if only briefly. From 1982 to 1983, CBS ran seventeen episodes of *Bring 'Em Back Alive*, starring Bruce Boxleitner as Frank Buck. The jungle was virgin again, and natives dressed in strange masks. Colonial Singapore had its smugglers and villains, and Buck called the suave sultan of Johore "HH," just as in his autobiography. Buck again wore his pith helmet and trademark suit in the jungle and evening wear at Raffles Hotel. Ali was once more the subservient but loyal sidekick. As Buck danced with an episode's heroine at Raffles, she asked him why he trapped the animals. "Oh I do it for the money," he said, as she nodded disbelievingly, "and for the thrill." Then he paused: "The thrill I get when I see the look in a little kid's eyes watching animals at the zoo or at the circus." In his rebirth after the death of the animal trade, Frank Buck was in it for the kids and for the animals. "You see what I do I hope in a small way preserves them."[84] And then she kissed him.

In the dramatic pause, Buck was reborn a modern conservationist and the lavish stage set was colonial nostalgia, like so many zoo-faris. This series never caught on and was soon canceled. Maybe it was network TVs way of saying goodbye to the animal trade.

CONCLUSION

Searching for the Yeti

MARLIN PERKINS LONGED to bring back alive one last rare animal. Or, maybe, he just wanted to travel deep into the Himalayas with the last great adventurer of the twentieth century. In his last days at the Lincoln Park Zoo, before moving to St. Louis and just after *Zoo Parade* ended its run, Perkins received the offer of a lifetime. *World Book Encyclopedia* was equipping an expedition to try to find and capture a yeti, the abominable snowman. Even better, the expedition would be led by Sir Edmund Hillary, who had scaled Mt. Everest as empires crumbled below him.[1]

Perkins had faced down rhinos in Africa, and in his upcoming *Wild Kingdom* television show, he would dive the deeps, cross deserts, and penetrate jungles. Yet even for the famous zoo man, a search for the yeti promised to be different. It was a last expedition, the final effort to bring back alive a truly rare prize. Over Perkins's life as he evolved from a childhood zoo visitor to a regular zoo worker to a director and television star, animals once rare had become commonplace in zoos. By the 1960s, everyone in America knew the anthropoid apes. Even children could tell the difference between a chimpanzee, a gorilla, and an orangutan. Some could now identify the differences between lowland and mountain gorillas. This was a remarkable change in what ordinary people knew about tropical animals: these same children's parents or grandparents grew up with only hazy ideas about what a chimpanzee or gorilla might look like. Once gorillas were shrouded in mist and myth; maybe the yeti was the next great discovery.

The yeti (if it lived anywhere) prowled the Himalayan heights. Perkins seemed skeptical. He had, after all, collected newspaper clippings about the conspiracy theorists and science fiction dreamers' most outlandish yeti.[2] Nonetheless, for Perkins, a zoo director who never felt contented sitting behind an office desk, a fantastical journey was in the offing. In 1960, *Zoo Parade,* comfortably set in the confines of the Lincoln Park Zoo, had been canceled. Perkins knew the old formula had gone stale. Viewers, whether they were watching television or visiting the zoo, craved more than just watching the animals through the bars of cages. They wanted their animals running through the wild or in open-air enclosures at the zoo. They were still searching for wildness, even in the age of extinction.

Television audiences demanded the same colorful natural splendor they saw on the big screen. In the waning months of his time as the Lincoln Park Zoo's director, Perkins pleaded with his employers to allow him to leave the zoo to search for the yeti. The remote chance of its discovery surely made his absence worthwhile. After all, the Lincoln Park Zoo owed its existence to the popularity of Bushman, its sullen and recently deceased gorilla. A new species could revitalize an urban zoo eager for good publicity. Maybe the abominable snowman was real and the expedition could capture a specimen. "Lincoln Park Zoo could have become the first Zoo in the world to secure and display an Abominable Snowman," Perkins tantalized. However small, it was too good a chance to ignore. "I cannot over-estimate the potentiality of such a stupendous possibility."[3]

Perkins posed for publicity pictures while loading tranquilizer darts into the capture rifle. He packed his expensive cameras and rolls of wire to make crates for the yeti. In retrospect, all that packing seemed farcical.[4] At first, Perkins, Hillary, and the rest of the expedition enjoyed comforts that harkened back to the days of British imperial influence in Nepal. They rested in Kathmandu where Perkins watched the exotic and strangely intoxicating Festival of the Little Living Goddess. Entranced if exhausted from following the ornately carved chariot of the exquisitely bejeweled six- or seven-year-old "Living Goddess" from her nunnery around the city, Perkins retired to his comfortable hotel. The open fire and fine food of an expat European chef beckoned. The hotel's "Yak and Yeti" bar offered a welcome relief from sightseeing.[5]

Once outside of the "fascinating city," the promising expedition unraveled. Perkins and the other "sahibs" struggled with blisters and the weight of their packs. Soon, Perkins shifted his pack to the shoulders of Ang Pemba, his personal sherpa. The sherpas bore their loads into the high hills where glaciers shimmered in tropical sun. Hillary and Perkins had planned and packed with the same seriousness of past animal adventurers. In the mountains, Perkins prepared sophisticated camera snares. He attached costly cameras to thin trip wires and mounted them on mountain paths and cave roofs. Perhaps, then, the last ape could be caught, caged, and brought home to Chicago where the zoo would have Bushman's worthy successor.[6]

Now to Perkins, the yeti appeared like more than a mirage, the desperate visions of a mountain climber blinded by the snow. Instead, he believed he was uncovering a vast native fraud. Of course, animals had long been part of hoaxes back at home in the United States. P. T. Barnum, the museum owner and circus impresario, had a made a fortune out of natural humbugs. He had stitched together a mermaid and mounted an elephant. Visitors had delighted in detecting his flimflam. Now, Perkins described humbug simply as the evidence of the miserable and degraded lives of those people who eked out an existence in the shadows of magical mountains. In their fetid shelters, around platters of rancid food, these native hosts promised proof of the yeti. Perkins was going to uncover fraud.

Perkins had little interest in a creature so important in local religious and cultural beliefs. Instead, he was focused on a massive footprint in the snow. Twice the size of Perkins's own foot, natives insisted it was proof that a folkloric creature was a long-lost species. Perkins was dubious. Before World War II, when Americans sought fortunes buying and selling Asian animals, natives like these had been essential allies. Their knowledge helped track down the perfect specimens. They knew the habits of the animals and their routes through the jungles as well as their commercial value. Frank Buck and so many other animal traders had boasted of the loyalty native peoples showed their bosses. Their faithfulness and knowledge had unlocked the secret animal treasures of the tropics.

Now Perkins and his fellow adventurers regarded native peoples with suspicion. Likely the yeti was just "stories told by the sherpas around camp fires," Perkins said.[7] The footprint, Perkins demonstrated, was merely from mountain dog whose small prints had in the intense sun melted into the yeti's spoor. If this mistake could be laughed off as honest ignorance, a mangy hide seemed a deliberate hoax. A generation before, Americans had eagerly accepted Buck's nature-faking and staged fights. Virtually everyone knew it was deception, but no one really minded. Now folklore seemed fraud, tradition revealed as deception. Locals assured Perkins that this scrap of fur removed from a mountain temple was really a yeti's scalp. After debunking the snowy footprint, Perkins and Hillary distrusted natives. Instead, they hurried the fur to a waiting plane bound for Chicago and its biology labs. After the fur returned a few weeks later, Perkins revealed it was merely an old, malodorous hide of a Himalayan serow (a small ungulate that scrambled up the mountain slopes like a goat). The *Encyclopedia* had cajoled two of the world's best-known adventurers into a science fiction chase. Hillary and his fellow mountaineers soon climbed into the heights to test whether they could last the winter in a specially built hut, while Perkins headed home to take up a new job at the St. Louis Zoo.[8]

Perkins remembered the yeti expedition with a hint of sadness and nostalgia. When zoo men were off chasing snowmelt and mangy fur—hoaxes—the era of dangerous safaris and wild cargo had truly ended. Perkins probably never believed he would actually find a yeti. After all, he had collected enough silly newspaper clippings on abominable snowmen. Like adventurers before him, he also received piles of letters from Americans begging to come along, but this time they seemed less like admirers eager for tropical adventure than like true believers in monsters and ghosts. "I was reading articles about the Snowman long before the more popular, or respectable magazines would dare print them," confessed a Chicago police officer. He promised to resign his job if Perkins offered him a place. An anthropologist worried partly about being overweight and out of shape and partly about the Russian, Chinese, and Japanese teams he claimed were also searching the mountains for the yeti.

In hopes of "victory" for Perkins's team, he promised he could leave on a moment's notice. More prosaically, a nurse offered to quit her job: "I know of no better reason for quitting than to help you look for the fuzzy man. I'm dead serious."[9]

A mountain blizzard of conspiracy theories and cross-cultural mistrust buried the adventurers' era of wild animal trades that defined zoos' first century. Some American observers were hopeful that the breeding zoo, the safari park, and conservation programs would define a new era. Others wanted to phase out the zoo. Some saw only the naked cage. Others saw Noah's Ark.

The Viet Cong and the Keeper Blues: Vietnam, Long Island

In 1968, Marvin Jones saw firsthand how the animal trade was devolving from manly tropical adventure in the colonies into postcolonial smuggling. Jones was a sergeant serving in Vietnam and leading soldiers out on patrol. In the tropical heat with full pack and rifles, Buck must have been the last person on most soldiers' minds. "We were hit last week and had our potable water supply knocked out," Jones reported. The dangers lurking in the jungle weren't tigers but guerrillas. As the soldiers peered into the gloomy underbrush for Viet Cong ambushes, Jones was also looking out for smugglers. His mind was never far from zoos back home, and at least one official was appalled by the news that the man who was keeping animal records for the nation's zoos was also out on jungle patrol: "You are too valuable to the world zoological community. . . . What a system!"[10]

Jones wrote home to zoo officials with bad news. The war was obviously affecting Vietnam's wildlife, and not just because of habitat destruction from Agent Orange incineration. The "natives are known to kill many of the animals for food." In the small towns, small animal markets had also sprung up offering birds and a few monkeys for sale, mostly to GIs. Out on patrol, Jones watched the Viet Cong and local wildlife. He saw too many of the former, too little of the latter. Even the birds seem to know "that GI's shoot them."[11]

Jones was on the trail of smugglers transporting rare douc langurs, gorgeous, multicolored primates today listed under CITES as endangered to critically endangered. "Sure wish I could pin down the export area of the

Doucs," he wrote home. Somehow these rare animals were getting out of Vietnam and into America. Before the war began, there were only three on display in American zoos. Now there were three dozen. One zoo received a pair from a San Diego dealer, but they soon died. The Tulsa Zoo also bought a pair of doucs, but they were in desperate conditions, infested with lice. Jones was furious with a dealer clearly selling smuggled animals and with the zoos who bought them. "Articles like Desmond Morris' may make zoo folks unhappy, but they sure need a swift kick in the pants," he fumed. He was convinced the Viet Cong or the North Vietnamese Army were actually smuggling douc langurs into Laos. From there, American dealers brought them home to the zoo.[12]

As Henry Trefflich posted his letters to the National Zoo, zoo directors, and President Jimmy Carter, he also knew the era of animal trading was limping to its end. Once Trefflich owned a sprawling animal emporium in Lower Manhattan, but it had been condemned to build the World Trade Center, a perfect irony about the transition from one type of global trade to another. He had tried to construct a holding compound on Long Island, but neighbors successfully fought his plan. The animal trade just didn't have the same allure anymore, and recovering from his heart attack, Trefflich didn't have the energy or money to fight back.

Trefflich had built a million-dollar business in the animal trade. He had quietly supplied some of the monkeys for Buck's Jungleland, and with a great deal more bluster, he had delivered monkeys, gorillas, and chimps to the nation's biggest zoos. "Many of you will remember Henry Trefflich," he wrote zoo directors, "in his prime from 1928–1965." This had been the heyday of the animal game, and he had hired Jungle Jenny and used her allure to help build his own animal empire. Now, in 1977, animal trade millionaires' fortunes became nickel penury, and some days Trefflich slept in the store just to save the money he would have spent driving home to New Jersey.[13] His letter to the National Zoo promised he would compensate the zoo for the money he still owed on a pygmy hippopotamus, if only he could get back on his feet. Maybe the zoo could share its list of surplus animals? Then he could fulfill the few remaining orders.[14] The animals weren't arriving from abroad anymore. He even wrote the president, begging for import permits to bring in elephants from India. Once his stock lists boasted of jungle adventure and animal cargo. Now, desperate and ill, he appealed for pity.[15] When he died in 1978, the National Zoo began proceedings against his bankrupt estate to recoup its losses.[16]

Fred Zeehandelaar faced the end of the era another way. He was proud of his still-busy days in his Long Island office, shuffling customs and quarantine forms from the inbox to the outbox. He ran to the airport to receive a wild cargo, then dashed home to phone a zoo to beg for past due payment. Along the way, he smoked a couple packets of cigarettes. The AAZK invited him to give a keynote to their 1972 annual conference where he lauded the hard conservation labor of keepers at the front lines of species survival. He shared a few inside jokes at the expense of zoo directors. What really did they do when all the husbandry, feeding, breeding, and shit shoveling work was done inside the enclosure?[17]

Yet zoo directors, keepers, and Zeehandelaar knew well there was nothing particularly clean about the animal trade, either. Sometimes illegally collected apes moved from one country to another—from Indonesia to Singapore, for example—and received a whole new set of papers.[18] It was animal laundering of the worst kind. At the end of the age of empire and at the dawn of a new era of mass extinctions, the animal trade had entered the shadows and, no matter how much Zeehandelaar, Trefflich, and others protested their innocence, they found themselves in trouble. In Zeehandelaar's case (it was soon before the courts), he was desperate to import cheetahs before the Endangered Species Act came into effect. In one of the first prosecutions under the new law, Zeehandelaar was accused of backdating the documents on the import order.

By 1973, Zeehandelaar faced indictment, accused of illegally transporting endangered species. Marvin Jones, for one, was relieved. "If he goes under of course many many zoos will have to come up with answers." He was found guilty, was fined $5,000, and spent three months in jail—ironically, an animal trader beyond bars.[19] Zeehandelaar's commodities were "living, breathing creatures destined to educate the public about life in the animal world," or so he justified his business.[20] The problem was that by the 1970s these were endangered commodities and the public was beginning to think about them less as the dangerous prizes of the abundant jungles than as the last lovable specimens of dying species.

The very next year Ed Roberts implored fellow keepers to recognize that animals, once replaceable specimens, had become endangered species. Roberts, a keeper at the Stone Memorial Zoo in Stoneham, Massachusetts, had been active in the fledgling AAZK for seven years when he offered advice to a new generation of keepers in the age of extinction. In his own career, Roberts experienced the dramatic transformation of zoo work. Keepers had gone abroad in large numbers as soldiers but returned to zoos just as zookeeping was demilitarized. Uniforms now looked less like military garb and more like safari-ware. Roberts had strong ideas about what it meant to care for endangered species. Love helped, but it wasn't enough. Sure, animal lovers filled out their applications and started work on the other side of the enclosure, thinking zookeeping was "just a lifetime of playing with attractive animals." Zookeeping had evolved over the first hundred years of American zoos, but some parts of the job never changed, like smells, feces, and "heavy physical labor."[21]

After decades of labor strife inside the zoo, the AAZK reassured directors and reminded its rank and file that it wasn't a union. Yet it represented its members who knew that their vision of conservation as care clashed with that of directors, trained to wheel and deal. Keepers knew the animals for their individual character, less for their genetic individuality. The keeper, Roberts insisted, was "a friend and confident of his animals."[22]

For this new generation, zookeeping was conservation by affection, but that too could create animosities. "Old timers," though, flaunted the rules by sneaking extra treats. Maybe they still looked the other way when kids tossed peanuts. As well, "direct contact" with endangered species didn't help keepers find common cause with locals in Africa and Asia. When keepers, like other ordinary people, could finally afford to travel, they came back firmly convinced they must save the animals by their hands alone. One keeper returned home to Chicago from a comfortable safari in Kenya with "new respect" for "the good old U.S.A. and . . . zoos and keepers."[23] Africa, though, troubled her. Tourists could land by plane right into the heart of game country, but the animals were skittish. They had become accustomed not to tourists but to poachers. After her African safari, zookeeping seemed that much more important as conservation work. She reassured her fellow keepers: "If, as a keeper, you have had endangered animals on your run reproduce," you accomplished "more to help his survival than has his native country."[24]

Keepers may have been at the front lines of conservation, but zoos remained disorderly places. Like a century before, animals never quite accepted confinement, whether in open enclosures or cages. Directors still imagined they knew better then keepers about how to care for and display animals. Visitors still wanted to enjoy the animals their own way despite the rules. Soon, zookeepers were singing the "Talking Zookeeper Blues." "Keepers Care!" But "we all suffer some frustration. . . . Animals that don't feed, species that won't breed, Public who don't read, and people who won't heed the Keepers' Advice and suggestions."[25]

The End of Game: Nepal, San Francisco

"Most of us can recall our first visits to the zoo," wrote one observer around the time Trefflich faced bankruptcy. When she was a child in the interwar period, a day at the zoo was still a wonderful outing, a chance to gasp at the elephant, cower before the lion, and giggle at the rhino. Those animals were there to amuse, and they had been "captured by some Frank Buckish super-hero in a vast jungle populated with millions more." Adults now, such visits all seemed so childish.[26] In the 1970s facing the end of the animal trade and a rising chorus of critics, zoos were forced more than ever before to justify their present and plan for the future. Such plans acquired an even greater urgency, as it was clear visitor favorites from cheetahs to tigers were rapidly disappearing in the wild.

It was the end of game. Wild animals had become endangered instead of trophies. In the 1970s, the American public was more divided than ever before in their opinion on zoos. Critiques of zoos coalesced, breeding new organizations that today continue to challenge the ethics of keeping animals in captivity. Many of the arguments for phasing out the zoo advanced in the 1970s ring familiar today. Zoos have responded, often with hyperbole about their role as the foundation for animal conservation. Advocates, like William Conway, now the director emeritus at the Bronx Zoo, claimed zoos in 1977 as the "repositories of some irreplaceable living treasures." Given changes in animals' "homelands," whether high in the Himalayas or deep in the Malaysian jungle, "there will be no other way of preserving them." The zoo wasn't a prison or even a naked cage, he argued, it was an asylum, and animals weren't specimens or commodities but refugees.[27]

Here was a vision of a world divided between those who conserved and those who encroached, poached, and threatened. Such a vision remains deeply entrenched in contemporary pleas for wildlife conservation. Animals depended on those American zoo visitors to survive because they were fleeing other humans. When Perkins went to the high mountains, he went to capture and display, with promises to conserve. He came home an empty-handed consumer filled with condemnations of native people. But a question nagged. Could the zoo be anything more than a consumer of wildlife? Could zoos, at the end of the century, use their existing stock to become producers of wildlife, and if so, what needed to change in visitors' expectations and exhibit design? Zoos could boast of their role as the linchpins of global animal conservation, but the challenge to breed the animals zoos needed for exhibition was hard to meet. It was another challenge entirely to meld species survival with visitors' demands to see particular animal "treasures." Zoos still struggle over which animals to breed, save, and exhibit and which to consign, regrettably, to the less-than-natural struggle for survival.

In 1975, David Simon had worked at the San Francisco Zoo for four years when he added a dose of sober reality to the debate about the future of animals in captivity. As zoos became animal producers, partly out of necessity and partly out of a genuine zeal to preserve the world's endangered species, adaptations to humans became a new factor in natural selection. Zoos, he realized, were a kind of evolutionary "pressure cooker," adding new stresses to animals' struggle to survive. Certainly humans had a great deal to learn about husbandry, but animals were going to have to learn as well about breeding and raising their young in captivity. In a few short generations, the living, breathing commodity had to adapt to the "new stressful situation of zoo living." If not, "it dies."[28] Pause for a moment on the yeti. Perkins realized the zoo would be tempted by a charismatic animal star to replace Bushman. The yeti expedition was the zoo equivalent of a televised talent search.

Some animals, like those cheetahs Zeehandelaar tried to import, breed only with difficulty in zoos. In the wild, notes one conservationist, they "breed like rabbits."[29] As well, particular species' appeal to humans became a factor in their struggle for captive survival. Fur that might once have camouflaged became striking beauty. Iridescent feathers to entrance a mate could also enthrall the human visitor. Venom that once paralyzed prey excited humans. History also matters in the struggle for captive survival. Over the past century, Americans have told endless stories about tigers to their kids. They know (or think they do) about their ferocious character, man-eating, and jungle regency. Those elephants that once carried children on their broad shoulders in America's first zoos are skinless ghosts that haunt our relentless desire to see elephants in our zoos. The wallowing rhino still evokes images of the dusty charge across the African veldt. Dying in the wild, such animals live everlasting in our collective culture. Such animal treasures still evoke the exotic, and we want them saved. But does the okapi, the douc langur, the Amazonian frog evoke the same historical memories? Only a small percentage of the earth's animals receive entry to the American zoo. Reptiles, insects, amphibians, even birds are immediately disadvantaged.

Zoos still spend resources on exhibits for charismatic animals poorly suited for their captive environment or, frankly, less in need of refuge than others that might bore visitors. The St. Louis Zoo, for example, despite its landlocked location in a city of long, fierce summers, has labored to display penguins and puffins and to rebuild its polar bear habitat and revive the bear pits, the famous pioneer examples of open enclosures.[30] It is not hard to argue that humid St. Louis is not the best place for polar bears. Wilderness conservation may be the better solution, critics argue. With endangered species lists lengthening but zoos ever more dependent on visitor admission fees, zoo officials face tough choices about which species to save and which to exhibit. Zoo directors with an eye on business models and bottom lines often must become impresarios like their ancestors George Vierheller or, even, Perkins. Sea lions bursting from the water to grab a tossed fish or a tiger embodying the fearsome jungle are infinitely more exciting than the somnolent snake. So zoos keep camels—children can still ride them at many zoos—but they hardly need a place on the ark.

Some animals become stuffed toys given out in adoption programs. Others slither, tread, walk, crawl, trot, flutter, and swim quietly toward extinction. Human love toward some creatures and indifference toward others become a factor in survival as zoos balance entertainment and conservation. Some like Conway described zoos as the last refuge for the world's last wildlife, a justification as common in 1977 as in 2016. Other zoo directors recognized the zoo's "primary function is recreation."[31] Zoo directors were (and still are) trying to figure out exactly why visitors actually go to the zoo. What do they expect to see? Visitors are opinionated. They enjoy certain animals but ignore others. Many visitors went and still go to zoos because they enjoy looking at cute, cuddly, ferocious, colorful, and funny animals.

It's worthwhile walking around the zoo looking at the animals but listening to the humans. Children whine, hungry, tired, or keen to bypass the sleeping rhino for the pacing tiger, which, at least is moving. Adults and kids alike clap their hands or beat their chests, desperate for the gorilla or orangutan to look back. Visitors have a lot to say about their zoos. Parents might talk to the children about the signs explaining global warming or habitat destruction. They might also walk silently on by. Visitors critique the enclosures: too small, too large, not enough privacy, too hard to see the animals. They wonder aloud about the pacing. They try to wake the sleeping. They wish they could feed or touch the animals. They mock other humans for how they treat other animals, like a visitor to the National Zoo who went in 1979 to watch other people. "I saw one guy whistling at a snake, trying to get its attention. Can you imagine?" he asked. "That's like trying to teach my dog calculus."[32]

Enclosing Animals: Washington, Seattle

Now turn the camera away from the captive animals for a moment and photograph zoo patrons. Some bring "zoo-gear": sunglasses, walking shoes, picnics, and, naturally, cameras. Others come empty-handed. *Homo zoo visitor* groups in families and feeds on purchased concessions. In 1979, the National Zoo, like a naturalist in the bush, studied its visitors. The zoo's study revealed that visitors came for a spectrum of reasons but with firm ideas of what the zoo needed: elephant rides, penguins, a petting zoo, and a monorail. At least one visitor asked for "egg-creams," a chocolaty, carbonated drink—like so many other zoo treats even today— that has very little to do with environmentalism. As visitors were ever more conscious of the naked cage at home and extinctions abroad, they not only articulated strong ideas about what captivity should be but also believed firmly they could tell when animals were happy, neurotic, natural, bored, or sad. More than ever before, zoos in the 1970s asked, "Where do we go from here?" and "How do we get there?" The future of zoos has been debated with similar fervor ever since.[33]

Some sightseers went to zoos, proud of zoos' conservationist advocacy. "I know the Zoo has been trying to breed some of the endangered animals," explained one visitor. Visitors wanted to train their kids to look at animals in new ways. Endangered animals were more than cuddly toys. They faced extinction, and visitors wanted to see them before they disappeared. Breeding programs had their own attraction. Rare births, like a new giraffe, could still draw crowds. "I wanted to say I saw it," explained a visitor about a new baby, "like people say they saw King Tut." Still, many came to the zoo simply for recreation.[34]

When animals were active, crowds lingered, but sleeping animals were tiresome. "When there is no movement there is no fun." Visitors also watched the keepers as they interacted with their charges. Visitors were eager to witness moments of cross-species tenderness such as when keepers called their animals by their names.[35] They wanted to see keepers care *for* their animals because they, too, had learned to care *about* the rare treasures they saw on display. As a result, visitors thought, and continue to think, about what made animals happy. But their evaluations were "affective," according to one visitor, feelings rather than biological knowledge. They knew very little about what the habitat of animals' ancestors would have been. They didn't know if mambas lived in jungles or deserts but cared if snakes could "stretch out" behind the glass. Visitors might not recall whether gorillas lived in warm jungles or cool mountaintops, but they still hated the enclosures' concrete floors because they were too hard "on the gorillas [sic] feet." Like those visitors decades before complaining about Gunda's imprisonment, they imagined that the chains on the elephants' legs must irritate their skin.[36]

A century of change and a decade condemning the naked cage had their effects. Zoo visitors came to hate looking at animals through metal bars. They abhorred concrete, empty cages, and plastic objects. "The animals shouldn't be in bare cages," said one tourist. "They should have things to play with." Just like patrons a century ago who had all kinds of suggestions for William T. Hornaday and the Bronx Zoo at the end of

Gunda's life, visitors in 1979 had a touch of the exhibit designer and architect in them. Some thought animals wanted more open space "because it is a more natural setting." Others would have planted trees and designed private nooks for the animals.[37]

In 1979, zoo visitors didn't hate captivity, but they wanted it to be comfortable. Visitors were convinced they could discern just by looking how animals felt about the enclosed life. Maybe it was simply by "watching his expressions," but zoo patrons tried to figure out when animals were wild or neurotic. To one visitor the monkeys looked like "nervous wrecks." Perhaps they needed padded cages.

Children (such as me) who began going to zoos in the 1970s have grown up. Today, we still care if animals are neurotic, an animal disease our parents first diagnosed in the 1960s. Just listen to visitors who trace the tiger's pacing to the inadequacy of its enclosure.[38] Zoos have taught us to experience animals with our eyes. A century ago, visitors were encouraged to examine the animals as specimens of species, but visitors have never truly behaved. Our parents and grandparents imagined those animals as characters from jungle pictures. Peer into the enclosure to imagine yourself in Buck's jungle empire. Over the years, zoos have become about sight more than any other sensory experience. Ventilation attempts to remove animal aromas and shit stench. We no longer feed, and except for the occasional domestic animal, we don't pet. We want animal enclosures to look natural, but we care much less if they don't feel natural to the animals in other sensory ways. We've been taught, above all, to look at the animals; so how can we begin to understand how animals sense the zoo from the other side of the enclosure? Visitors even write letters of complaint if the carnivores' food looks like natural kill, carcasses with bones and viscera, rather than unobtrusive, tidily cut meat chunks.

We have learned to accept the enclosure as long it looks natural. Zoos have responded, tearing down cages or restoring New Deal–era projects. As a child, I was drawn to the new ape house in St. Louis. The fake trees were certainly an improvement over the cold bars and fetid cages they replaced. The chimpanzee that habitually spat on visitors was gone (not before it wet my mother). The apes seemed to be playing, and there were enough visual cues, painted murals, for example, for me to imagine myself encountering a chimpanzee family in full jungle. Of course, I couldn't begin to imagine what the apes might have felt. Could they engage in the same flights of fancy?

But what *is* enclosure? Visitors still have very clear ideas of how they want exhibits designed; we vote with our feet and our entrance fees. The problem for zoo officials is that visitors have contradictory priorities. Should enclosures provide easy access to viewers? Or should they feature private spaces for animals to connect with some simulation of the environment and even with other animals, including those their ancestors would have encountered in the wild?

Today, some zoo directors and designers have encouraged a greater simulation of nature, even if it means favorite animals are harder to see. David Hancocks entered zoo work around the time Marvin Jones patrolled the jungles and Henry Trefflich began sleeping in the back of his store. Hancocks directed Seattle's Woodland Park Zoo and, in Tucson, the Arizona-Sonora Desert Museum. (He has also headed zoos in his native Australia.) From inside the zoo gates, Hancocks has offered some harsh critiques of zoos, their architecture, and the AZA. Just before he entered the zoo world, zoos could still grab many of the animals they needed from the wild. As he moved through the hierarchy of zoo work toward a directorship, he saw the hard labor of husbandry and listened to the chorus of publicity that announced each new birth.[39]

In Seattle, Hancocks helped design a lush enclosure filled with real plants and a managed ecosystem for the gorillas. Hancocks's immersion exhibit was a deliberate critique of zoo design. No more concrete floors. No plastic. No night cages. No winters trapped indoors. Too many zoos, he scolds, keep the wrong animals. Why house large tropical animals in a cold northern zoo? Animals should have access to clear skies, real dirt, and living plants. If humans have an innate affinity to nature, so, too, do captive animals.[40]

Zoo design still has all the artificiality of a movie set, Hancocks argues. He's more right than he probably imagines; after all, zoos have long struggled to match the naturalism of first Buck's animal adventure films and later television. Contemporary exhibits seem to him "Tarzanesque." They look like artificial stage sets designed to help visitors imagine they are peering into the primitive jungle. Of course, replace resin trees with real ones, murals with leafy plants, and expand the enclosure, and it's still an illusion. At the zoo, naturalism is not nature, but visions of what the tropics are meant to look like: jungly, Edenic, flush with game, and utterly devoid of native people.

Management is very much present in the immersion display. It's there in the work of the directors and registrars who move animals from zoo to zoo to meet the demands of husbandry and exhibit design. It's there in the careful selection of animals that can coexist. Sometimes hidden glass or a well-designed waterfall strategically separates predator from prey. In other enclosures, different species occupy the same spaces. We are not meant to see the human hand. Even keepers today are more restrained in touching their charges. The caresses, rides, pokes, slaps, cuddles, and commands are touches of the past.[41]

We have heard promises of naturalism before in the evolution of the zoo. Gone with the old and uncomfortable, the smelly and the cramped and in with the modern, the natural, and the expansive. Bars gave way to moats and monkey islands and African veldts (whose descriptions can often sound a great deal like immersion exhibits). Zoos debate the next stage in their evolution, asking, for example, if the twenty-first-century zoo is something we watch only on hidden cameras. As we have those urgent conversations, it is worth pausing to think about what exactly is an enclosure.

With the benefit of history, I have come to believe zoos are not somehow prisons suggesting that animals are punished. They are enclosures, and *enclosure* is a word of multiple meanings: at once architectural space, habitat, and a long historical process that has profoundly reshaped life on earth from pre-modernity to modernity and from empire to independence. At zoos, enclosure, instead of bars, suggests light and naturalism, but in global history, enclosures evoke something much darker. Enclosure, historians have argued, is a human process in which common land is fenced, passing from common hands into private ownership. Historian Peter Linebaugh describes enclosure as privatization that built "barriers between people and land." Enclosure began centuries ago, turning grazing land into estates and jungles into farms, then plantations. Plantations became the economic engines of colonies. Colonies become new nations, and with independence, zoo directors, like animal traders, realized there were new pressures to return land to the commons. Zoo directors were pessimistic. They envisioned only permanent conflict between peoples who wanted land upon which to grow and hunt food and the animals who sought habitat.

A Pilgrimage to the Zoo: Congo, Chicago

Look beyond the informational plaque, the ornamented gateways, and the verdant plantings, and here's the zoo's history lesson: enclosure has shaped class and imperial relations and, just as much, affected animal life and propelled extinction. Enclosure also builds a barrier between animals and land and, then, places animals into bounded exhibits. Enter the animal trade, turning animals into extractable resources, removed from the commons and transported to enclosures. Enclosure also constructs new barriers between peoples and animals, but these barriers looked different at opposite ends of the globe. In the colonized tropics, barriers turned animals into private property, game for aristocratic sport or traders with permits to buy and export. Later, these animals became valuable attractions drawing tourists with cameras. In the United States, barriers between people and animals became the singular problem of zoo design. In the long history of enclosure, zoos have become our most cherished cultural institutions, even if workers, visitors, and animals have challenged zoo officials' visions of orderly display.

Over the course of the past century, Americans have searched for wildness in all the captive places. Zoos were produced out of a global system of enclosure. It's hard to take "dominion" out of zoo design. One of the profound ironies of the twenty-first century is we've come to accept that animal enclosures are the only place where endangered animals can live. Some animals live in game parks and reserves—their own form of enclosure. They also survive as species in American zoos. The global history of enclosure built empires. It also turned the African gorilla into American gorillas with habitats in the heart of cities or in sprawling suburbs. Over a century and a quarter American zoos have evolved from rows of cages—specimen boxes—into a patchwork habitat of enclosures. In this habitat, animals depend on thousands of keepers for care and food, on registrars and directors for migrations, on veterinarians for health care, on visitors' love and donations for species survival.[42] More than ever before, colorful fur, cuteness in infancy, perceived character in storybooks and film, and, of course, the ability to adapt to captivity have become factors in the struggle for survival.

Hancocks calls the zoo enclosures "a different nature." I call it a "third nature." If "first nature," as the environmental historian William Cronon defines it, is the nature before human encroachment and "second nature" is the nature of the human, enclosed, and exploited environment, then zoos represent third nature.[43] This built, managed, husbanded habitat of human care represents nature not as animals know it or how it really exists but how it should be in a fantasy of tropical lushness. Third nature is history, a product of how we built, lived, and contested empires. It is wildness and wilderness suspended at the moment of their initial enclosure when there were still plenty of animals for the taking. Guns became cameras. Heads and horns became holiday snaps and cross-species family pictures. Safaris begat zoo-faris and ecotourism. New construction, like the Bronx Zoo's Congo exhibit, is directly linked to supporting conservation projects in central Africa. The zoo remains a portal for virtual ecotourism from urban (or suburban) temperate America to a tropical world, now ravaged by its peoples. If the Bronx Zoo connects Americans to Congolese nature, it describes a great human rift, layered over older divides of race, class, and empire, between those who conserve and those who kill. Third nature also articulates a vision of what the Third World should be. The lushness of the Bronx Zoo's Congo exhibit heightens condemnations of the Congo as a postcolonial dystopia of civil war, disease, poaching, genocide, unregulated mining, and bushmeat.

As visitors, we've been trained to think a lot about the ethics of the zoo. "Should I let my child go to the zoo on the school trip?" I'm often asked by friends and neighbors. They know their kids will love the zoo and its animals, but they are not sure they want them to. We also celebrate what zoos and their patrons do or can do for animals. It is soothing to think that seeing an animal in the enclosure somehow contributes to its survival or that when we send a donation we have truly become animal parents. Does seeing the sleeping tiger and buying its stuffy doppelganger make our kids and us more aware of conservation? Or does it shape that struggle as a conflict between those encroaching upon habitats and the cats themselves without much attention to those long histories of enclosure that put different peoples and different animals into competition over land?

In a strange way, zoo enclosures are shared space. Zoo visitors, the Brookfield Zoo discovered, spent as much time evaluating the displays as they did examining the animals. They wanted exhibits that were "good for the animal(s)" but just as "good" for visitors.[44] Whether it was the outrage over Gunda's plight or, decades later, Oofy's hug for his human mother, we like to think our zoo animals not only thrive in captivity but also can learn to like us. Or, even, to be like us. Perhaps this is why Binti-jua became *Newsweek* magazine's 1996 hero of the year and one of the "world's most fascinating people"—even if she was a gorilla living at the Brookfield Zoo.

Binti-jua was a typical American gorilla, the offspring of two zoo gorillas, one parent from the Bronx and another from across the country in San Francisco. She has done her part for species survival, giving birth to two babies, including one with Ramar, a gorilla who came to the zoo after a circus career. The jungle was a long way in her genetic past when in August 1996, a three-year-old boy (just about Julie Ann Vogt's age) tumbled into the gorilla enclosure. Julie Ann's case revolved around whether zoo animals were, or even should be, wild beasts. This little boy's case revived the idea of the family of animals. The gorilla proved even more humane than humans.[45]

As the crowd gasped and screamed and the keepers came running, Binti-jua reached over and picked up the unconscious boy and stood her ground against the other gorillas who wanted to inspect the toddler who entered the enclosure. That night, the nation watched a visitor's film of the accident. The gorilla seemed to rock the young boy gently in her arms before she handed him over to her keepers. As the young human recovered anonymously in the hospital, the gorilla lounged with her own gorilla baby in front of camera crews. The zoo's phones rang and its mailbox filled. A few people worried about the safety of the exhibit, but the zoo resisted any changes. "We don't want to put an 8-foot high, chain-linked fence across." The National Zoo had done just that after Vogt's death, and it made the zoo seem like a slum.

Most people, though, were enthralled. Some cried. Others made a "pilgrimage to the zoo just to see her." A local grocer offered twenty-five pounds of bananas. Dozens called the zoo offering to adopt the gorilla. The zoo reminded the public that it had an "adoption program" with donations starting at $25.

Binti-jua has "been a very good mother," said Craig Demitros, her keeper. What made her so maternal? Demitros and the zoo had one answer: she was an American gorilla. Her name in Swahili meant "daughter of sunlight," but, like so many other zoo animals, her real mother was neither ape nor sunlight but human. After her 1988 birth, her natural mother rejected her and she was removed to human care. Humans cuddled and cradled her and fed her bottles. She was lucky enough to be reintroduced successfully to her captive gorilla family, but she knew very little about how to nurture. Zookeepers taught her to mother by giving her a stuffed animal, just like the ones donors could receive if they adopted an animal. Keepers trained Binti-jua in the fundamentals of motherhood. She learned to carry, cradle, and breast-feed her stuffed animal. Crucially, they also taught her to retrieve her baby, just what she did when the boy fell into the enclosure, the zoo said. They "taught her to be a mom."[46]

Other observers were not so sure. Maybe this was an important insight into the very animals kept in captivity and described in story and film for a century as fearsome apes. Gorillas were naturally humane and maybe we could learn a lesson, wrote admirers. Some primatologists agreed. "Compassion is a natural tendency in animals," wrote Frans de Waal, a primatologist from the Yerkes Primate Center. Animals can care for each other not simply within their families but also across species. When the boy fell into the enclosure, he crossed the species line, and for many, the drama that followed proved the zoo worked. Endangered or not, Binti-jua showed "primal compassion."[47]

Here at last was the yeti, across town and several decades removed from where Perkins began his expedition. Anthropoid, certainly, Binti-jua was trapped between the human and ape. Unlike the stories of the Himalayan snowman that proved to be frauds, this gorilla, raised in American arms, cuddled, cradled, and cared. Off in the last unknown wilderness, the snowman was rumored to plunder and rape, but here, in Chicago, in the heart of civilization, living in the most modern enclosure, the gorilla proved humane.

NOTES

Abbreviations

AAZPA-04-191 American Zoo and Aquarium Association Records, 1961–1995, Accession 04-191, Smithsonian Institution Archives

AAZPA-96-024 American Zoo and Aquarium Association Records, 1963–1992, Accession 96-024, Smithsonian Institution Archives

AZA American Zoo and Aquarium Association Records, 1961–1995, Accession 04-191, Smithsonian Institution Archives

BJB Bernard and Jewel Bellush Papers, WAG 030, Tamiment Library / Robert F. Wagner Labor Archives, New York University

DC 37 American Federation of State, County, and Municipal Employees (AFSCME), District Council 37 Records, WAG 265, Tamiment Library / Robert F. Wagner Labor Archives, New York University

MG Office of the President (Madison Grant, W. Redmond Cross) Records. Wildlife Conservation Society Archives

MJ Marvin Jones Papers (unprocessed), Milwaukee Zoo Archives

MP Marlin Perkins Papers (sl 516), Western Historical Manuscript Collection, University of Missouri–St. Louis

NPD National Zoological Park, Office of the Director, Records, circa 1946–1973, Accession 96-139, Smithsonian Institution Archives

NYWF New York World's Fair 1939 and 1940 Incorporated Records 1935–1945, MssCol 2233, Manuscripts and Archives Division, New York Public Library

NYZS New York Zoological Society Archives, Wildlife Conservation Society Archives

NZA   National Zoological Park, Office of the Director, Animal Information Files, 1968–1972, Accession 03-020, Smithsonian Institution Archives

NZOP   National Zoological Park, Office of Public Affairs, Records, 1899–1988 and undated, with material from 1805, Smithsonian Archives, Record Unit 365

NZP   National Zoological Park Records, 1877–1966, Record Unit 74, Smithsonian Institution Archives

NZPA   National Zoological Park, Office of Public Affairs, Records, 1899–1988

OD   Office of the Director (W. T. Hornaday, W. R. Blair), 1899–1939. Wildlife Conservation Society Archives

OP   Office of the President (Fairfield Osborn) Records, Wildlife Conservation Society

PZ   Philadelphia Zoo Records, Philadelphia Zoo Library and Archives

SG   Richard D. Sparks Gorilla Papers, University of Arizona Special Collections

STZ   St. Louis Zoological Park Records, 1910–1941, Western Historical Manuscripts, University of Missouri-St. Louis, Record Group 194

WLM   William M. Mann and Lucile Quarry Mann Papers, circa 1885–1981, Record Unit 7293, Smithsonian Institution Archives

WPA   National Archives, R669 Records of the WPA, Division of Information Service (Primary), File 1936–1942 333a, Entry 678

WPALC   Collection: W.P.A.—Special Projects, Library of Congress

WPC   Records of the WPA, Central Files—State, 1935–1944

WPD   National Archives, R669 Records of the WPA, Division of Information Service Division of Information Service, 333b

WPN   National Archives, R669 Records of the WPA, Division of Information – Newspapers Clippings File, 1935–1942—Project Zoo, Entry 150

WPS   National Archives, WPA History and Archives, State Reports of Research, Publications, 1935–1943, HM 1999

WTH   William Temple Hornaday Papers, Manuscript Division, Library of Congress

ZSM   Zoological Society of Milwaukee County: Records, 1910–2000, University of Wisconsin-Milwaukee Libraries Archives Department

Introduction

1. "Trapping Wild Animals," *Los Angeles Times* (November 18, 1906): VII7.

2. On attendance to accredited American zoos (as well as the process of accreditation), see the data collected by the Association of Zoos and Aquariums (AZA): https://www.aza .org/about-aza/ as well as the more detailed 2009 report https://www.aza.org/uploaded Files/About_Us/aza-economic-impacts-zoos-aquariums-final-2009.pdf (both accessed October 14, 2014).

3. Bob Mullan and Gary Marvin, *Zoo Culture: The Book about Watching People Watch Animals* (Urbana, IL: University of Illinois Press, 1998).

4. http://www.stlzoo.org/visit/thingstoseeanddo/riversedge/ (accessed December 9, 2015).

5. http://www.worldwildlife.org/species/tiger and http://www.worldwildlife.org/publications /tiger-conservation-landscape-data-and-report (accessed October 14, 2014).

6. Nicole Shukin, *Animal Capital: Rendering Life in Biopolitical Times* (Minneapolis: University of Minnesota Press, 2009); Susan McHugh, *Animal Stories: Narrating across Species Lines* (Minneapolis: University of Minnesota Press, 2011); Stephanie Rutherford, *Governing the Wild: Ecotours of Power* (Minneapolis: University of Minnesota Press, 2011); Akira Mizuta Lippit, *Electric Animal: Toward a Rhetoric of Wildlife* (Minneapolis: University of Minnesota Press, 2008); Matthew Brower, *Developing Animals: Wildlife and Early American Photography* (Minneapolis: University of Minnesota Press, 2010).

7. Peter Ryhiner (as told to Daniel P. Mannix), *The Wildest Game* (Philadelphia: J.B. Lippincott, 1958), 16.

8. "Raja at 20: Still a Special Place in Our Hearts," *St. Louis Post-Dispatch* (December 27, 2012), http:// www.stltoday.com / entertainment / raja-at-still-a-special-place-in-our-hearts /article_9259c4a6-5111-5451-b343-8798dda5c785.html (accessed October 14, 2014).

9. Matthew Calarco, *Zoographies: The Question of the Animal from Heidegger to Derrida* (New York: Columbia University Press, 2008); Cary Wolfe, ed., *Zoontologies: The Question of the Animal* (Minneapolis: University of Minnesota Press, 2003); Erica Fudge, *Animal* (London: Reaktion, 2002); Harriet Ritvo, *The Animal Estate: The English and Other Creatures in the Victorian Age* (Cambridge, MA: Harvard University Press, 1987).

10. Heidi Dahles, "Game Killing and Killing Games: An Anthropologist Looking at Hunting in a Modern Society," *Society and Animals* 1, no. 2 (1993): 169–189; John M. MacKenzie, *The Empire of Nature: Hunting, Conservation, and British Imperialism* (Manchester: Manchester University Press, 1988).

11. Eric Ames, *Carl Hagenbeck's Empire of Entertainments* (Seattle: University of Washington Press, 2009); Vicki Croke, *The Modern Ark: The Story of Zoos, Past, Present, and Future* (New York: Scribner, 1997); Susan G. Davis, *Spectacular Nature: Corporate Culture and the Sea World Experience* (Berkeley: University of California Press, 1997); Elizabeth Hanson, *Animal Attractions: Nature on Display in American Zoos* (Princeton: Princeton University Press, 2002); R. J. Hoage and William A. Deiss, eds., *New Worlds, New Animals: From Menagerie to Zoological Park in the Nineteenth Century* (Baltimore: Johns Hopkins University Press, 1996); Nigel Rothfels, *Savages and Beasts: The Birth of the Modern Zoo* (Baltimore: Johns Hopkins University Press, 2002).

12. Eric Baratay and Elisabeth Hardouin-Fugier, *Zoo: A History of Zoological Gardens in the West* (London: Reaktion, 2002), 71–264; Vernon N. Kisling Jr., ed., *Zoo and Aquarium History: Ancient Animal Collections to Zoological Gardens* (Boca Raton, FL: CRC, 2001), 147–180.

13. Ralph R. Acampora, ed., *Metamorphoses of the Zoo: Animal Encounter after Noah* (Lanham, MD: Lexington Books, 2010).

14. Harriet Ritvo, *The Platypus and the Mermaid, and Other Figments of the Classifying Imagination* (Cambridge, MA: Harvard University Press, 1998).

15. Andrea Friederici Ross, *Let the Lions Roar! The Evolution of the Brookfield Zoo* (Chicago: Chicago Zoological Society, 1997), 255.

16. Mary Delach Leonard, *Animals Always: 100 Years at the Saint Louis Zoo* (Columbia: University of Missouri Press, 2009), 186.

17. Linda Kroebner, *Zoo Book: The Evolution of Wildlife Conservation Centers* (New York:

Doherty, 1994); David Hancocks, *A Different Nature: The Paradoxical World of Zoos and Their Uncertain Future* (Berkeley: University of California Press, 2001).

18. Clyde Hill, "Baseball . . . Football . . . Zoos," *Zoonooz* (April 1965): 13–14.

1. The Elephant's Skin

1. "R. S. Tompkns to Ned Hollister," (August 2, 1923), NZP, Box 101, Folder 4.

2. "Souvenir: The Elephant Show for the Benefit Washington Park's Zoo," (February 12–13, 1906) and *Bulletin of the Washington Park Zoological Society of Milwaukee* IV (May 1933): 20 in ZSM, Box 21, Folders 6 and 11; Elizabeth Hanson, *Animal Attractions: Nature on Display in American Zoos* (Princeton: Princeton University Press, 2002), 44–45.

3. Eric Baratay and Elisabeth Hardouin-Fugier, *Zoo: A History of Zoological Gardens in the West* (London: Reaktion, 2002); Clark DeLeon, *America's First Zoostory: 125 Years at the Philadelphia Zoo* (Virginia Beach, VA: Donning, 1999); Mary Delach Leonard, *Animals Always: 100 Years at the Saint Louis Zoo* (Columbia: University of Missouri Press, 2009); David Ehrlinger, *The Cincinnati Zoo and Botanical Garden, From Past to Present* (Cincinnati: Cincinnati Zoo and Botanical Garden, 1993); William Bridges, *Gathering of Animals: An Unconventional History of the New York Zoological Society* (New York: Harper and Row, 1974).

4. *Seventh Annual Report of the New York Zoological Society* (April 1903), 46.

5. Susan Nance, *Entertaining Elephants: Animal Agency and the Business of the American Circus* (Baltimore: The Johns Hopkins University Press, 2013); Eric Scigliano, *Love, War, and Circuses: The Age-Old Relationship between Elephants and Humans* (Boston: Houghton Mifflin, 2002); Jennifer L. Mosier, "The Big Attraction: The Circus Elephant and American Culture," *Journal of American Culture* 22 (Summer 1999): 7–18; Dan Wylie, *Elephant* (London: Reaktion Books, 2009); Shana Alexander, *The Astonishing Elephant* (New York: Random House, 2000); Nigel Rothfels, "Why Look at Elephants?" *Worldviews: Global Religions, Culture, and Ecology* 9, no. 2 (2005): 166–183.

6. "Philadelphia 'Zoo' Elephant Dead," *New York Times* (January 19, 1898): 3.

7. "Wallet Made from Elephant Hide—Jennie—at the Philadelphia Zoo for 23 Years," (ca. 1898), PZ, Box ACAA1; Williams Biddle Cadwalader, "Bears, Owls, Tiger and Others! Philadelphia's Zoo, 1874–1949," PZ, Box PUB A3; A. E. Brown, "The First Zoological Gardens in America and Their History," *Philadelphia Public Ledger* (September 4, 1903), PZ, Box ZH A1.

8. Samuel J. M. M. Alberti, ed., *The Afterlives of Animals: A Museum Menagerie* (Charlottesville: University of Virginia Press. 2011); Rachel Poliquin, *The Breathless Zoo: Taxidermy and the Cultures of Longing* (University Park: Pennsylvania State University Press, 2012); David Maddon, *The Authentic Animal* (New York: St. Martin's Press, 2011); Stephen T. Asma, *Stuffed Animals and Pickled Heads: The Culture and Evolution of Natural History Museums* (New York: Oxford University Press, 2001); Clifton D. Bryant and Donald J. Shoemaker, "Dead Zoo Chic: Some Conceptual Notes on Taxidermy in American Social Life," *Free Inquiry in Creative Sociology* 16, no. 2 (1988): 195–202.

9. William T. Hornaday, "The Elephant in Jungle, Zoo, and Circus," *The Mentor* 12 (June 1924), author's collection.

10. Amy Louise Wood, "'Killing the Elephant': Murderous Beasts and the Thrill of Retribution, 1885–1930," *Journal of the Gilded Age and Progressive Era* 11 (July 2012): 405–444.

11. Ewin Hulfish, *Illustrated Guide and Hand-Book of the Zoological Garden of Philadelphia* (Philadelphia, 1875), PZ, Box PUB A4, 64–65.

12. "Children's Elephant at Forest Park with Mayor and Head of School Board in the Howdah," *St. Louis Post Dispatch* (ca. 1910), STZ, Series II, Scrapbook, Vol. 1.

13. "Roars of Zoo Beasts Awaken Schoolgirls," (n.d., ca. 1913), STZ, Series II, Scrapbook, Vol. 4. For a typical letter of complaint about night noises from the zoo, see: "Cecil French to Frank Baker," (November 8, 1908), NZP, Box 102, Folder 1.

14. "The Circus," *Arkansas Gazette* (January 8, 1878): 2; "Saw the Elephant," *Cleveland Plain Dealer* (April 29, 1895): 1. On circus histories and menageries, see Bluford Adams, "'A Stupendous Mirror of Departed Empires': The Barnum Hippodromes and Circuses, 1874–1891," *American Literary History* 8 (Spring 1996): 34–56; George L. Chindahl, *A History of the Circus in America* (Caldwell, ID: Caxton, 1959); Janet M. Davis, *The Circus Age: Culture and Society under the American Big Top* (Chapel Hill: University of North Carolina Press, 2002).

15. *Bulletin of the Washington Park Zoological Society of Milwaukee* 4 (May 1933): 5.

16. "Story of the Philadelphia Zoo," *Harper's New Monthly Magazine* (April 18, 1879), PZ, Box ZH A1; Vernon N. Kisling Jr., "The Origin and Development of American Zoological Parks to 1899," in R. J. Hoage and William A. Deiss, eds., *New Worlds, New Animals: From Menagerie to Zoological Park in the Nineteenth Century* (Baltimore: Johns Hopkins University Press, 1996), 109–125; DeLeon, *America's First Zoostory,* 33–48; Nigel Rothfels, *Savages and Beasts: The Birth of the Modern Zoo* (Baltimore: Johns Hopkins University Press, 2002).

17. William Temple Hornaday, "The London Zoological Society and Its Gardens—An Object Lesson for New York," *Second Annual Report of the New York Zoological Society* (March 15, 1898), 43; Harriet Ritvo, *The Animal Estate: The English and Other Creatures in the Victorian Age* (Cambridge: Harvard University Press, 1987), 205–242.

18. "F. A. Crandall Jr. to Frank Baker," (October 5, 1903), NZP, Box 100, Folder 6.

19. "Public Fund Is Advocated for an Extension of Zoo," *St. Louis Post-Dispatch* (August 28, 1910); "Many Cities See Value in Good Zoo," (ca. 1910); "De Vry Says City Lacks Only Zoo," (ca. 1911), all in STZ, Series II, Scrapbook, Vol. 1; Leonard, *Animals Always,* 5–36.

20. "Assistant Superintendent, Franklin Park Zoo, Boston to S. S. Flower," (April 16, 1912), NZP, Box 100, Folder 10.

21. "M. O. Stone to Frank Baker," (October 7, 1903) and "Arthur B. Brayton to Frank Baker," (March 26, 1902), NZP Records, Box 100, Folder 6. For a survey of American animal collections by 1904, see "Statistics of Zoological Collections," (ca. 1903), NZP, Box 100, Folder 6; Hanson, *Animal Attractions,* 11–17.

22. "L. M. DeLaussure to Frank Baker," (January 18, 1910), NZP, Box 100, Folder 8.

23. "J. T. Bethel to Ned Hollister," (October 1 and October 4, 1923), NZP, Box 101, Folder 4.

24. "Edward Bailey to Ned Hollister," (November 2, 1923), NZP, Box 101, Folder 4.

25. Hulfish, *Illustrated Guide and Hand-Book of the Zoological Garden of Philadelphia,* 4.

26. Ibid.; Harriet Ritvo, *The Platypus and the Mermaid, and Other Figments of the Classifying Imagination* (Cambridge: Harvard University Press, 1998), 1–50; Patrick H. Wirtz,

"Zoo City: Bourgeois Values and Scientific Culture in the Industrial Landscape," *Journal of Urban Design* 2 (February 1997): 61–82; Carla Yanni, *Nature's Museum's: Victorian Science and the Architecture of Display* (Baltimore: Johns Hopkins University Press, 1999).

27. *Eighth Annual Report of the Zoological Society. Cincinnati. For the Year 1881* (1882), 5–7; Ehrlinger, *The Cincinnati Zoo*, 3–56.

28. *Twenty-Second Annual Report of the Board of Directors of the Zoological Society of Philadelphia* (1894), 2–8.

29. Gregory J. Dehler, *The Most Defiant Devil: William Temple Hornaday and His Controversial Crusade to Save American Wildlife* (Charlottesville: University of Virginia Press, 2013); Bridges, *A Gathering of Animals*, 20–31.

30. "Report of the Director of the Zoological Park," *Third Annual Report of the New York Zoological Society* (May 1899), 39–41.

31. *Second Annual Report of the New York Zoological Society*, 32; Baratay and Hardouin-Fugier, *Zoo*, 113–130; Ritvo, *The Animal Estate*, 205–242.

32. Hornaday, "The London Zoological Society and Its Gardens," 67. See also Kurt Koenigsberger, *The Novel and the Menagerie: Totality, Englishness, and Empire* (Columbus: Ohio State University Press, 2007).

33. "Report of the Director of the Zoological Park," *Third Annual Report of the New York Zoological Society*, 49; Helen L. Horowitz, "Animal and Man in the New York Zoological Park," *New York History* 56 (October 1975): 426–455; William T. Hornaday, "The New York Zoological Park," *The Century* 61 (November 1900), 89–90.

34. "Report of the Executive Committee," *Fourth Annual Report of the New York Zoological Society* (May 1900), 27. On class, social order, and public institutions like museums, see, for example, Sven Beckert, *The Monied Metropolis: New York City and the Consolidation of the American Bourgeoisie* (Cambridge: Cambridge University Press, 2003), 172–236; Julia B. Rosenbaum, "Ordering the Social Sphere: Public Art and Boston's Bourgeoisie," in Sven Beckert and Julia B. Rosenbaum, eds., *The American Bourgeoisie: Distinction and Identity in the Nineteenth Century* (New York: Palgrave Macmillan, 2010), 193–208; Tony Bennett, *Pasts beyond Memory: Evolution, Museums, Colonialism* (London: Routledge, 2004).

35. "The Philadelphia Zoo," *Harper's New Monthly Magazine*, 719–720.

36. *Eighth Annual Report of the New York Zoological Society* (April 1904), 50.

37. "Publications," *Fifth Annual Report of the New York Zoological Society* (May 1901), 41.

38. "The Philadelphia Zoo," *Harper's New Monthly Magazine*, 719.

39. Arthur Erwin Brown, *Guide to the Garden of the Zoological Society of Philadelphia, Fourth Edition* (1888), 7, PZ, Box PUB A4.

40. "J. C. Thompson to Arthur E. Brown," (July 15, 1906); "W. W. Stiles to Arthur E. Brown," (February 2, 1909), PZ, Box RES A1.

41. Brown, *Guide to the Garden of the Zoological Society of Philadelphia*, 3; William T. Hornaday, *Official Guidebook to the New York Zoological Park* (New York Zoological Society, 1928; first ed. 1899).

42. "Statement of John Lover," (August 31, 1902), PZ, Box ZSP A1.

43. "The National Zoological Park," (ca. 1894), NZP, Box 8.

44. *Twenty-Fifth Annual Report of the Board of Directors of the Zoological Society of Philadelphia* (1897), 7–9.

45. Brown, *Guide to the Garden of the Zoological Society of Philadelphia*, 31.

46. "Philadelphia's Dead Elephant," *Charlotte Observer* (January 21, 1898): 3. Paola Cavalieri, with Matthew Calarco, J. M. Coetzee, Harlan B. Miller, and Cary Wolfe, *The Death of the Animal: A Dialogue* (New York: Columbia University Press, 2009). John F. Kasson, *Amusing the Million: Coney Island at the Turn of the Century* (New York: Hill and Wag, 1978); Lawrence Levine, *Highbrow/Lowbrow: The Emergence of Cultural Hierarchy in America* (Cambridge: Harvard University Press, 1990).

47. "J. B. Quinlan to Frank Baker," (December 13, 1909) and "Baker to Quinlan," (December 14, 1909), both in NZP, Box 102, Folder 1.

48. "Vicious Vandals Try to Poison Animals," *The St. Louis Zoo* (September 1933): 1, STZ, Folder 62.

49. "To the Employees of the N.Y. Zoological Park," (January 30, 1900), OD, Box 131, Folder: Circulars/Notices to Park Staff, 1900–1920.

50. *Twelfth Annual Report of the New York Zoological Society* (January 1908), 84.

51. *Twentieth Annual Report of the New York Zoological Society* (January 1916), 64.

52. Hornaday, *Official Guidebook*, 94; "Barbary Lion, 'Sultan,'" (Postcard, 1905).

53. *Illustrated Guide and Hand-Book of the Zoological Garden of Philadelphia* (Philadelphia, 1875), 64, in PZ, Box PUB A4; *Bulletin of the Washington Park Zoological Society of Milwaukee* 4 (May, 1933): 10.

54. "Philadelphia's Dead Elephant."

55. Ibid.

56. *Thirty-First Annual Report of the Board of Directors of the Zoological Society of Philadelphia* (1903), 14.

57. "Philadelphia's Dead Elephant."

58. "Jennie Is Dead," *Philadelphia Inquirer* (January 19, 1898): 13.

59. *Ninth Annual Report of the New York Zoological Society* (January 1905), 61, 69–71; *Twelfth Annual Report of the New York Zoological Society*, 78. Debra L. Forthram, Lisa F. Kane, David Hancocks, and Paul F. Waldau, eds., *An Elephant in the Room: The Science and Well-Being of Elephants in Captivity* (North Graftan, MA: Tufts Center for Animals and Public Policy, 2009).

60. William Temple Hornaday, *Two Years in the Jungle: The Experiences of a Hunter and Naturalist in India, Ceylon, the Malay Peninsula and Borneo* (New York: Charles Scribner's Sons, 1885), 161–166, 200–214; Stefan Bachtel, *Mr. Hornaday's War: How a Peculiar Victorian Zookeeper Waged a Lonely Crusade for Wildlife That Changed the World* (Boston: Beacon, 2012).

61. Hornaday, *Two Years in the Jungle*, 214–217.

62. "Frank Baker to H. C. Moore," (August 15, 1907), NZP, Box 76, Folder 6.

63. "Gifts to the American Museum of Natural History," *Seventh Annual Report of the New York Zoological Society*, 48; William S. Walker, *A Living Exhibition: The Smithsonian and the Transformation of the Universal Museum* (Amherst: University of Massachusetts Press, 2013), 11–43.

64. William T. Hornaday, *A Wild-Animal Round-Up: Stories and Pictures from the Passing Show* (New York: Charles Scribner's Sons, 1925), 300, 305.

65. "Animal Collections," *Seventh Annual Report of the New York Zoological Society*, 59.

66. Andrew C. Isenberg, *The Destruction of the Bison: An Environmental History, 1750–1920*

(Cambridge: Cambridge University Press. 2000); William Beinart and Peter Coates, *Environment and History: The Taming of Nature in the USA and South Africa* (New York: Routledge, 1995), 20–33.

67. Madison Grant, "History of the Zoological Society," *Zoological Society Bulletin* 37 (January 1910): 590–91; Jonathan P. Spiro, *Defending Master Race, Conservation, Eugenics, and the Legacy of Madison Grant* (Burlington: University of Vermont Press, 2009); Matthew Pratt Guterl, *The Color of Race in America, 1900–1940* (Cambridge: Harvard University Press, 2001), 100–153.

68. Daniel Justin Herman, "From Farmers to Hunters: Cultural Evolution in the Nineteenth-Century United States," in Kathleen Kete, ed., *A Cultural History of Animals in the Age of Empire* (Oxford: Berg, 2007), 47–72. It was, in fact, Roosevelt who wrote down some rules for the "fair chase" in Roosevelt, *The Wilderness Hunter: An Account of the Big Game of the United States and Its Chase with Horse, Hound, and Rifle* (New York: G.P. Putnam's, 1893), esp. 27–29.

69. Madison Grant, "Article for Natural History" (n.d.), MG, Box 1, Folder 9; John M. MacKenzie, *The Empire of Nature: Hunting, Conservation, and British Imperialism* (Manchester: Manchester University Press, 1988); Madison Grant, *Passing of the Great Race, or the Racial Basis of European History* (New York: Charles Scribner's Sons, 1916).

70. William T. Hornaday, "Our Last Buffalo Hunt, Part I and II," in Hornaday, *A Wild Animal Round Up*, 5–53.

71. *Tenth Annual Report of the New York Zoological Society* (January 1906): 42, 181–200.

72. Ibid., 41.

73. *First Annual Report of the New York Zoological Society* (March 1897), 19–21.

74. *The National Collection of Heads and Horns, Part I* (New York Zoological Park, May 1907); *Eleventh Annual Report of the New York Zoological Society* (January 1907), 41–42; *Twelfth Annual Report of the New York Zoological Society*, 35.

75. "The Philadelphia Zoo," *Harper's New Monthly Magazine*.

76. "Selection of Animals," *Fourth Annual Report of the New York Zoological Society*, 50.

77. "Frank Baker to I. A. Martin," (October 7, 1910), NZP, Box 20.

78. "Crandall to Baker"; Helen L. Horowitz, "The National Zoological Park: 'City of Refuge' or Zoo?" *Records of the Columbia Historical Society, Washington, D.C.* 49 (1973/1974): 405–429.

79. "The National Zoological Park," (ca. 1894); "J. T. McCaddon to Frank Baker," (July 5, 1895), NZP, Box 38, Folder 64.

80. "Animal Collections," 60.

81. "60 Years of Chains Still before Gunda," *New York Times* (June 28, 1914): 13.

82. Ibid.

83. "Selection of Animals," *Fourth Annual Report of the New York Zoological Society*, 50.

84. "Report of the Director of the New York Zoological Park," *Fifth Annual Report of the New York Zoological Society*, 43.

85. Raymond L. Ditmars, "Department of Reptiles," *Fifth Annual Report of the New York Zoological Society*, 52.

86. "Open to the Public," *The Press* (July 1, 1874), Philadelphia Zoo Records, Box ZH A1.

87. "When the Animals Escape from the Zoo," *The Ladies' Home Journal* 18 (May 1901): 3.

88. "Gunda Tries to Kill Keeper: Bronx Zoo Elephant Barely Misses Crushing Walter Thurman," *New York Times* (July 29, 1909): 4; "Gunda, the Good Elephant," *Washington Post* (October 15, 1905): JJ6. James Turner, *Reckoning with the Beast: Animals, Pain, and Humanity in the Victorian Mind* (Baltimore: The Johns Hopkins University Press, 1980).

89. *Nineteenth Annual Report of the New York Zoological Society* (January 1915): 40; "Gunda Breaks Tusk in Fight to Be Free," *New York Times* (July 14, 1914): 1.

90. "Hornaday Defends Chaining of Gunda," *New York Times* (January 24, 1914): 22; "The Electrocution of Topsy the Elephant" (Edison Film Company, 1903).

91. "Gunda Breaks Tusk in Fight to Be Free."

92. "Gunda Drops a Chain and Swats a Poem," *New York Times* (July 12, 1914): 3.

93. "Times Readers Protest against Gunda's Imprisonment," *New York Times* (July 19, 1914): SM6. Jason Hribal, *Fear of the Animal Planet: The Hidden History of Animal Resistance* (Oakland: AK Press, 2011); Susan J. Pearson, *The Rights of the Defenseless: Protecting Animals and Children in Gilded Age America* (Chicago: University of Chicago Press, 2011), 17–97.

94. "Hornaday Defends Chaining of Gunda," 22.

95. "Hornaday on Gunda: Call Pleases for Chained Elephant Attacks on the Zoological Park," *New York Times* (June 27, 1914): 6.

96. "Gunda as an Agent of 'Uplift'," *New York Times* (June 28, 1914): 14; "Vetoes $1,000 Plan for Relief of Gunda," *New York Times* (July 1, 1914): 5.

97. W. R. Hotchkin, "Gunda's Jungle: How a Home Might Be Constructed for Him in the Park," *New York Times* (June 28, 1914): 14.

2. The Voyage of the *Silverash*

1. William M. Mann, *Ant Hill Odyssey* (Boston: Little, Brown, 1948); William M. Mann, *Wild Animals in and out of the Zoo* (Washington, DC: Smithsonian, 1930); Lucile Q. Mann, *From Jungle to Zoo: Adventures of a Naturalist's Wife* (New York: Dodd, Mead, 1934).

2. Lucile Mann, "Diary of East Indies, National Geographic-Smithsonian Expedition," 1937), WLM, Box 7, Folder 1, 118–127; Lucile Mann, "Ark from Asia" (1937), WLM, Box 8, Folder 3, 189–280. On William and Lucile Mann's different expeditions, see: Elizabeth Hanson, *Animal Attractions: Nature on Display in American Zoos* (Princeton: Princeton University Press, 2002): 100–129.

3. For the production of commodities and the rise of global empires, see, for example, Mona Domosh, *American Commodities in an Age of Empire* (New York: Routledge, 2006); Jonathan Curry-Machado, ed., *Global Histories, Imperial Commodities, Local Interactions* (New York: Palgrave Macmillan, 2013); Arjun Appadurai, ed., *The Social Life of Things: Commodities in Cultural Perspective* (Cambridge: Cambridge University Press, 1986). On the ways the economic value of animals is determined, see Jennifer Price, *Flight Maps: Adventures with Nature in Modern America* (New York: BasicBooks, 2000); Susan D. Jones, *Valuing Animals: Veterinarians and Their Patients in Modern America* (Baltimore: Johns Hopkins University Press, 2003); Susan R. Schrepfer and Philip Scranton, eds., *Industrializing Organisms: Introducing Evolutionary History*

(New York: Routledge, 2004), 167–260; Clay McShane and Joel Tarr, *The Horse in the City: Living Machines in the Nineteenth Century* (Baltimore: Johns Hopkins University Press, 2007), 84–126; Nicole Shukin, *Animal Capital: Rendering Life in Biopolitical Times* (Minneapolis: University of Minnesota Press, 2009).

4. The animal trade was not simply the buying the selling of animals but part of the confrontation of species. See Cary Wolfe, *Animal Rites: American Culture, the Discourse of Species, and Posthumanist Theory* (Chicago: University of Chicago Press, 2003), 1–94; Donna J. Haraway, *When Species Meet* (Minneapolis: University of Minnesota Press, 2007). For other examples of moments of the species encounter, see Jon T. Coleman, *Vicious: Wolves and Men in America* (New Haven: Yale University Press, 2004), 37–51; Jennifer Mason, *Urban Animals, Sentimental Culture, and American Literature, 1850–1900* (Baltimore: Johns Hopkins University Press, 2005), 52–94; Akira Mizuta Lippit, *Electric Animal: Toward a Rhetoric of Wildlife* (Minneapolis: University of Minnesota Press, 2000), 101–135; Harriet Ritvo, *The Animal Estate* (Cambridge: Harvard University Press, 1987), 205–242. Steve Baker, *Picturing the Beast: Animals, Identity, and Representation* (Urbana: University of Illinois Press, 1993); Randy Malamud, *Poetic Animals and Animal Souls* (New York: Palgrave, 2003), 3–48.

5. Carl Trocki, *Singapore: Wealth, Power, and the Culture of Control* (New York: Routledge, 2005), 7–94; C. M. Turnbull, *A History of Modern Singapore, 1819–2005,* rev. ed. (Singapore: National University of Singapore Press, 2009); Tim Harper and Sunil Amrith, eds., *Sites of Asian Interaction: Ideas, Networks, and Mobility* (Cambridge: Cambridge University Press, 2014).

6. "Trapping for the Zoo," *Washington Post* (September 5, 1894): 5; "Western Trappers Ready," (1910), STZ, Series II, Scrapbook, Vol. 1. The newly formed St. Louis Zoo also looked to western American animals as it tried to build its collection. See also Hanson, *Animal Attractions,* 71–99; Mark V. Barrow Jr., "The Specimen Dealer: Entrepreneurial Natural History in America's Gilded Age," *Journal of the History of Biology* 33 (Winter 2000): 493–534. For a contemporary study of the animal trade, see: Jacqueline L. Schneider, *Sold into Extinction: The Global Trade in Endangered Species* (New York: Praeger, 2012).

7. "Frank Baker to Brigadier General H. C. Corbin," (July 12, 1899), NZP, Box 11; "C. C. Todd to Commanding Officer, USS Wilmington," (June 21, 1899), NZP, Box 40, Folder 109; "Are Asked to Help the Zoo," *Chicago Daily Tribune* (November 26, 1899): 58; "Sailors Asked to Assist in Stocking Local Zoo," *Washington Post* (July 19, 1925): S9. On nature, environment, and the U.S. empire, see Greg Bankoff, "Conservation and Colonialism: Gifford Pinchot and the Birth of Tropical Forestry in the Philippines" and Stuart McCook, "'The World Was My Garden': Tropical Botany and Cosmopolitanism in American Science, 1898–1935," both in Alfred W. McCoy and Francisco A. Scarano, eds., *Colonial Crucible: Empire in the Making of the Modern American State* (Madison: University of Wisconsin Press, 2009), 479–488, 499–507.

8. "D. C. Worster to George P. Merrill," (April 14, 1900); "Richard Rathbun to Quartermaster's Department," (August 4, 1900), both in NZP, Box 88, Folder 17.

9. "A. H. Pinney to Frank Baker," (January 22, 1908); "Baker to Pinney," (January 31, 1908); "Chief of Police and Marshal of the Canal Zone to Secretary of Agriculture," (April 18, 1907); "L. C. Wilson to National Zoological Park," (May 24, 1908); "Surgeon, U.S. Navy,

Colon Hospital to Curator of Zoology, Smithsonian Institute," (April 30, 1910); "Baker to Surgeon," (May 19, 1910), all in NZP, Box 88, Folder 16.

10. "Frank Baker to A. C. Robison," (June 28, 1900); "Robison to Baker," (July 5, 1900), NZP, Box 94, Folder 10.

11. Owen J. Lynch, "The U.S. Constitution and Philippine Colonialism: An Enduring and Unfortunate Legacy," in McCoy and Scarano, eds., *Colonial Crucible*, 353–364; Department of Agriculture, "Special Order Prohibiting the Landing of Animals from the Philippine Islands at Any of the Ports of the United States or of the Dependencies Thereof," (December 13, 1901), NZP, Box 89, Folder 1.

12. "Frank Baker to Edward Dietrich," (July 2, 1904), NZP, Box 15; Amy Kaplan, *The Anarchy of Empire* (Cambridge, MA: Harvard University Press, 2005).

13. "Grocery Stores Produce Many Snakes," *St. Louis Zoo* 2 (November, 1930), STZ, Folder 61.

14. "National Geographic Society—Smithsonian Institution Sumatra Expedition," WLM, Box 5, Folder 5; Mann, "Ark from Asia," 3–32; "H. M. Smith to W. M. Mann" (March 27, 1937) and "G. M. Khan and Sons to W. M. Mann" (1937), WLM, Box 5, Folder 5; William and Lucile Mann, "Around the World for Animals" *National Geographic* 73 (June 1938): 665–714

15. "A. B. Baker to Henry P. Bridges," (October 6, 1916) and "N. Hollister to Henry P. Bridges," (January 3, 1917), both in NZP, Box 24.

16. Charles F. Brownell, "Marketing Wild Animals," *Leslie's Monthly Magazine* 15 (July 1905): 288.

17. Eric Ames, *Carl Hagenbeck's Empire of Entertainments* (Seattle: University of Washington Press, 2008); Nigel Rothfels, *Savages and Beasts: The Birth of the Modern Zoo* (Baltimore: Johns Hopkins University Press, 2002); Karl Hagenbeck, "List of Animals Now for Sale," (January 27, 1898), NZP, Box 36, Folder 41; David Ehrlinger, *The Cincinnati Zoo and Botanical Garden: From Past to Present* (Cincinnati: Cincinnati Zoo and Botanical Garden, 1993), 42–43; Lorenz Hagenbeck, *Animals Are My Life* (London: Bodley Head, 1956), 54.

18. Hagenbeck, *Animals Are My Life,* 102–118; T. R. Ybarra, "'Hagenbeck's' Closes Its Doors," *New York Times* (November 28, 1920), 49. For a typical example of how the Hagenbeck firm sold and bought animals with American zoos, see "Carl Hagenbeck to Arthur Brown," (June 13, 1898) and "Hagenbeck to Brown," (April 28, 1899), PZ, Box ACAA1; "Letter from C. L. Williams," (September 10, 1902), PZ, Box PER AA2. See also Richard W. Flint, "American Showman and European Dealers: The Commerce in Wild Animals in 19th Century America," in R. J. Hoage and William A. Deiss, eds., *New Worlds, New Animals: From Menagerie to Zoological Park in the Nineteenth Century* (Baltimore: Johns Hopkins University Press, 1996), 97–108.

19. J. Olin Howe, "Another Monopoly Lost to Germany—The Tiger Trade" *New York Tribune* (January 26, 1919) 5; "Frank Baker to Richard Rathbun," (July 3, 1916) and "W. T. Hornaday to Arthur B. Baker," (May 12, 1916), both in NZP, Box 105, Folder 1; J. Alden Loring, "Trip to South Africa by Mr. J. Alden Loring," (January 10, 1917), NZP, Box 105, Folder 2; J. Alden Loring, *African Adventure Stories* (New York: Charles Scribner's Sons, 1914); William Bridges, *Gathering of Animals: An Unconventional History of the New York Zoological Society* (New York: Harper and Row, 1974), 60–68.

20. Frank Buck with Ferrin Fraser, *All in a Lifetime* (New York: Robert M. McBride, 1941), 113–114.

21. "Singapore. The Port for Wild Cargoes," *Singapore Free Press and Mercantile Advertiser* (July 30, 1914): 12; Karel Willem Dammerman, *The Agricultural Zoology of the Malay Archipelago* (Amsterdam: J.H. de Bussy, 1929); Greg Bankoff and Peter Boomgaard, eds., *A History of Natural Resources in Asia* (New York: Palgrave Macmillan, 2007); Yeoh Seng Guan, Loh Wei Leng, Khoo Salma Nasution, and Neil Khor, eds., *Penang and Its Region: The Story of an Asian Entrepôt* (Singapore: National University of Singapore Press, 2009).

22. "250 Crocodile Skins for Singapore," *Straits Times* (August 25, 1937): 16; "Vessel Brings Wild Animals," *Los Angeles Times* (November 2, 1922): II1; Frank Buck, "A Jungle Business," *Asia: The American Magazine on the Orient* 22 (August 1922): 633–638.

23. Mann, "Diary of East Indies Trip" (1937), 16–18.

24. K. W. Dammerman, *Preservation of Wild Life and Nature Reserves in the Netherlands Indies* (Fourth Pacific Science Congress, Java, 1929), appendix IV.

25. Buck, *All in a Lifetime,* 101. See also: Mahesh Rangarajan, *India's Wildlife History: An Introduction* (Delhi: Permanent Black, 2005), 46–79; Rangarajan, *Fencing the Forest: Conservation and Ecological Change in India's Central Provinces 1860–1914* (Oxford: Oxford University Press, 1999); "Wild Animals Hotel," *Singapore Free Press and Mercantile Advertiser* (February 2, 1923): 5; "Chinese Enterprise in Singapore," *Singapore Free Press and Mercantile Advertiser* (October 8, 1935): 4.

26. Buck, *All in a Lifetime,* 101.

27. *Report of the Wild Life Commission* (Singapore: Govt. Print. Office, 1932); Theodore R. Hubback, *Wild Animals and Wild Birds Committee* (Singapore: Govt. Print. Office, 1934); Hubback, *Elephant and Seladang Hunting in the Federated Malay States* (London: Rowland Ward, 1905); William George Maxwell, *In Malay Forests* (Edinburgh: Blackwood, 1907); Arthur Locke, *The Tigers of Trengganu* (New York: Scribner, 1954); Fiona L. P. Tan, "The Beastly Business of Regulating the Wildlife Trade in Colonial Singapore," in Timothy P. Barnard, ed., *Nature Contained: Environmental Histories of Singapore* (Singapore: NUS Press, 2014): 145–178.

28. "Wild Life In Malaya," *Straits Times* (December 2, 1932): 14; "Singapore's Bird and Animal Shops May Disappear," *Straits Times* (April 17, 1934): 18; "Disgrace to Singapore," *Straits Times* (January 28, 1939): 12.

29. Buck, *All in a Lifetime,* 75, 231.

30. Mann, "Diary of East Indies Trip" (1937), 55.

31. Mann, "Diary of East Indies Trip" (1937), 16–118.

32. "Wild Animals Hotel," *Singapore Free Press and Mercantile Advertiser* (February 2, 1923): 5.

33. Mann, "Diary of East Indies Trip" (1937), 16–118.

34. Mann, "Diary of East Indies Trip" (1937), 31–65.

35. Hagenbeck, *Beasts and Men,* 70–71.

36. W. M. Mann, "Smithsonian-Chrysler Expedition to Africa to Collect Living Animals, Field Season of 1926," WLM, Box 14, Folder 6.

37. "John Tate to William Mann," (May 27, 1926); "John L. Burns to Mann," (March 3,

1926); "Raymond F. Briggs to Mann," (March 1, 1926); "Charles Day to Mann," (March 1, 1926); "Jason Stuart to Mann," (March 8, 1926), all in NZP, Box 194, Folder 5.

38. "William Mann to Austin Clark," (n.d., ca. July, 1926), WLM, Box 12, Folder 11.

39. "Wild Ape Is Meek at Museum Show," *New York Times* (November 1, 1936): N7.

40. William Mann, "Our recent trip to Africa . . ." (n.d., ca. 1927), WLM, Box 12, Folder 10.

41. Brian Herne, *White Hunters: The Golden Age of Safaris* (New York: Henry Holt, 1999); Edward I. Steinhart, *Black Poachers, White Hunters: A Social History of Hunting in Colonial Kenya* (Athens: Ohio University Press, 2006).

42. "William Mann to A. B. Baker," (June 1, 1926), NZP, Box 195, Folder 1; Hanson, *Animal Attractions,* 109.

43. "William Mann to A. B. Baker," (June 1, 1926).

44. "In a letter written from Umbugwe . . . ," (n.d., ca. 1926), WLM, Box 10, Folder 10.

45. Mann, "Our recent trip to Africa . . ."

46. "William Mann to Austin Clark and Carl Getz," (August 26, 1926), NZP, Box 195, Folder 1.

47. "Names Submitted in Giraffe Naming Contest," (n.d., ca. 1926), NZP, Box 195, Folder 1; Mann, "Our recent trip to Africa . . ."; "Giraffe Is Captured for Washington Zoo," *Washington Post* (August 25, 1926): 11.

48. "Eric Reid, District Officer, to William Mann," (August 19, 1926), NZP, Box 195, Folder 1; "Tanganyika Territory, Foreign Contract of Service," (September 6, 1926), WLM, Box 15, Folder 11.

49. Karl Hagenbeck, *Beasts and Men: Being Carl Hagenbeck's Experiences for Half a Century among Wild Animals,* Hugh S. R. Elliot and A. G. Thacker, trans. (New York: Longman's Green, 1912), 43–90.

50. Hagenbeck, *Beasts and Men,* 43–90.

51. Mann, "Smithsonian-Chrysler Expedition to Africa to Collect Living Animals."

52. Raymond Blathwayt, "Wild Animals. How They Are Captured, Transported, Trained, and Sold," *McClure's Magazine* 1 (June, 1893): 26–32.

53. "John Camp to Frank Baker," (November 2, 1901), NZP, Box 85, Folder 15.

54. Dammerman, *Preservation of Wild Life and Nature Reserves in the Netherlands Indies;* Robert Cribb, "Conservation in Colonial Indonesia," *Interventions* 9 (March, 2007): 49–61; Jane Carruthers, *The Kruger National Park: A Social and Political History* (Pietermaritzburg: University of Natal Press, 1995); Roderick Neumann, "Africa's 'Last Wilderness': Reordering Space for Political and Economic Control in Colonial Tanzania," *Africa* 71, no. 4 (2001): 641–665; Richard Grove, *Green Imperialism: Colonial Expansion, Tropical Island Edens and the Origins of Environmentalism* (Cambridge: Cambridge University Press, 1995).

55. Mark Cioc, *The Game of Conservation: International Treaties to Protect the World's Migratory Animals* (Athens: Ohio University Press, 2009), 40–45; John M. MacKenzie, *The Empire of Nature: Hunting, Conservation and British Imperialism* (Manchester: Manchester University Press, 1988).

56. Steinhart, *Black Poachers, White Hunters,* esp. 42–173.

57. Mann, "Our recent trip to Africa . . ."

58. "William Mann to Austin Clark," (n.d., ca. July, 1926), WLM, Box 12, Folder 10; "Reid to Mann," (August 19, 1926).

59. "Affidavit, Saidi bin Kisanda," (November 11, 1926), NZP, Box 194, Folder 5.

60. "Affidavit, Sultanti Chanzi Machafula," (November 11, 1926); "Affidavit, Jumbe Kondo," (November 11, 1926); "An Extract from a Letter from the District Officer, Morogoro," (November 29, 1926), all in NZP, Box 194, Folder 5.

61. "An Extract from a Letter from the District Officer, Morogoro."

62. Ellen Velvin, *From Jungle to Zoo* (New York: Moffat, Yard, 1915), 125–169.

63. Gerald G. Gross, "Dr. Mann Back with Rare Beasts," *Washington Post* (August 11, 1940), 25.

64. Henry Trefflich and Edward Anthony, *Jungle for Sale* (New York: Hawthorn, 1967), 12–20; Benjamin Burbridge, *Gorilla, Tracking and Capturing the Ape-man of Africa* (New York: Century, 1928), 22.

65. J. L. Buck, "Horace—My Pigmy Hippo," *Asia Magazine* (January 27), STZ, Series II, Scrapbook, Vol. 6; J. L. Buck, "The Most Thrilling Adventure of My Life," *Washington Post* (June 19, 1927): SM2; J. L. Buck, "Dramas of the Jungle," *Los Angeles Times* (August 26, 1928): L16. For almost exactly the same drama of meat versus specimen, see Wynant Davis Hubbard, "A Fight with an Angry Hippo," *Los Angeles Times* (September 26, 1926): K21.

66. "Confidential, Urgent, Eastern Province, Dar-es-salaam," (January 17, 1927), NZP, Box 194, Folder 5; "Prices of Animals Sold to Mr. William Mann," (1926) and "William Mann to Carl Getz," (November 13, 1926), NZP, Box 195, Folder 1.

67. "W. M. Mann to Layang Gaddi," (n.d., ca. 1937), WLM, Box 5, Folder 5; Elizabeth Oldfield, "Zoo Animal Cargo Arrives in Good Shape," *Washington Times* (September 29, 1937), NZPA, clippings.

68. Mann, "Diary of East Indies, National Geographic-Smithsonian Expedition," 127–129.

69. "He's a Modern Noah," *Washington Post* (April 3, 1910): E14.

70. "Rough 'Zoo' Voyage Ends," *New York Times* (July 18, 1949): 33.

71. Buck, *All in a Lifetime,* 190–199.

72. Buck, *All in a Lifetime,* 146–148; Blythwayt, "Wild Animals."

73. Buck, *On Jungle Trails,* 109–128.

74. Mann, "Diary of East Indies Trip" (1937), 143; "Zoo Animal Cargo Arrives in Good Shape;" Gerald G. Gross, "Dr. Mann's 'Ark' Docks Today; 'Keep Fingers Crossed,' He Says," *Washington Post* (September 27, 1927): 13.

75. "Zoo Animal Cargo Arrives in Good Shape."

76. Lucile Mann, "The Passengers Were Wild," (n.d., ca. 1937), WLM, Box 9, Folder 6, 3.

77. Mann, "Diary of East Indies Trip" (1937), 143

78. "Mary E. Slavin to Lucile Mann," (October 5, 1939) and "E. H. Dodd to Lucile Mann," (January 23, 1939), both in WLM, Box 8, Folder 3.

3. Jungleland

1. "Annual Report," (1934), STZ, Folder 18.

2. "Annual Report," (1933), STZ, Folder 17; "Giant Tapir to Join Bronx Zoo Family, Wild Animal Captured in India to Arrive Next Month," *New York Times* (March 25, 1923): E2.

3. "Zoo Pays $8800 for Rhino, Will Show It Friday," *St. Louis Post* (January 19, 1934), STZ, Series II, Scrapbook, Vol. 8; L. C. Rookmaaker et al., *The Rhinoceros in Captivity: A List of 2439 Rhinoceroses Kept from Roman Times to 1994* (The Hague: Kugler, 1998), 40–41.

4. "Zoo Pays $8800 for Rhino."

5. "Annual Report (1934)," STZ, Folder 18.

6. For other studies of the place of the animals in American culture, see Cary Wolfe, *Animal Rites: American Culture, the Discourse of Species, and Posthumanist Theory* (Chicago: University of Chicago Press, 2003), 1–94; Donna J. Haraway, *When Species Meet* (Minneapolis: University of Minnesota Press, 2007); Jon T. Coleman, *Vicious: Wolves and Men in America* (New Haven: Yale University Press, 2004), 37–51; Jennifer Mason, *Urban Animals, Sentimental Culture, and American Literature, 1850–1900* (Baltimore: Johns Hopkins University Press, 2005), 52–94.

7. Lillian G. Genn, "Animals DANGEROUS Only When AROUSED," *Washington Post* (August 28, 1932): SM4; "'Emperor of the Jungle' Tells How," *New York Times* (April 1, 1934), X4.

8. Wynant Davis Hubbard, *Wild Animals: A White Man's Conquest of Jungle Beasts* (New York: D. Appleton, 1926); Hubbard, "I Catch My Wild," *Popular Science Monthly* (April 1927): 46–47, 139.

9. Margaret Carson Hubbard, *No One to Blame: An African Adventure* (New York: Minto, Balch and Company, 1934).

10. Hubbard, *Wild Animals*, 81.

11. Ibid., 75.

12. Charles Mayer, "How I Capture Jungle Animals," *Boy's Life* 15 (January 1925): 7, 37, 44.

13. "List and Prices of Animal Birds Snakes Etc." (December 15, 1898), NZP, Box 37, Folder 61.

14. Charles Mayer and Morgan Stinemetz, *Jungle Beasts I Have Captured* (New York: Doubleday, Page and Company, 1924), 2.

15. Charles Mayer, "Trapping an Elephant Herd in Trengganu," *Boston Daily Globe* (August 7, 1921): SM8.

16. Lawrence Buell, *Writing for an Endangered World: Literature, Culture, and Environment in the U.S. and Beyond* (Cambridge: Belknap, 2003); Graham Huggan and Helen Tiffin, *Postcolonial Ecocriticism: Literature, Animals, Environment* (New York: Routledge, 2010).

17. Charles Mayer, "Recruiting for the Menagerie, Thrilling Fight to Capture a Jungle Snake Thirty-Two Feet Long," *Boston Daily Globe* (July 24, 1921): SM6; Mayer, *Trapping Wild Animals in Malay Jungles* (Garden City: Garden City Publishing Co., 1920), frontispiece.

18. "The Jungle's Vengeance," *Singapore Free Press and Mercantile Advertiser* (September 29, 1924): 6.

19. Mayer and Stinemetz, *Jungle Beasts*, 95.

20. Charles Mayer, "A Tiger, a Rhino, and the Ghost Mountain," *Asia: Journal of the American Asiatic Association* 23 (February 1923): 93–97.

21. Mayer, "A Tiger, a Rhino, and the Ghost Mountain," 93–97.

22. Mayer and Stinemetz, *Jungle Beasts*, 128–140.

23. Ibid., 141–144.

24. Ibid., 144–148.

25. "200 Tigers Deaths a Year, when Terror Stalked in Singapore," *Singapore Press and Mercantile Advertiser* (June 19, 1933): 8.

26. "Amazing Case of Plagiarism Is Exposed Locally," *Straits Times* (October 15, 1933): 1; G. P. Sanderson, *Thirteen Years among the Wild Beasts of India* (London: William Allen, 1878).

27. Hubbard, "I Catch My Wild," 46.

28. "'Emperor of the Jungle' Tells How."

29. John Scott, "Mr. Buck Tells All About Animals in Four Minutes," *Los Angeles Times* (August 7, 1932): B5.

30. Theodore Roosevelt, *On African Game Trails: An Account of the African Wandering of an American Hunter-Naturalist* (New York: Charles Scribner, 1910); Edgar Rice Burroughs, *Tarzan of the Apes* (New York: A.C. McClurg, 1912).

31. "Frank Buck Says . . . A Black Panther Isn't Half as Treacherous as a Blow-Out [Ad for Goodrich Tires]," *Palm Beach Post* (April 23, 1935): 5; "Use Your Frank Buck Watch . . . Jungle Fun? [Ad for Wheaties]," *Los Angeles Times* (March 27, 1949): F14; "It Takes Healthy Nerves for Frank Buck to Bring-Em-Back Alive [Ad for Camel Cigarettes]," *Eugene Register-Guard* (September 11, 1933): 3; Sally Joy Brown, "Kids, You Will See Frank Buck at Next Party," *Chicago Daily Tribune* (August 5, 1934): D2.

32. Frank Buck with Ferrin Fraser, *All in a Lifetime* (New York: Robert M. McBride, 1941), 3. Buck's retreat to the jungle in order to escape the confines of an American class system stands in comparison to Donna Haraway's reading of Carl Akeley's earlier jungle expeditions. Haraway suggests that the photographer and taxidermist Akeley's move from impoverished youth to intimate ties to elites, especially Theodore Roosevelt, was enabled by his vision of "jungle peace" and his expressed identity as scientist and artist. Haraway, *Primate Visions: Gender, Race, and Nature in the World of Modern Science* (New York: Routledge, 1989), 35–46.

33. "Frank Buck to Emerson Brown," (January 30, 1924), PZ, Box PER AA2; "Frank Buck to Bring Back an Exciting Film," *Washington Post* (November 26, 1933: JP2.

34. Mae Tinne, "'Wild Cargo' Gives Inside Dope on Explorers' Derring-Do Deeds," *Chicago Daily Tribune* (April 8, 1934): D1.

35. Cynthia Erb, *Tracking King Kong: A Hollywood Icon in World Culture* (Detroit: Wayne State, 1998); David Stymeist, "Myth and the Monster Cinema," *Anthropologica* 51, no. 2 (2009): 395–406; James Snead, "Spectatorship and Capture in King Kong: The Guilty Look," *Critical Quarterly* 33 (Spring 1991): 53–69; Gregg Mitman, *Reel Nature: America's Romance with Wildlife on Film* (Cambridge, MA: Harvard University Press, 1999), 59–84; Cynthia Chris, *Watching Wildlife* (Minneapolis: University of Minnesota Press, 2006), 1–44; Matthew Brower, *Developing Animals: Wildlife and Early American Photography* (Minneapolis: University of Minnesota Press, 2011). For details on Singapore and Malaya as the centers of the jungle film industry, see "Singapore Becomes Equatorial Hollywood," *Straits Times* (January 8, 1936): 12.

36. "Wild Animal Business Flops; Hunter Turns Film Advisor," *Los Angeles Times* (March 4, 1934): A5.

37. "Making a Living in Odd Ways: The Orient Produces Some Strange Jobs," *New York Times* (February 28, 1932), 16; Buck, *All in a Lifetime*, 101; "Wild Animals Hotel," *Singapore Free Press and Mercantile Advertiser* (February 2, 1923): 5; "Chinese Enterprise in Singapore," *Singapore Free Press and Mercantile Advertiser* (October 8, 1935): 4.

38. Frank Buck and Ferrin Fraser, *Tim Thompson in the Jungle* (New York: D. Appleton,

1935), 11–12. The premise of the novel was to bring to life the fantasy of those who wrote to Buck asking to come on an animal capture expedition. The fourteen-year-old and white Thompson stowed away in order to join Buck in Malaya.

39. Buck, *All in a Lifetime,* 174–179.

40. Peter Linebaugh, "The Law of the Jungle," *Capitalism Nature Socialism* 18 (December 2007): 38–53; Akira Mizuta Lippit, *Electric Animal: Toward a Rhetoric of Wildlife* (Minneapolis: University of Minnesota Press, 2000), 162–198.

41. "How Are Your Nerves?" *Evening Standard* (May 14, 1934): 3.

42. Buck, *All in a Lifetime,* 97–99.

43. Ibid., 102.

44. Frank Buck, "Story of Lal," *Collier's Weekly* (May 5, 1934): 21, 63–64. My reading of Buck and his representation of native assistants is indebted to Toni Morrison, *Playing in the Dark: Whiteness and the Literary Imagination* (New York: Vintage, repr. 1993); Wolfe, *Animal Rites,* 165–166.

45. Frank Buck, "Clouded Leopard," *Collier's Weekly* (March 24, 1934): 9, 54–55; Frank Buck, *Classics Illustrated: On Jungle Trails* (September 1957), cover.

46. Buck, *On Jungle Trails,* 59.

47. Ibid., 109–128.

48. Philip K. Scheuer, "Hunter's Diary Exciting Fare," *Los Angeles Times* (May 14, 1934): 6.

49. "Frank Buck to Talk at Hillstreet," *Los Angeles Times* (May 7, 1934): 9.

50. Burns Mantle, "Jungle Animal Film Best Show of N.Y. Week," *Chicago Daily Tribune* (June 26, 1932): F9; "John Scott, "Jungle's Thrills Captured," *Los Angeles Times* (August 2, 1932): A7; "Sultan to View Film at R-K-O," *Los Angeles Times* (May 10, 1934); "Jungle Film Retained at Keith House," *Washington Post* (July 3, 1932): A1; "Frank Buck's Bring 'Em Back Alive," *Singapore Press and Mercantile Advertiser* (May 18, 1933): 16.

51. "Bring 'Em Back Alive Opens Fox This Friday: Picture Is Jungle Talkie Film," *Atlanta Daily World* (August 11, 1932): 6A.

52. *Jungle Cavalcade* (1941).

53. Buck, *All in a Lifetime,* 230–235.

54. *Chang: A Drama of the Wilderness* (1927).

55. Ibid.; Mitman, *Reel Nature,* 36–57; Fatimah Tobing Rony, *The Third Eye: Race, Cinema, and Ethnographic Spectacle* (Durham: Duke University Press, 1996), 135–137.

56. "Frank Buck's Jungle Thriller," *Rochester Evening Post and the Post Express* (June 24, 1932): 20; "Animal Film Liked at Mission," *San Jose Evening News* (August 10, 1932): 11.

57. "When Jungle Gangsters Meet," *Singapore Free Press and Mercantile Advertiser* (October 3, 1932): 3; "Frank Buck to 'Shoot Bandits'," *Singapore Free Press* (July 2, 1949): 5; Lee Grieveson, *Mob Culture: Hidden Histories of the American Gangster Film* (New Brunswick: Rutgers University Press, 2005). On tiger hunting, see Heather Schell, "Tiger Tales," in Deborah Denenhold Morse and Martin A. Danahay, eds., *Victorian Animal Dreams: Representations of Animals in Victorian Literature and Culture* (Burlington, VT: Ashgate, 2007), 229–248; Peter Boomgaard, *Frontiers of Fear: Tigers and People in the Malay World, 1600–1950* (New Haven: Yale University Press, 2001), esp. 107–144.

58. *Fang and Claw* (1934); *Jungle Cavalcade.*

59. Frank Buck, "Jungle Trails," *Collier's Weekly* (May 19, 1934): 22, 59–60.

60. Frank Buck, "Never a Dull Moment: Wild Life in Johore," *Collier's Weekly* (January 26, 1935): 28, 54–55.

61. Buck, *All in a Lifetime*, 209–211; Alfred Albelli, "How Frank Buck Filmed His Tiger-Python Battle," *Modern Mechanix and Inventions* (November 1932): 34–39, 160–161.

62. Armand Denis, *On Safari: The Story of a Man's Life in Search of Adventure* (New York: E.P. Dutton, 1963), 55–68. Buck recreated the same story in a number of different sites, including *Tim Thompson*, 144–187. See also "Buck Lives Up to His Billing out in Jungle," *Washington Post* (February 10, 1935): X2.

63. "Bring 'Em Back Alive Opens Fox This Friday: Picture Is Jungle Talkie Film," *Atlanta Daily World* (August 11, 1932): 6A; Frank Buck with Edward Anthony, "Striped Fury: Mr. Buck Meets a Bad Actor," *Collier's Weekly* (April 9, 1932): 28–35.

64. "Telegram: Frank Buck to C. Emerson Brown," (August 8, 1932), PZ, Box FIN AA1.

65. "Official Handbook for Members of Frank Buck's Adventurer's Club" (1934), author's collection.

66. Frank Buck with Edward Anthony, *Bring 'Em Back Alive* (New York: Pocket Books, 1930), 31.

67. William Cronon, "The Trouble with Wilderness," in Cronon, ed., *Uncommon Ground: Toward Reinventing Nature* (New York: W.W. Norton, 1995), 79–84; Candace Slater, "Amazonia as Edenic Narrative," in *Uncommon Ground*, 117; Enright, *The Maximum of Wilderness*, 9–34; Laura Wright, *Wilderness into Civilized Spaces: Reading the Post-colonial Environment* (Athens: University of Georgia Press, 2010), 6–13; Ramachandra Guha, *Environmentalism: A Global History* (New York: Longman, 2000), 39–46.

68. Frank Buck, *Capturing Wild Elephants!* (Chicago: Merrill Publishing Co., 1934); "Sakais to Be Filmed," *Straits Times* (June 12, 1935): 13.

69. Cary Wolfe, ed., *Zoontologies: The Question of the Animal* (Minneapolis: University of Minnesota Press, 2003), xx; Catherine A. Lutz and Jane L. Collins, *Reading National Geographic* (Chicago: University of Chicago Press, 1993), 115–216.

70. "Joe Cook and Frank Buck Are Variety Stars," *Chicago Daily Tribune* (February 23, 1936), E3; "Buck's 'Jungle Cavalcade an Exciting Film," *Chicago Daily Tribune* (July 20, 1941), E4; "Frank Buck Will Tell Experiences along with Film," *Los Angeles Times* (May 6, 1934), A2; "Malayan Beasts for New York World Fair," *Straits Times* (January 31, 1939): 14; "Malay Village at World Fair," *Straits Times* (December 18, 1938): 17.

71. Buck, *Bring 'Em Back Alive*, 10; "Edgar Rice Burroughs, Inc. to Grover Whalen," (1938), NYWF, Box, 536, Folder 6; "John T. Benson to Grover Whalen," (June 30, 1937); "Carl Hagenbeck to Grover Whalen," (October 19, 1937); and "George P. Smith to Director of Concessions," (May 3, 1938), all in NYWF, Box 537, Folder 12.

72. "Frank Buck's Jungle Camp" (n.d., ca. 1938), NYWF, Box 537, Folder 5; "Frank Buck's Jungleland, Press Release," (1939) and "News Release: Frank Buck's Jungleland," (April 8, 1939), NYWF, Box 978, Folder 7; Raymond C. Shindler to George P. Smith, Jr.," (February 11, 1938) and "Frank Buck Show Application," (May 25, 1938), NYWF, Box 1546, Folder 2.

73. "Buck to Whalen," (July 20, 1939), NYWF, Box 537 Folder 4; "Frank Buck's Jungle Camp," (n.d., ca. 1938), NYWF, Box 537, Folder 5.

74. Buck, *All in a Lifetime*, 274–276; "The Bathyspherium," (July 5, 1938) and "Maurice Mermey to Amusements Area Board," (August 5, 1938), NYWF, Box 617, Folder 16.

4. The Monkeys' Island

1. *The Brookfield Zoo, 1934–1954* (Brookfield, IL: Chicago Zoological Society, 1954), 8–19; Eric Ames, *Carl Hagenbeck's Empire of Entertainments* (Seattle: University of Washington Press, 2009), 141–197; Nigel Rothfels, *Savages and Beasts: The Birth of the Modern Zoo* (Baltimore: Johns Hopkins University Press, 2002), 161–177.

2. "Children Ask for Zoo 'Pets' as Closing Is Threatened," *Record* (October 12, 1932) and "Boy Wants Lion, but Zoo Objects," *Evening Ledger* (October 17, 1932), both in PZ, Box FAC A1.

3. "Zoological Garden Attendance since Opening," (n.d.), PZ, Box ZH1 A1.

4. "Zoo May Board Animals That Philadelphia Cannot Feed," *Washington Post* (October 10, 1932): 1; "Children Have Raised $89 to Help Feed Zoo Animals," *Ledger* (August 19, 1932): "Zoo Head Threatens to Destroy Animals unless Council Appropriates $100,000," *Record* (October 9, 1932), "Failure of City to Give Funds Will Mean End of the Philadelphia Zoo," *Evening Ledger* (October 8, 1932), all in PZ, Box FAC A1. See also "A Report on Zoos," (1935–36); "Survey, Analysis, and Plan of Procedure for the Zoological Society of Philadelphia," (January 1936); "Suggestions for Improving Grounds and Animals," (1936–37), all in PZ, Box CON B1. See also Jesse C. Donahue and Erik K. Trump, *American Zoos during the Depression: A New Deal for Animals* (Jefferson, NC: McFarland, 2010); Samuel Redman. "The Hearst Museum of Anthropology, the New Deal, and a Reassessment of the 'Dark Age' of the Museum in the United States," *Museum Anthropology* 34, no. 1 (2011): 43–55.

5. David M. Kennedy, *Freedom from Fear: The American People in Depression and War, 1929–1945* (New York: Oxford University Press, 1999), 43–130.

6. "San Diego Zoo Auction Fails to Find Buyer," *Los Angeles Times* (August 31, 1932): 10.

7. "San Diego Zoo Sale Due Anew," *Los Angeles Times* (September 11, 1932): 10.

8. "Complete Zoo Possible Quickly and Cheaply," *Cleveland Plain-Dealer* (July 21, 1941) in WPA, Box 19.

9. "Contented Polar Bears," *Pittsburgh Post-Gazette* (September 10, 1937), WPN, Box 193.

10. "Como Park Zoo," (August 14, 1936), WPA, Box 19.

11. "Who'll Save Zoo?" *Record* (October 14, 1932), PZ, Box FAC A1.

12. "Failure of City."

13. "Zoo Sells Lion Cubs to Obtain Funds to Feed Their Parents," *Evening Ledger* (October 12, 1932), PZ, Box FAC A1.

14. "An Animal Garden in Fairmount Park," (Zoological Society of Philadelphia, 1988), PZ, Box ZH A1, 21

15. "The Jungle Comes to Philadelphia," (June 19, 1936), PZ, Box ZSP A1.

16. "Philadelphia Wants a Free and Modern Zoo," (1936), PZ, Box ZSP A1.

17. "A FREE and MODERN ZOO FOR PHILDEPHIA," *Bulletin of the Zoological Society of Philadelphia* (September 1936): 1, PZ, Box ZSP A1.

18. "The Civic Importance of Modern Zoos in the United States," (1936), 8; and "The Jungle Comes to Philadelphia," (June 19, 1936), both in PZ, Box ZSP A1.

19. "Annual Report," (1934), STZ, Folder 18; "The Civic Importance of Modern Zoos," 3.

20. Timothy Messer-Kruse, *Banksters, Bosses, and Smart Money: A Social History of the Great Toledo Bank Crash of 1931* (Columbus: Ohio State University Press, 2005).

21. "W. R. Blair to Officers of the New York Zoological Park," (July 11, 1934) and "W. R. Blair to All Maintenance Employees," (July 27, 1932), OD, Box 131, Folder: Circulars/ Notices to Park Staff, 1920–1940.

22. Roger Conant, *A Field Guide to the Life and Times of Roger Conant* (Toledo: Toledo Zoological Society, 1997), 41; "An Animal Garden in Fairmount Park," 21,

23. "Wm. B. Schmuhl to Col. F. C. Harrington," (June 21, 1937), WPC, Box 2265, File, 651–109; "Toledo (Ohio) Zoological Park," (ca. 1937), WPD, 13–15; Toledo Zoological Park, ". . . given work and wages . . . ," (1937), WPD.

24. "Excerpt from Ohio Press Release," (August 6, 1937); "Speech Given by William B. Schmuhl, Director, District #1, WPA in Ohio," (May 30, 1939), WPA, Box 19.

25. "Drive for Enlarged Columbus Zoo Is Launched by Public-Spirited Citizens," *Columbus Evening Dispatch* (April 1, 1938), WPN, Box 193; "News Release from WPA in Ohio," (May 14, 1941), WPA, Box 19.

26. "Supplementary List of Zoo Projects for Operation," (May 31, 1936), WPA, Box 19.

27. "No Boondoggle for Bronx Zoo," *New York Sun* (August 1, 1936), WPN, Box 193; Donahue and Trump, *American Zoos during the Depression*, 7–35; Roy Rosenzweig and Barbara Melosh, "Government and the Arts: Voices from the New Deal Era," *Journal of American History* 77 (September 1990): 596–608; Jason Scott Smith, *Building New Deal Liberalism: The Political Economy of Public Works, 1933–1956* (Cambridge: Cambridge University Press, 2006); Barbara Melosh, *Engendering Culture: Manhood and Woman-hood in New Deal Public Art and Theater* (Washington: Smithsonian Institution Press, 1991); Susan Quinn, *Furious Improvisation: How the WPA and a Cast of Thousands Made High Art out of Desperate Times* (New York: Walker, 2008).

28. "Start Work on Paddock for Tillie at Nay Aug," *Scranton Times* (July 26, 1937), WPN, Box 193.

29. "New Deal Ends Slum Clearance Project at the Audubon Zoo," *New Orleans Item-Tribune* (September 6, 1936); "Sunday Release," (May 18, 1938); and "Excerpt from LA. Narrative Report," (June 20, 1936), all in WPA, Box 19.

30. "American Association of Zoological Parks and Acquariums [*sic*]," (September 29, 1931), AZA, Roll 100; "William Temple Hornaday to C. Emerson Brown," (December 5, 1936), WTH, Box 11, Folder Brown-BU.

31. "New Deal Ends Slum Clearance Project at the Audubon Zoo," *New Orleans Item-Tribune* (September 6, 1936), WPN, Box 193; "Sunday Release"; "Excerpt from LA. Narrative Report."

32. "Zoo Director Awaits WPA Shovels," *Washington Times* (January 22, 1936), WPN, Box 193; "Zoological Bldgs.," (January 16, 1936), WPA, Box 19; David Hancocks, *Animals and Architecture* (New York: Praeger, 1971); Bob Mullan and Gary Marvin, *Zoo Culture: The Book about Watching People Watch Animals,* 2nd ed. (Urbana: University of Illinois Press, 1998); Ralph R. Acampora, "Extinction by Exhibition: Looking In and At the Zoo," *Human Ecology Review* 5, no. 1 (1998): 1–4.

33. "WPA Building New-Fangled Fenceless Enclosures at Zoo," *Little Rock Gazette* (February 9, 1939), WPN, Box 193.

34. "WPA Zoo Building Program Finished at $92,189 Cost," *San Antonio Express* (June 28, 1936); "Extent of Modernization Program at Buffalo Zoo," *Buffalo Evening News* (June

30, 1936); "Seals to Move in New Home in Modernized Buffalo Zoo," *Buffalo Courier-Express* (August 9, 1936), 2, all in WPN, Box 193.

35. "Mesker Zoo, 'Out of Mud,' Now Has National Rating," *Evansville Courier* (May 30, 1937), WPN, Box 193.

36. "San Francisco Zoological Garden," (November 23, 1940), WPA, Box 19.

37. "More than $800,000 Spent on Zoo in Last Five Years," *Cincinnati Enquirer* (February 9, 1938); "Zoo Receives New Animals," *Jackson Daily News* (November 25, 1937), WPN, Box 193.

38. "Excerpt from WPA News Release," (May 8, 1936), WPA, Box 19.

39. Earl Chapin May, "To the Zoos Has Come a New Deal," *Parks and Recreation* 18 (April 1935): 330–333.

40. "Toledo (Ohio) Zoological Park."

41. "Excerpt from New York State Press Releases," (November 21, 1937), WPA, Box 19.

42. "Work on Barless Bear Pits to Be Started July 1," *Kansas City Journal Post* (June 16, 1937); "Buffalo Gets New $842,133 Grant for Zoo," *Buffalo Courier Express* (April 30, 1939), both in WPN, Box 193; "Release Immediately" (June 2, ca. 1939), WPA, Box 19.

43. "Zoo and Smithsonian Seek More Money," *Washington News* (August 15, 1940); "Work Is Completed on Zoo's Aviary for Tropical Birds," *Toledo Blade* (March 18, 1937), WPN, Box 193.

44. "Excerpt from WPA News Release."

45. May, "To the Zoos Has Come a New Deal," 330–333.

46. "William Mann to Charles Knight," (December 12, 1933); "Memorandum of Projects for Zoo," (December 7, 1933), in NZP, Box 215, Folder 17.

47. "WPA Helps Zoo Toward Ideal," *Parks and Recreation* 20 (September 1936): 42–45.

48. T. E. Powers, "A Four-Star WPA Project," *New York Evening Journal* (July 22, 1936), WPN, Box 193.

49. "Speech Given by William B. Schmuhl."

50. "Third District, Ramsey County, St. Paul, Minnesota," (August 14, 1936), WPA, Box 19.

51. "WPA to Provide Animal Lectures for Zoo Visitors," *Milwaukee Sentinel* (August 18, 1936), WPN, Box 193; Donahue and Trump, *American Zoos during the Depression*, 36–67. See also: Holly Allen, *Forgotten Men and Fallen Women: The Cultural Politics of New Deal Narratives* (Ithaca, NY: Cornell University Press, 2015).

52. "Monkey Island Opens as 25,000 Visit Zoo," *Memphis Commercial Appeal* (September 14, 1936), WPN, Box 193. For more contemporary studies of visitors and animal interactions, see Michael D. Kreger and Joy A. Mench, "Visitor Animal Interactions at the Zoo," *Anthrozoos* 8, no. 3 (1995): 143–143; John Knight, "Monkey Mountain as a Megazoo: Analyzing the Naturalistic Claims of 'Wild Monkey Parks' in Japan," *Society and Animals* 14, no. 3 (2006): 245–245.

53. "Excerpt from the I.A. Narrative Report."

54. "Monkey Business," *Pittsburgh Press* (July 20, 1941), 12, WPN, Box 193.

55. *Illustrated Zoo Book, Guide to the Columbus Zoo* (Columbus, 1940), 11, WPS, Box 63.

56. "Excerpt from *Free Press,* Detroit," (January 19, 1936), WPA, Box 19.

57. "WPA Helps Zoo Toward Ideal," 42–45.

58. Ibid.

59. Douglas Deuchler and Carla W. Owens, *Brookfield Zoo and the Chicago Zoological Society* (Charlestown, SC: Arcadia Publishing, 2009), 51; *The Brookfield Zoo 1934–1954,* 69.

60. "Memphis Allots $14,000 for Monkey Playground," *Richmond Times Dispatch* (February 10, 1936); "Monkey Mountain Major Attraction at Zoo," *Lansing Journal* (June 26, 1936); "$65,000 Builds Monkey House," *New York Sun* (July 18, 1936), all in WPN, Box 193.

61. Federal Writers Project, "Highland Park Zoological Gardens," *Zoo Views* (June 8, 1939), 3, WPALC, Box A-816, File, "Pennsylvania, "Zoo Views."

62. Hugh S. Davis, "Humanizing the Monkey Island," *Parks and Recreation* 15 (June 1932): 669–671

63. Henry Trefflich and Edward Anthony, *Jungle for Sale* (New York: Hawthorn Books, 1967).

64. "Lewis Wirch to William Mann," (June 15, 1939), NZP, Box 152, Folder 10; "William Mann to Jack Supp," (February 16, 1942), NZP, Box 152, Folder 12.

65. "Prevue of Monkey Island—Frisky Monks Like New Home," *Memphis Press Scimitar* (September 10, 1936), WPN, Box 193.

66. Mary R. Lewis, *At the Zoo* (New York: Thomas Nelson and Sons, 1927).

67. W. W. and Irene Robinson, *At the Zoo* (New York: Macmillan, 1940).

68. "Don't Feed the Animals," *St. Louis Star* (August 21, 1935), STZ, Series II, Scrapbook, Vol. 8.

69. "What Should Zoo Animals Eat?" (n.d.), WPD.

70. Leonard Outhwaite, "A Report for the New York Zoological Society," (March 1938), NYZS, Small collection file, Consultant's reports, Box 3.

71. "Zoo Encourages Natural History Study by New Attractions for School Children," *New York Times* (August 24, 1941): 44.

72. Helen Louise Thorndyke, *Honey Bunch: Her First Visit to the Zoo* (New York: Grosset & Dunlap, 1932); Helen Thurston Monahan, *Puss in the Zoo* (New York: Greystone Press, 1940).

73. Ken Kawata, "Zoo Animal Feeding: A Natural History Viewpoint," *Der Zoologische Garten* 78, no. 1 (2008): 17–42; Nicholas Sammond, *Babes in Tomorrowland: Walt Disney and the Making of the American Child, 1930–1960* (Durham: Duke University Press, 2005); Steven Mintz, *Huck's Raft: A History of American Childhood* (Cambridge, MA: Harvard University Press, 2004), 213–253.

74. Outhwaite, "A Report for the New York Zoological Society," 127.

75. News Realease No. 786," (April 28, 1939), NYWF, Box 537, Folder 5.

76. "Fair Attendance is 25, 927 in Spite of Rain Threats," *Chicago Tribune* (June 5, 1934): 2; "Sally's Guests to See Monkey Colony at the Fair," *Chicago Tribune* (August 7, 1934): 15.

77. "Mally Wild de Villareal to Grover Whalen," (February 24, 1937);"J. Frank Hatch to Maurice Mermey," (April 1, 1937); "Telegram, to Grover Whalen," (October 19, 1937); "Carl Hagenbeck to Grover Whalen," (October 19, 1937); "Memorandum, Albin K. Johnson to W. H. Standley," (November 11, 1937); "George P. Smith to Director of Concessions," (May 3, 1938), all in NYWF, Box 537, Folder 12.

78. "Frank Buck's Jungle Camp," (n.d., ca. 1938), NYWF, Box 537, Folder 5.

79. "Frank Buck's Jungleland," (flyers, ca. 1939–1940), author's collection.

80. "Stage and Screen Happenings," *Pittsburgh Post-Gazette* (August 3, 1932): 4; Ames, *Carl Hagenbeck's Empire of Entertainments,* 198–229.

81. "Thousands Swarm into Open as Mercury Mounts to 90," *Philadelphia Record* (May 8, 1939), 1, WPN, Box 193.

82. "Will Yolen to Lee Pollack" (August 12, 1939), NYWF, Box 537, Folder 5; "Frank Buck to George Smith" (May 17, 1940), NYWF, Box 1546, Folder 2.

83. "Alfonso Carvajal to M. V. Little," (March 21, 1939); "John Wolfe to W. Franklin Dove," (December 23, 1938), both in NYWF, Box 536, Folder 6 Animal Shows (1939); "Mable Brown to Paul M. Massman," (November 16, 1936); "George P. Smith, Dept. of Exhibits and Concessions, to G. C. Allen," (April 28, 1938); "Wilson Baguly to George P. Smith," (February 17, 1939); and "William Cenoire to Grover Whalen" (May 26, 1939), all in NYWF, Box 547, Folder 9.

84. Robert Bogden, *Freak Show: Presenting Human Oddities for Amusement and Profit* (Chicago: University of Chicago Press, 2014), 54.

85. "News Release No. 687," (1939) and "American Consulate, Java, to E. W. Cobb," (March 27, 1939), NYWF, Box 537, Folder 5

86. "Mermaid," (1939), NYWF, Box 537, Folder 5.

87. Outhwaite, "A Report for the New York Zoological Society," 125.

88. Ibid., 21–22.

89. Fairfield Osborn, "World's Fair Experiences," *New York Zoological Society Bulletin* 42 (November–December 1939): 183–186; Osborn, "Our Building at the World's Fair," *New York Zoological Society Bulletin* 42 (January–February 1939): 3–7; William Bridges, "Zoology Goes to the World's Fair," *New York Zoological Society Bulletin* 33 (September–October 1939): 149–153; John Tee-Van, "Our Building at the World's Fair in 1940," *New York Zoological Society Bulletin* 43 (May–June 1940): 70–71.

90. Outhwaite, "A Report for the New York Zoological Society," 116.

91. "The Jungle Comes to Philadelphia." Italics are original.

5. Aping

1. *Detroit Zoo Log* (Detroit: Detroit Zoological Society, 2003), 30–32; Walter D. Stone, "Detroit's Chimpanzee Shows," *Parks and Recreation* (April 1957), 40–41; "Detroit Zoo Buys Elephants: Draws 2 Million in Year," *Billboard* (November 15, 1952), 71. Of those 2 million, 233,000 paid to watch a show at the Jo Mendi arena.

2. C. Emerson Brown, "Notes on Some Anthropoid Apes," in "Meeting of the American Association of Zoological Parks and Aquariums," (September 29, 1931), AZA, Roll 100.

3. Paul Du Chaillu, *Explorations and Adventures in Equatorial Africa* (New York: Harper and Brothers, 1861). See also Stuart McCook, "'It May Be Truth, but It Is Not Evidence': Paul du Chaillu and the Legitimation of Evidence in the Field Sciences," *Osiris* 11 (1996): 177–197; Georgina M. Montgomery, "'Infinite Loneliness': The Life and Times of Miss Congo," *Endeavour* 33 (September 2009): 101–105; Richard Kerridge, "Ecologies of Desire: Travel Writing and Nature Writing as Travelogue," in Steve Clark, ed., *Travel Writing and Empire: Postcolonial Theory in Transit* (New York: St. Martins, 1999): 164–182; John Miller, *Empire and the Animal Body: Violence, Identity, and Ecology in Victorian Adventure Fiction* (London: Anthem, 2012).

4. "Today and, Perhaps, Tomorrow, Those Wonders the Male and Female Chimpanzee," *Cincinnati Commercial* (July 17, 1878): 8.

5. "Frank Baker to F. E. Kitzmiller," (1890), NZP, Box 7, Folder 2.

6. "Sixteenth Annual Report of the New York Zoological Society," (January 1912), 69; Alyse Cunningham, "A Gorilla's Life in Civilization," *New York Zoological Society Bulletin* 25 (September 1921): 118–124.

7. Carl E. Akeley, *In Brightest Africa* (New York: Doubleday, Page, 1923).

8. "Chimpanzees and Gorillas," *Philadelphia Inquirer* (February 28, 1915): 4; "Darwin's Theory in Far Liberia," *Philadelphia Inquirer* (February 14, 1909): 3; Robert E. Kohler, *All Creatures: Naturalists, Collectors, and Biodiversity, 1850–1950* (Princeton: Princeton University Press, 2006), 227–245.

9. "Consul, Chimpanzee, Mr. Belmont's Guest," *New York Times* (July 27, 1907): 7; " 'Consul' Comes to America," *Washington Post* (June 21, 1909): 1.

10. Donna J. Haraway, *Primate Visions: Gender, Race, and Nature in the World of Modern Science* (New York: Routledge, 1989), 4–5 helps locate primatology as storytelling or, as she writes, "elaborate narratives about origins, natures, and possibilities." As a production of a narrative about taxonomies of animals and people, the case of the popularity of the chimp show identifies the inherent entanglement of entertainment, primatology, and zoo display.

11. "J. S. Edwards to William T. Hornaday," (1897), WTH, Box 11, Folder E.

12. "If Darwin Only Knew, a Dressed Up Chimpanzee Taken for a Negro Girl," *The Sunday Oregonian* (May 11, 1980): 13; Constance Areson Clark, *God—or Gorilla: Images of Evolution in the Jazz Age* (Baltimore: Johns Hopkins University Press, 2008).

13. "J. S. Edwards to William Temple Hornaday," (1897), WTH, Box 11, Folder E; W. Henry Sheak, "Disposition and Intelligence of the Orang-Utan," *Journal of Mammalogy* 3 (February 1922): 47–51.

14. William Ballantine, *Wild Tigers and Tame Fleas* (New York: Rinehart, 1958), 105.

15. "For Vaudeville's Patrons," *New York Times* (April 19, 1914): X7.

16. Ibid. On blackface performance, see Louis Chude-Sokei, "The Uncanny History of Minstrels and Machines, 1835–1923," in Stephen Johnson, ed., *Burnt Cork: Traditions and Legacies of Blackface Minstrelsy* (Amherst: University of Massachusetts Press, 2012), 104–132; W. T. Lhamon, *Raising Cain: Blackface Performance from Jim Crow to Hip Hop* (Cambridge, MA: Harvard University Press, 1998); Eric Lott, *Love and Theft: Blackface Minstrelsy and the American Working Class* (New York: Oxford University Press, 1993): Matthew Frye Jacobson, *Whiteness of a Different Color: European Immigrants and the Alchemy of Race* (Cambridge, MA: Harvard University Press, 1998), 28–29.

17. "Smith & Murray, Saturday's Extraordinary Price Attractions," *Springfield Republican* (June 4, 1910): 5.

18. "Some Chase Prospects" *Washington Post* (September 19, 1909): SM2; David A. H. Wilson, "Racial Prejudice and the Performing Animals Controversy in Early Twentieth-Century Britain," *Society and Animals* 17, no. 2 (2009): 149–165.

19. "Chase's Polite Vaudeville" (Advertisement), *Washington Post* (January 5, 1910): 4.

20. "Vaudeville Houses Open New Bills," *Morning Oregonian* (October 26, 1909): 11.

21. "Monkey Shows Amazing Ways," *Chicago Daily Tribune* (January 3, 1904): 15.

22. "Keiths, Here Is a Great Show" (Advertisement), *Washington Post* (December 19, 1925): 10.

23. "Consul's Life Story," *Washington Post* (April 25, 1911): SM3; "Consul Jr., Acting Ape, Dead," *New York Times* (March 17, 1910): 4.

24. "$75,000 Monkey Killed," *Oregonian* (June 17, 1910): 5.

25. "'Consul' Comes to America" *Washington Post* (June 21, 1909): 1; Chude-Sokei, "The Uncanny History of Minstrels and Machines."

26. Ann Stoler, "Tense and Tender Ties: The Politics of Comparison in North American History and (Post) Colonial Studies," *Journal of American History* 88 (December 2001): 829–865; Harriet Ritvo, *The Platypus and the Mermaid and Other Figments of the Classifying Imagination* (Cambridge, MA: Harvard University Press, 1997); Michael Lundblad, "Archeology of a Humane Society: Animality, Savagery, Blackness," in Marianne DeKoven and Michael Lundblad, eds., *Species Matters: Humane Advocacy and Cultural Theory* (New York: Columbia University Press, 2012), 75–102; Donna J. Haraway, *Primate Visions: Gender, Race, and Nature in the World of Modern Science* (New York: Routledge, 1989), 19–25.

27. "How Sammy Green Got His Name," *The St. Louis Zoo* 4 (June 1932): 3, STZ, Folder 61. There are similarities in the humanoid display of African apes and the exhibition of African humans, especially at zoos: see Sadiah Qureshi, *Peoples on Parade: Exhibitions, Empire and Anthropology in Nineteenth Century Britain* (Chicago: University of Chicago Press, 2011); Robert W. Rydell, "Darkest Africa: African Shows at America's World's Fairs, 1893–1940" and Harvey Blume, "Ota Benga and the Barnum Perplex," in Bernth Lindfors, ed., *Africans on Stage: Studies in Ethnological Show Business* (Bloomington: University of Indiana Press, 1999), 137–155, 188–202; Harvey Blume and Phillips Verner Bradford, *Ota Benga: The Pygmy in the Zoo* (New York: St. Martins, 1992); Nicolas Bancel et al., eds., *Human Zoos: From the Hottentot Venus to Reality Shows* (Chicago: University of Chicago Press, 2009).

28. "Massa, Dean of Captive Gorillas," (n.d.), PZ, Box AC B1.

29. "Only One of His Kind in America," *Inquirer* (October 4, 1927), PZ, Box AC B1.

30. "Chimpanzee's Friend Mourns; Negro Playmate of Consul Would Rather Have Lost a Relative," *New York Times* (May 6, 1904): 7.

31. "Chimpanzees Guests of New York Hotel," *Lexington Herald* (December 19, 1909): 5.

32. "L. A. Camacho to William T. Hornaday," (May 5, 1910), WTH, Box 12, Folder FA; William T. Hornaday, *The Mind and Manners of Wild Animals* (New York: Charles Scribners, 1922), 86–90.

33. Hornaday, *The Mind and Manners of Wild Animals,* 74–75.

34. "Baldy Learns to Skate," *New York Times* (June 13, 1910): 13; New York Zoological Park, "Chimpanzee 'Baldy'" (1910), postcard in author's collection.

35. William T. Hornaday, *Official Guide Book to the New York Zoological Park* (New York: New York Zoological Society), 109.

36. Lightner Witmer, "A Monkey with a Mind," *Psychological Clinic* 3 (1909): 169–178; "A Monkey with a Mind," *New York Times* (January 22, 1910): BR1: "Trick Chimpanzee Fulfills Mind Test," *New York Times* (December 18, 1909): 7; "Chase's Polite Vaudeville" (Advertisement), *Washington Post* (January 5, 1910): 4; Gregg Mitman and Lorraine Daston, eds., "Introduction: The How and Why of Thinking with Animals," in *Thinking with Animals: New Perspectives on Anthropomorphism,* eds. Gregg Mitman and Lorraine Daston (New York: Columbia University Press, 2005), 8.

37. Hornaday, *The Minds and Manners of Wild Animals,* 82–83; W. T. Shepherd, "Some Observations on the Intelligence of the Chimpanzee," *Journal of Animal Behavior* 5 (September–October 1915): 391–396.

38. Clive D. L. Wynne, "Rosalià Abreu and the Apes of Havana" *International Journal of Primatology* 29 (2008): 289–302.

39. "Philadelphia to Get Some of Madame Abreu's Collection," *Parks and Recreation* (May, 1931), 519.

40. "Abreu to Yerkes" (December 27, 1916), Robert Yerkes Papers, Yale University Manuscripts and Archives, Box 1, Folder 6.

41. Robert M. Yerkes, *Almost Human* (New York: The Century Co., 1925), 49–74.

42. Ibid., 128, 244–255.

43. Ibid., 267–274.

44. Ibid., 260–263; "Sixth Annual Report of the New York Zoological Society," (April 1, 1902), 61. See also James L. Newman, *Encountering Gorillas: A Chronicle of Discovery, Exploitation, Understanding, and Survival* (Lanham, MD: Rowman and Littlefield, 2013); Donald A. Dewsbury, *Monkey Farm: A History of the Yerkes Laboratories of Primate Biology, Orange Park, Florida, 1930–1965* (Lewisburg, PA: Bucknell University Press, 2006).

45. "Two Apes and 'Link' Arrive at Dayton," *New York Times* (July 15, 1925): 1.

46. "Exhibits on Evolution Draw Museum Throngs" *Washington Post* (July 26, 1925): 2.

47. William T. Hornaday, *The American Natural History* (New York: Scribner's, 1904), 11.

48. "An African Monkey," *Kansas City Journal* (February 17, 1895): 7.

49. "Royal Theater," *Tulsa World* (February 5, 1921): 13.

50. "Joe Mendi, Best Paid Chimpanzee, Is Dead," *Toledo New-Bee* (July 24, 1930): 15.

51. "Amusements," *Boston Daily Globe* (October 12, 1925): 5; "Irene Franklin in New Songs at Keith's," *Boston Daily Globe* (October 13, 1925): 23A; "Joe Mendi, Famous Chimp, Coming to Davis Theater," *Pittsburgh Press* (January 2, 1926): 3.

52. *14th Annual Report of the Washington Park Zoological Society* (1924).

53. Reginald William Thompson, *Wild Animal Man: Being the Story of the Life of Reuben Castang* (New York: W. Morrow, 1934), 288.

54. "Thirty-Seventh Annual Report of the New York Zoological Society" (January 1933), 41

55. "Skippy, Zoos Gentlemanly Chimp, Is Dead," *Chicago Daily Tribune* (August 23, 1939): 9.

56. "Chimpanzee Lord and Lady Are Here," (July 18, 1923) in STZ, Series II, Scrapbooks, 1921–26.

57. *Guide Book, the Cincinnati Zoo* (Zoological Society of Cincinnati, 1942), 5; "Zoo Show to Present Chimpanzee and Pony," *Los Angeles Times* (February 21, 1954): B1.

58. "Animal Acts Whet Public's Interest in Zoo," *The St. Louis Zoo* 5 (May 1933): 2, STZ, Folder 62; "Chimpanzees perform at St. Louis Zoo . . . ," *Fox Movietone News,* Volume 30, Release 40 (May 13, 1957), National Archives and Records Administration, Series: Motion Picture Newsreel Films, Collection: Fox Movietone News Collection.

59. Gary Ferguson, "Mr. Moke," *St. Louis Post-Dispatch* (December 23, 1959) in STZ Series III, Newsclippings, 1959–1960.

60. "Lock at Ape House Broken—Warrant against Man Who Sold the Animal," *St. Louis Post-Dispatch* (November 8, 1959), STZ, Newsclippings, 1959–1960.

61. Pete Goldman, "Just the Facts, Moke," *St. Louis Globe-Democrat* (December 27, 1959), Goldman, "The Man Who Fell in Love with a Chimp," in *St. Louis Globe-Democrat* (January 3, 1960), Sid Ross, "For the Love of Moke," *St. Louis Post-Dispatch* (February

7, 1960), all in St. Louis Zoological Park Records, Newsclippings, 1959–1960; "Tells How He Stole Bosom Pal from Zoo," *Chicago Daily Tribune* (Dec 30, 1959): 1; "Mr. Moke, Trainer Close as Father, Son, State Is Told," *St. Petersburg Times* (February 4, 1960: 1

62. "Contentment," *St. Louis Post-Dispatch* (December 31, 1959), "Tomarchin Still Has Mr. Moke after Dramatic Court Scene," *St. Louis Post-Dispatch* (January 7, 1960), "Vierheller Says Moke Won't Miss Tomarchin," *St. Louis Globe-Democrat* (February 4, 1960), all in STZ, Newsclippings, 1959–1960.

63. Ross, "For the Love of Moke."

64. "Special Show on Saturday," *Miami News* (August 7, 1958): 10B; "Tomarchin Makes Most out of Moke," *St Louis Globe-Democrat* (March 1, 1960); Goldman, "The Man Who Fell in Love with a Chimp," all in STZ, Newsclippings, 1959–1960.

65. "Moke Is No.1 Attraction at Zoo," *St. Louis Post-Dispatch* (September 5, 1961); "Tomarchin Placed on 18 Months' Parole," *St. Louis Globe-Democrat* (September 7, 1961), in Newsclippings, 1960–61; Ernest Havemann, "Mr. Zoo: Nobody since Noah Has Known So Many Animals So Well," *Life* 30 (January 8, 1951): 39–42, 47–49.

66. "The Private Life of Mr. Moke," *St. Louis Post-Dispatch* (October 22, 1961), STZ, Newsclippings, 1961–63; "Goodbye Kiss for Mr. Zoo," *Life* 52 (May 25, 1962): 45.

67. *The St. Louis Zoo Album* (1961), MP, Series 7, Box 7, Folder 157.

68. Belle Benchley, *My Friends, the Apes* (Boston: Little, Brown, 1942), 125, 161–162. See also Rebecca Bishop, "Several Exceptional Forms of Primates: Simian Cinema," *Science Fiction Studies* 35 (July 2008): 238–250; Fatimah Tobing Rony, *The Third Eye: Race, Cinema, and Ethnographic Spectacle* (Durham: Duke University Press, 1996), 162.

69. On race and domestic images of intimacy, see Laura Wexler, *Tender Violence: Domestic Visions in the Age of U.S. Imperialism* (Chapel Hill: University of North Carolina Press, 2000).

70. "All around Detroit," *Luddington Daily News* (September 8, 1932): 7.

71. "Playmate for Young Gorilla at Zoo Sought," *St. Louis Post-Dispatch* (ca. 1931), STZ, Series II, Scrapbook, Vol. 7.

72. "Yonnah, 6-Year Old Gorilla, Is a 'Stooge' in Zoo Stage Debut," *St. Louis Globe-Democrat* (ca. 1935), STZ, Series II, Scrapbook, Vol. 8.

73. "Richard D. Sparks to Mary L. Jobe Akeley," (October 28, 1932), Series 1, Folder 2; "Yonnah Cuddles Up to Her Old Pal after Separation of Three Years," *St. Louis Star* (February 21, 1936), STZ, Series II, Scrapbook, Vol. 9.

74. "Richard D. Sparks to Harold C. Bingham," (October 10, 1931), SG, Series 1, Folder 4.

75. "Belle Benchley to Richard Sparks," (November 6, 1931), SG, Series 1, Folder 3.

76. "Massa, Dean of Captive Gorillas."

77. "Only One of His Kind in America."

78. "Gorilla Mauls Keeper at Zoo," *Philadelphia Inquirer* (April 23, 1947); "Gorilla Bamboo Breaks Arms of Zoo Keeper; 2d Attack in 2 Months," *Philadelphia Inquirer* (August 11, 1947), PZ, Box AC B1.

79. "Congorilla," *Variety* (July 26, 1932): 17, 46

80. M. E. Buhler, "Ota Benga," *New York Times* (September 19, 1906).

81. "Still Stirred about Benga; Negro Communities Heard From—He Is Happy Now He Is Free," *New York Times* (September 23, 1906): 9; "Negro Ministers Act to Free the Pygmy;

Will Ask the Mayor to Have Him Taken from Monkey Cage," *New York Times* (September 11, 1906): 2.

82. "African Pygmy's Fate Is Still Undecided," *New York Times* (September 18, 1906): 9.

83. "Who's Who in the New Films," *New York Times* (July 24, 1932), x2.

84. "Richard D. Sparks to Martin Johnson," (September 18, 1931), SG, Series 1, Folder 7.

85. "Richard D. Sparks to Harold C. Bingham," (August 12, 1931), SG, Series 1, Folder 4.

86. "Tordis Graim to Harold C. Bingham," (September 22, 1931), SG, Series 1, Folder 4.

87. "Mary L. Jobe Akeley to Richard Sparks," (September 28, 1931), SG, Series 1, Folder 2.

88. "Mary L. Jobe Akeley to Richard Sparks," (December 4, 1931) and (September 28, 1931), SG, Series 1, Folder 2; "Harold C. Bingham to Richard Sparks," (September 1, 1931), SG, Series 1, Folder 4.

89. "Richard D. Sparks to Mary L. Jobe Akeley," (September 11, 1931), SG, Series 1, Folder 2 and "Richard D. Sparks to Martin Johnson," (September 11, 1931), SG, Series 1, Folder 7.

90. "Richard D. Sparks to Robert N. Yerkes," (August 12, 1933), SG, Series 1, Folder 11.

91. "Big City Startles Two African Boys," *New York Times* (July 5, 1931): 3.

92. Ibid.

93. "Reading Guidance," *Miami Daily News* (October 11, 1931), 4.

94. "Two African Boys Find Harlem Odd," *New York Times* (July 6, 1931): 15; "Africans Glad to Quit America," *The Spokesman-Review* (October 2, 1931): 9; "Two African Youths Enjoy Harlem Visit," *Chicago Defender* (July 18, 1931): 10.

95. "Measuring Gorillas," *Amsterdam News* (May 29, 1929): 20.

96. "Score Housing of Native Tribesmen," *Amsterdam News* (July 22, 1931): A20; "Garvey Women Rescue African Natives from Stable of New York Zoo," *Afro-American* (July 25, 1931): 5.

97. "The Man in the Street," *Amsterdam News* (August 19, 1931): 8.

98. Ibid.

99. "Two African Boys Here as Attendants of Huge Gorilla," *The Pittsburgh Courier* (August 1, 1931): A6.

100. "Americanize Quickly," *Reading Eagle* (September 13, 1931): 32.

101. "Richard D. Sparks to Harold C. Bingham," (August 28, 1931), SG, Series 1, Folder 4; "Richard D. Sparks to Mary L. Jobe Akeley," (October 28, 1932) and "Richard D. Sparks to Mary L. Jobe Akeley," (December 10, 1931), both in SG, Series 1, Folder 2.

102. "Harlem Sharpens Uganda Boys' Wits," *New York Times* (July 20, 1931): 17.

103. "African Natives Here, Homesick, Will Sail," *New York Times* (October 1, 1931): 20.

104. Romeo L. Doughbery, "Our Safari Returns from Africa," *Amsterdam News* (October 5, 1932): 8.

105. "Richard D. Sparks to Mary L. Jobe Akeley," (October 12, 1931), SG, Series 1, Folder 2; "Richard D. Sparks to John C. Merriam," (September, 17, 1931), SG, Series 1, Folder 4; "Two Live Gorillas Purchased for Zoo," *Berkeley Daily Gazette* (October 1, 1931): 2.

106. Benchley, *My Friends, the Apes*, 173.

107. Ibid., 125.

108. Ibid., 161–162, 187–188; "Belle Benchley to Richard Sparks," (September 25, 1931), SG, Series 1, Folder 3.

109. Benchley, *My Friends, the Apes*, 186–189.

110. "Richard D. Sparks to Belle Benchley," (October 8, 1931), SG, Series 1, Folder 3.

111. "Belle Benchley to Richard Sparks," (October 14, 1931), SG, Series 1, Folder 3.

112. Benchley, *My Friends, the Apes*, 206.

113. Ibid., 207.

114. "Richard D. Sparks to Mary L. Jobe Akeley," (February 4, 1932), SG, Series 1, Folder 2; Benchley, *My Friends, the Apes*, 207–210.

115. Benchley, *My Friends, the Apes*, 224; "Belle Benchley to Richard Sparks," (October 28, 1932), SG, Series 1, Folder 3.

116. Benchley, *My Friends, the Apes*, 278–281.

117. Ibid., 187–188.

118. "Belle Benchley to Richard Sparks," (February 13, 1950), SG, Series 1, Folder 3.

119. "Jean Lowman to Saunders," (February 10, 1954), Museum of History and Industry, Lowman Family Collection on Bobo the Gorilla, Folder 10.

120. "Simian Society of America," (November 6, 1959), Lowman Collection, Folder 10; "Picture from the Family Album," *Whidbey News-Times* (February 29, 1968) in Lowman Collection, Folder 15.

121. "Bobo Mobil Gas," Photograph, (1954), Lowman Collection, Folder 7 and "Expenses for IRS records," (1951–1954), Folder 18.

122. "Gertrude Lintz to Jean Lowman," (June 17, 1961), Lowman Collection, Folder 10.

123. "Bobo's 'Mother' Recalls: Gorilla Was 'Dear Child'," *Seattle Times* (May 22, 1983), Lowman Collection, Folder 16.

124. George P. Vierheller, "Behind the Scenes with the Chimps" (October, 1948), National Zoological Park Records, Box 173, Folder 11; Wayne Whittaker, "3 Ring Zoo" *Popular Mechanics* 90 (August 1948): 92–97, 260–264.

125. "Bath room . . . ," (ca. 1953), Lowman Collection, Folder 10.

126. "Itemized Damages," (1951–1953), Lowman Collection, Folder 19.

127. Steven Berentson, "Bobo," *Skagit Argus* (November 30, 1982), Lowman Collection, Folder 16.

128. "Vallos Zirbel to Raymond and Jean Lowman," (November 30, 1953) and "Sam and Lola to Raymond and Jean Lowman," (February, 1968), Lowman Collection, Folder 10.

129. John J. Reddin, "Bobo: Big, Ugly, Irresistible," (1968), Lowman Collection, Folder 15; "Gorilla Newly-Weds," *Universal Newsreel*, Volume 30, Release 1 (December 27, 1956), National Archives and Records Administration, Series: Motion Picture Releases of the Universal Library, Collection: MCA/Universal Pictures Collection.

130. Tom Robbins, "Bobo and His Keeper," *Seattle Times* (June 24, 1962), Lowman Collection, Folder 10.

131. Reddin, "Bobo: Big, Ugly, Irresistible."

6. Don't Feed the Keepers

1. "Frank Baker to Charles D. Walcott," (October 29, 1906), NZP, Box 150, Folder 8.

2. "W. H. Blackburne to Frank Baker," (February 28, 1905); "Blackburne to Baker," (May 30, 1905); "N.Z.P. Permanent Roll," (ca. 1905); "William T. Hornaday to Frank Baker," (April 2, 1905), all in NZP, Box 147, Folder 6; "Payroll of Employees, Length of Service," (1916); "Note on Salaries for Keeper," (March 9, 1917), both in PZ, BOX PER B1.

3. "M. C. Hunter to Frank Baker," (January 28, 1905), NZP Box 142, Folder 9.

4. "Keepers to Frank Baker," (October 30, 1906), NZP, Box 147, Folder 6.

5. "Report on the Employees' Committee on Positions," (November 24, 1919), NZP, Box 148, Folder 1.

6. "Notice to Keepers and Janitors," (October 10, 1907), OD, Box 131, Folder: Circulars/ Notices to Park Staff, 1900–1920,

7. William Bridges, *A Gathering of Animals: An Unconventional History of the New York Zoological Society* (New York: Harper and Row, 1974), 364–370.

8. "To All Officers of the Zoological Park," (August 4, 1917), OD, Box 131, Folder: Circulars/Notices to Park Staff, 1900–1920; Jonathan P. Spiro, *Defending Master Race, Conservation, Eugenics, and the Legacy of Madison Grant* (Burlington: University of Vermont Press, 2009), 143–166.

9. "J. E. Dean et al to Frank Baker," (October 30, 1906) and "Petition to Charles D. Walcott," (January 27, 1908), both in NZP, Box 147, Folder 6.

10. "Petition to Walcott." On manly codes of labor, see Ava Baron, "An 'Other' Side of Gender Antagonism at Work: Men, Boys, and the Remasculinization of Printers; Work, 1830–1920," in Baron, ed., *Work Engendered: Toward a New History of American Labor* (Ithaca, NY: Cornell University Press, 1991), 47–69; David Montgomery, *Workers Control in America* (Cambridge: Cambridge University Press, 1979), esp. 1–31; Paul Michel Taillon, "To Make Men Out of Crude Material: Work Culture, Manhood, and Unionism in the Railroad Running Trades, c. 1870–1900," in Roger Horowitz, ed., *Boys and their Toys: Masculinity, Class and Technology in America* (New York: Routledge, 2001): 13–33. On notions of caring work, see Eileen Boris and Jennifer Klein, *Caring for America: Home Health Care Workers in the Shadow of the Welfare State* (New York: Oxford University Press, 2012).

11. "Regulations for Keepers of Primate House" (May 2, 1918), OD, Box 131, Folder: Circulars/Notices to Park Staff, 1900–1920.

12. "Frank Baker to S. P. Langley," (March 11, 1905), NZP, Box 147, Folder 6.

13. "Long Hours at the Zoo," *Washington Herald* (March 16, 1910) in NZP, Clippings: Labor Union (1910); "Petition to Walcott."

14. "William T. Hornaday to Frank Baker," (April 2, 1906), NZP, Box 147, Folder 6.

15. "Keepers to Arthur Brown," (Petition, June 24, 1902); "Petition from Keepers and Groundsmen," (November 22, 1906); "Monthly Wages of Employees," (March 1911); "Frank Baker to Robert Carson," (January 30, 1914); "Robert Carson to William T. Hornaday," (February 2, 1914); "Robert Carson to Sol Stephan," (February 2, 1914); "Keepers Wages, Length of Service, and Grievance," (February 2, 1916); "Letter from C. Emerson Brown to Keepers," (May 12, 1919), in PZ, BOX PER B1.

16. "To the Employees of the N.Y. Zoological Park," (January 30, 1900); "Notice to Keepers and Janitors," (October 10, 1907); "Warning against Gossiping," (November 5, 1915); "William Temple Hornaday to All Keepers," (July 17, 1918), all in OD, Box 131, Folder: Circulars/Notices to Park Staff, 1900–1920.

17. "W. R. Blair to Employees," (October 15, 1931), OD, Box 131, Folder: Circulars/Notices to Park Staff, 1920–1940.

18. "J. W. Buchanan to William H. Mann," (July 19, 1925), NZP, Box 146, Folder 7.

19. "Arthur Clark to A. B. Baker," (February 4, 1916), NZP, Box 144, Folder 1.

20. "C. A. Williams to Frank Baker," (May 11, 1916), NZP, Box 144, Folder 1.

21. C. E. Matz to Frank Baker," (January 9, 1919), NZP, Box 144, Folder 1.

22. "Table Showing Age of Employees, Length of Service in the Park," (January 31, 1908), NZP, Box 147, Folder 6.

23. Dee Burque and Gilbert Laurence, "Revue in the Zoo," (n.d.), Records of the Federal Theater Project Playscripts, National Archives, R669, Box 319.

24. Florenz K. Buschmann, "Personages to Be Met at the Zoo," *Washington Post* (October 26, 1930): SM4.

25. "W. H. Blackburne to Frank Baker," (June 23, 1914) and "Ned Hollister to Rep. Sydney E. Mudd," (January 29, 1917), both in NZP, Box 142, Folder 11.

26. "Memorandum of Understanding between the New York Zoological Society and Its Employees in the Zoological Park," (ca. 1907), NZP, Box 141, Folder 9.

27. "Report of Personal Injury," (December 13, 1926), NZP, Box 147, Folder 7.

28. A. W. Rolker, "The Rogues of a Zoo," *McClure's Magazine* 23 (May 1904).

29. "Zoo Elephant Killed with Big Rifle after It Goes on War Path," (1933), STZ, Series II, Scrapbook, Vol. 7.

30. Rolker, "The Rogues of a Zoo."

31. George "Slim" Lewis and Byron Fish, *I Loved Rogues* (Seattle: Superior Publishing Company, 1978), 86.

32. Ibid., 3, 89–90.

33. Ibid., 90–93.

34. "William T. Hornaday to Department of Mammals," (October 2, 1920), OD, Box 131, Folder: Circulars/Notices to Park Staff, 1900–1920.

35. "General Order," (July 11, 1914); "General Notice," (April 4, 1906); "Notice to Keepers and Janitors," (October 10, 1907), all in OD, Box 131, Folder: Circulars/Notices to Park Staff, 1900–1920. "General Order," (September 28, 1937); "W. Reid Blair to the Uniformed Employees," (February 28, 1936); "Backsliding from the Rules," (July 10, 1922), all in OD, Box 131, Folder: Circulars/Notices to Park Staff, 1920–1940.

36. "General Notice," (August 21, 1916) and "Rotten Fish," (August 14, 1916), OD, Box 131, Folder: Circulars/Notices to Park Staff, 1900–1920.

37. "Fairfield Osborn to the Employees," (February 9, 1942), OD, Folder: Employee Notices, ca. 1943–1952.

38. "Allen to Osborn," (June 25, 1942).

39. "Memorandum of Meeting between Daniel Allen and Fairfield Osborn and Harold J. O'Connell, Secretary," (November 28, 1941), OP.

40. "All Members of the Board of Trustees," (January 26, 1942) and "J. Watson Weebb to Osborn," (January 29, 1942), OP.

41. "Memorandum Telephoned by Mr. O'Connell," (December 3, 1941), "Untitled List," (ca. December 9, 1941), and "John Tee-Van to All Members of Personnel Committee," (February 25, 1942), OP. On the emergence of the CIO and trade unionism in the 1930s and in the postwar period, see Lizbeth Cohen, *Making a New Deal: Industrial Workers in Chicago, 1919–1939* (Cambridge: Cambridge University Press, 1990); George Lipsitz, *Rainbow at Midnight: Labor and Culture in the 1940s* (Urbana: University of Illinois Press, 1994).

42. "William Spahn, Organizational Director to Fairfield Osborn," (November 24, 1941), OP.

43. "Memo from Driscoll to J. Tee-Van," (January 31, 1942); "Tee-Van to Fairfield Osborn," (February 1, 1942), both in OP.

44. "Memo from Driscoll to J. Tee-Van," (January 31, 1942).

45. "Interview with John Tee-Van with William Cully, Head Keeper, Mammals, relating to conduct of certain Mammal Department Keepers," (February 1, 1942); "CIO Meeting Held February 2nd," (n.d., ca. 1942); "Wage Increase Promised!" (February 16, 1942), OP.

46. "Fairfield Osborn to Ely, Kelly, O'Connell," (July 8, 1942), OP.

47. Daniel Allen to Fairfield Osborn," (February 16, 1942); "Spring Pickets Will Sprout at Bronx Zoo," *PM* (March 29, 1942), OP.

48. "Meet Mickie Wolfe of Bronx, N.Y.," *State County and Municipal Workers News* (March 23, 1943), OP.

49. "Employees," (August 25, 1944); "Those Rumblings . . . ," *Daily Worker* (September 19, 1944), both in OP. On strikes and labor relations during World War II, see Nelson Lichtenstein, *Labor's War at Home: The CIO in World War II* (Cambridge: Cambridge University Press, 1982).

50. "San Diego Zoo Reopens after Strike," *Los Angeles Times* (July 3, 1965): A12; "Brookfield Workers Back on the Job," *Chicago Tribune* (August 28, 1966): 16; "Philadelphia Zoo Closes during Strike," *Chicago Tribune* (October 23, 1971): A27.

51. "Pickets Carry Signs Stressing Dignity Quest," *Washington Post* (July 10, 1974), C1.

52. "New York Zoological Society, Union Activities," (September 7, 1944), OP.

53. "Choate, Mitchell, and Ely to Laurence Rockefeller," (September 13, 1944); "Choate, Mitchell, and Ely to Fairfield Osborn," (September 14, 1944); "Osborn to Tee-Van," (August 3, 1946), OP.

54. "Memo, Conway, Cuyler, Driscoll, and Gandal to John Tee-Van," (October 22, 1959), OP.

55. "Jerry Wurf to Fairfield Osborn," (Telegram, March 30, 1961); "John Tee-Van to Jerry Wurf," (August 31, 1959); "Telegram, Jerry Wurf to John Tee-Van," (June 16, 1960), all in BJB, Series V, Box 5, Folder 38.

56. "No selfish goal . . . ," (December 12, 1959), OP.

57. "William Conway to William Evans," (March 10, 1961), BJB, Series V, Box 5, Folder 38.

58. "Catherine C. Huntington to Fairfield Osborn," (April 27, 1961); "Warren Kinney to Fairfield Osborn," (April, 1961), both in OP.

59. "John Tee-Van to Wyle," (December 1, 1959); "William Conway to the Negotiating Committee for District Council 37 and Local 1501," (November 4, 1960), both in OP.

60. "Pickets Go to the Zoo," *The Economist* (May 6, 1961), OP.

61. William Conway and Christopher Coates to All Employees," (April 6, 1961); "William G. Conway and Charles B. Driscoll to All Employees," (April 21, 1961), BJB, Series V, Box 5, Folder 38.

62. "Monkeys on the Picket Line," *New York World-Telegram and Sun* (April 11, 1961), OP.

63. For typical photos of strikers and their animals, see Photos, (March 18, 1961), (March 13, 1961), and (April 23, 1961), American Federation of State, County and Municipal Employees (AFSCME), District Council 37 Photographs, Tamiment Library/Robert F. Wagner Labor Archives, New York University, Series II, Box 31, Folder 1. For typical press coverage of animals on the picket lines, see "Strikes at Zoo Bring Pet Snakes," *New York Times* (April 10, 1961); "A Snake Graces a Picket Line," *New York Tribune* (April 10, 1961), DC 37, Series V, Box 54, Folder 23; as well as "Gertrude Weil Klein, Councilman,

Bronx, to Fairfield Osborn," (September 13, 1944), OP; "Animals Lonely in Zoo Strike," Volume 34, Release 29 (April 6, 1961), National Archives and Records Administration, Series: Motion Picture Releases of the Universal Library, Collection: MCA/Universal Pictures Collection.

64. "Silence of the Jungle Reigns at the Zoo," *Newsday* (April 3, 1961); "Bronx Zoo Elephants Settle into Strikes Popcornless Daze," *New York World-Telegram and Sun* (May 4, 1961), OP.

65. "Striking Pose," *New York Journal American* (April 3, 1961); "Grub's Tight at Bronx Zoo," *Newark Evening News* (May 9, 1961), both in OP.

66. "Pittsburgh Zoo Strike Leaves Animals Hungry," *Washington Post* (July 14, 1951): B2.

67. "Zoo Animals Miss the People," *St. Louis Globe-Democrat* (February 18, 1959) in Newsclippings, 1958–59.

68. "Zoo's Employes [*sic*] Still on Strike" and "Duke Lends a Hand," *St. Louis Globe Democrat* (February 17, 1959); "Picketing of Closed Zoo Threatened," *St. Louis Globe Democrat* (February 17, 1959), all in STZ, Series III, Newsclippings, Vol. 11.

69. "Up at Zoo, All Mourn for You," *New York Journal American* (May 9, 1961), OP.

70. "Otter Despair," *Newsweek* (April 17, 1961), OP.

71. "Hey Kids, Zoo Strike Is Over!" *New York Journal-American* (May 19, 1961), OP.

72. "Bronx Elephants Really Do Unpack Their Trunks," *New York Herald-Tribune* (May 21, 1961), OP.

73. "Fairfield Osborn to Members" (May, 1961), OP; Leon Fink, and Brian Greenberg, *Upheaval in the Quiet Zone: 1199 SEIU and the Politics of Health Care Unionism, 2nd edition* (Urbana: University of Illinois Press, 2009).

74. Joseph A. McCartin, "'A Wagner Act for Public Employees': Labor's Deferred Dream and the Rise of Conservatism, 1970–1976," *Journal of American History* 95 (June 2008): 123–148.

75. "John McKew to Robert O. Wagner," (September 30, 1982), and John McKew, "Labor and Management: Then, Now, and Tomorrow," (1982), AAZPA-96-024, Box 27.

76. McKew, "Labor and Management." For a discussion of the women zookeepers, see Chapter 8.

77. *AAZK Newsletter* 1 (November 1968): 3 in American Zoo and Aquarium Association Records, Accession 04-191, Smithsonian Archives, Box 4, Folder "AAZK Newsletter, 1968–73"; Rachel Watkins Rogers, *Zoo and Aquarium Professionals: The History of AAZK* (Topeka, KS: American Association of Zoo Keepers, 1992).

78. *AAZK Newsletter* 1 (November 1968): 3–9, AAZPA-04-191, Box 4, Folder "AAZK Newsletter, 1968–73."

79. Ed Roberts, "Keeper's Korner," *The Keeper* (September/October, 1973): 111–112 in AAZPA-04-191, Box 9, Folder "AAZK, The Keeper."

80. Lawrence Curtis, "Animal Technicians," *The Keeper* (September/October, 1973): 109–110, AAZPA-04-191, Box 4, Folder "AAZK Newsletter, 1968–73."

81. "Directive," (January 14, 1969); Richard Sweeney, "The AAZK Rejects Indifference," *AAZK Bulletin* 4 (November–December 1971): 2, both in AAZPA-04-191, Box 4, Folder "AAZK Newsletter, 1968–73."

82. Roberts, "Keeper's Korner."

7. The Zoo Man's Holiday

1. "The Zoo's African Safari Is Underway," *Zoonooz* (August 1956): 14.
2. "Five Days out of Nairobi," *Zoonooz* (September 1956): 3–7.
3. Ibid.
4. "Zoos," (June 1954), AZA, Roll 100.
5. George Pournelle, "In the African Bush," *Zoonooz* (October 1956): 5–7.
6. Ibid., 7.
7. Ibid.
8. http://cincinnatizoo.org/events/zoofari/ and http://www.lowryparkzoo.com/zoofari/. Both accessed August 9, 2014.
9. http://www.philadelphiazoo.org/Explore/Attractions/Camel-Safari.htm, http://www .philadelphiazoo.org/Explore/Attractions/Camel-Safari.htm, http://www.milwaukeezoo .org/visit/attractions/, http://zoofari.bronxzoo.com/, and http://www.stlzoo.org/visit /thingstoseeanddo/safari-tours/. All accessed August 9, 2014.
10. C. Emerson Brown, *My Animal Friends* (New York: Blue Ribbon Books, 1931), 192–193; 229–231; "Orders, Please," *Zoonooz* (June 1954): 6.
11. William Bridges, *Zoo Expeditions* (New York: William Morrow, 1945), 21.
12. Ibid., 23–24.
13. "Meeting for Monday," (September 14, 1953), AZA, Roll 100; William Bridges, *Zoo Careers* (New York: William Morrow, 1971), esp. 13–25
14. Quoted in Bridges, *Zoo Expeditions,* 24.
15. Theodore Roosevelt, "Foreword," in William Beebe, *Jungle Peace* (New York: The Modern Library, 1918), ix, 3–4.
16. Ibid., 4; Carol Grant Gould, *The Remarkable Life of William Beebe: Explorer and Naturalist* (Washington, DC: Island Press, 2004), 203–212.
17. Beebe, *Jungle Peace,* 3–4, 201–211.
18. William Beebe, *High Jungle* (New York: Duell, Sloan and Pearce, 1949), 319–335; William Beebe, *Jungle Days* (Garden City, NY: Garden City Publishing, 1923), 64–68.
19. William Beebe, *Tropical Wild Life in British Guiana* (New York: New York Zoological Society, 1917), 27; Beebe, "The Tropical Research Station, Bartica District, British Guiana," *Twenty-First Annual Report of the New York Zoological Society* (January 1917), 113–119.
20. Krista A. Thompson, *An Eye for the Tropics: Tourism, Photography, and Framing the Caribbean Picturesque* (Durham: Duke University Press, 2006); Kelly Enright, *The Maximum of Wilderness: The Jungle in the American Imagination* (Charlottesville: University of Virginia Press, 2012).
21. Beebe, *Jungle Peace,* 4.
22. Ibid., 147.
23. Beebe, *High Jungle,* 185–197.
24. Charles B. Driscoll, "New York Day by Day," *Reading Eagle* (May 19, 1940): 6; "Scientist Describes Weird Creatures at Lecture," *Milwaukee Sentinel* (January 31, 1931): 19.
25. Beebe, *High Jungle,* 185–197.
26. Beebe, *Jungle Peace,* v–ix, 94–95, 162, 168.
27. Ibid., 27–29.
28. "William Beebe's Jungle Days, Romance in Tropical Forests," *Milwaukee Journal* (July

17, 1925): 5; "Beebe Reports Unrestrainedly on Fantastic Andean Jungle," *Pittsburgh Post-Gazette* (May 22, 1949): 18.

29. Beebe, *High Jungle,* 138–145.

30. "Left Bright Lights for Jungle Trail, Ruth Rose, Ex-Broadway Star, Tells of Exploration in S. America," *Philadelphia American* (November 30, 1924), MG, Box 1, Folder 2.

31. "Herbert L. Satterlee to Madison Grant," (October 3, 1932), MG, Box 1, Folder 6.

32. "William Beebe to Madison Grant," (August 19, 1925), MG, Box 1, Folder 1.

33. "NYZoS Corporation," (November 10, 1940); "Concession Agreement," (November 19, 1938), both in World's Fair Committee, New York Zoological Society. Wildlife Conservation Society, Box 1.

34. "Left Bright Lights for Jungle Trail."

35. "Buys Large Yacht to Further Science," *New York Times* (May 20, 1925): 3; "H. D. Whiton Dean, Backed Beebe Trip," *New York Times* (November 1, 1930) 19; "Scientists Prepare Big Sea Expedition," *New York Times* (October 13, 1924): 17.

36. David Binney Putnam, *David Goes Voyaging* (New York: G.P. Putnam's Sons, 1928).

37. Raymond L. Ditmars, *The Forest of Adventure* (New York: Macmillan, 1933), 4–5.

38. Raymond L. Ditmars, "In Quest of the Bushmaster," *Zoological Society Bulletin* 35 (November–December 1932): 216–220.

39. "For MEN 10 to 60," *Boy's Life* (December 1933): 48.

40. Ditmars, *The Forest of Adventure.*

41. Ditmars, *Strange Animals I Have Known,* 319–326.

42. Anita Moffett, "A Jungle Romance by Dr. Ditmars: *The Forest of Adventure,*" *New York Times* (November 19, 1933): BR6; Beebe, *High Jungle,* 55–59.

43. Beebe, *High Jungle,* 23, 104–114.

44. Fairfield Osborn, "We 'Import' a Spot of Africa," *Zoological Society Bulletin* 44 (May–June 1941): 67–69; Harmon Goldstone, "The Zoological Park of the Future," *Zoological Society Bulletin* 43 (March–April 1940): 60–62; Lee S. Crandall, "Four Years of Africa-in-the-Bronx," *Animal Kingdom* 48 (May–June 1945): 59–61; L. H. Robbins, "Bringing Africa to the Bronx," *New York Times* (February 23, 1941): SM16–17.

45. Osborn, "We 'Import,'" 67; Allyn R. Jennings, "Our Future Zoo," *New York Zoological Society Bulletin* 44 (March–April 1941): 35–37.

46. Robbins, "Bringing Africa"; John O'Reilly, "Mechanized Zoo," *Popular Science Monthly* (January 1941): 104–106.

47. Fairfield Osborn, "The Opening of the Africa Plains," *Zoological Society Bulletin* 44 (May–June 1941): 72; Robbins, "Bringing Africa."

48. "Helene Carter to Fairfield Osborn," (June 2, 1942), New York Zoological Society Archives, RG 2, Control Number 1029, Fairfield Osborn Correspondence, 1935–1942, Box 4.

49. Brown, *My Animal Friends,* 233–234.

50. Osborn, "We 'Import,'" 67; Osborn, "The Opening of the Africa Plains," 72.

51. Goldstone, "The Zoological Park of the Future," 60–61.

52. Crandall, "Four Years of Africa-in-the-Bronx."

53. Osborn, "The Opening of the Africa Plains," 67.

54. "Osborn Promises Bigger Bronx Zoo," *New York Times* (June 17, 1943): 23.

55. William Mann, "Problems of Zoo in War Time," (ca. 1940), WLM, Box 12, Folder 7;

William Bridges, *Gathering of Animals: An Unconventional History of the New York Zoological Society* (New York: Harper & Row, 1974), 459.

56. "Roger Conant to William M. Mann," (December 9, 1941, and March 9, 1942); "Mary A. Bessemer to William M. Mann," (January 2, 1942); "Jay Gould to National Zoological Park," (January 12, 1942); "William M. Mann to Roger Conant," (March 10, 1942); "New York Zoological Society Information Service, Release Thursday, December 18, 1941," all in NZP, Box 191, Folder 2.

57. Mark Rosenthal, Carol Tauber, and Edward Uhlir, *The Ark in the Park: The Story of the Lincoln Park Zoo* (Urbana: University of Illinois Press, 2003), 73.

58. "The Zoo and Its Defense Activities," *Zoonooz* (February 1942): 7.

59. Clark Deleon, *America's First Zoostory: 125 Years at the Philadelphia Zoo* (Virginia Beach: Donning Company, 1999), 113–116.

60. "The Zoo and Its Defense Activities," 7.

61. "Bathysphere Used in War Research," *New York Times* (January 12, 1944): 25.

62. "Meeting for Wednesday," (September 16, 1953), AZA, Roll 100; "Bathysphere Used in War Research."

63. William Beebe, "The Pacific Is Before You: An Introduction to the Men of the Armed Services," in Fairfield Osborn, ed., *The Pacific World: Its Vast Distances, Its Lands and the Life upon Them, and Its Peoples* (New York: W.W. Norton, 1944), 13–20.

64. Osborn, *The Pacific World,* 17; "Genesis of a Book," *Animal Kingdom* (September–October 1996): 27–28.

65. "Genesis of a Book."

66. "Zoo Guarded Military Secrets," *Suburban Times* (August 23, 1945), PZ, Scrapbooks, PR R9.

67. Osborn, *The Pacific World,* 17.

68. Ibid., 13–16.

69. "Foreign Pets Find Homes," *Germantown Telegraph* (July 7, 1944); "GI News Item! Okinawa Pest Arrives at Zoo," *Philadelphia Record* (November 25, 1945), both in PZ, Scrapbooks, PR R9.

70. "Common Crisis in the Average Zoo," *Zoonooz* (December 1943): 6–7.

71. Ken Stott Jr., "With a Service Man in the Philippine Jungle," *Zoonooz* (October 1945): 6–7.

72. Ken Stott Jr., "Beginner's Africa, Part IX," *Zoonooz* (October 1950): 5–8; Stott, "Beginner's Africa, Part V," *Zoonooz* (March 1950): 7–8.

73. Ken Stott Jr., "Beginner's Africa, Conclusion," *Zoonooz* (August 1951): 6–8.

74. Ken Stott Jr., "Beginner's Africa, Part VII," *Zoonooz* (July 1950): 7–8.

75. "African Safaris, Tall Fish Stories Flow into Chicago," unnamed Chicago newspaper (August 12, 1955), MP, Series 3, Box 2, Folder 30.

76. Stott, "Beginner's Africa, Part IX," 6.

77. Ibid., 6.

78. Ken Stott Jr., "Beginner's Africa, Part XIV," *Zoonooz* (July 1951): 7–8.

79. Ken Stott Jr., "Beginner's Africa, Part XII," *Zoonooz* (January 1951): 7–8.

80. Ibid.

81. Ken Stott, Jr., "Beginner's Africa, Part X," *Zoonooz* (November 1950): 6–8.

82. Ken Stott Jr., "Beginner's Africa, Part I," *Zoonooz* (January 1950): 7–8; Stott, "Beginner's Africa, Part V," 7–8.

83. Stott, "Beginner's Africa, Conclusion," 8.

84. Ibid.

85. Ibid.

86. Ibid.

87. Charles Faust, "Sketchbook of East Africa," *Zoonooz* (February 1970): 4–17.

88. "Zoo Safari?" *Zoonooz* (February 1968): 13–14.

89. Mary Ellen Honsaker, "Africa's Presence," *Zoonooz* (June 1968): 8–9.

90. "Expedition to India," *Zoonooz* (July 1968): 4–7.

91. Raymond B. Cowles, "Vanishing Wildlife," *Zoonooz* (January 1956): 10–15.

92. John C. Devlin, "Zoo Head Finds African Big Game Holding Own: Osborn Returns from Tour Hopeful on Conservation," *New York Times* (April 7, 1960): 37; John S. Akama, Shem Maingi, and Blanca A. Camargo, "Wildlife Conservation, Safari Tourism, and the Role of Tourism Certification in Kenya: A Postcolonial Critique," *Tourism Recreation Research* 36, no. 3 (2011): 281–291; Colin Michael Hall and Hazel Tucker, eds., *Tourism and Postcolonialism: Contested Discourses, Identities and Representations* (London: Routledge, 2004).

93. Deleon, *America's First Zoostory,* 115.

94. "W. M. Mann to American Export Line," (June 5, 1941), NZP, Box 159, Folder 11.

95. "Modern Noah's Ark Reaches Pacific Coast with Elephants," *New York Times* (November 16, 1948): 31.

96. "Hunter, 'Babies,' Arrive in States," *Spokesman Review* (January 5, 1955): 1.

97. "Rosefelt Leaves Again," *Catalina Islander* (September 29, 1955): 2.

98. "Animal Exchange List," (May 17, 1956), AZA, Roll 100; "Conservation of Wildlife: Orang-Utans," (September 1962), AZA, Roll 101.

99. *Rhino! Rhino!* (Chicago: Chicago Zoological Society, 1949).

100. "Ralph Graham to Robert Bean," (March 14, 1948), in *Rhino! Rhino!*

101. "Graham to Bean," (April 18, 1948), in *Rhino! Rhino!*

102. "Graham to Bean," (February 20, 1948, and February 29, 1948), in *Rhino! Rhino!*

103. "Graham to Bean," (March 21, 1948, and March 28, 1948), in *Rhino! Rhino!*

104. "Graham to Bean," (May 9, 1948, and May 23, 1948), in *Rhino! Rhino!*

105. Bill Brinkley, "Zoo to Get 2 Baby Elephants as Nehru's Gift to U.S. Children," *Washington Post* (February 26, 1950): M1.

106. Department of State, "For the Press," (April 14, 1950); "T. N. Kaul to William Mann," (January 19, 1950), both in NZP, Box 159, Folder 9.

107. "For Immediate Release," (ca. 1962); "J. Lear Grimmer to Walter D. Stone," (January 31, 1962), both in NPD, Box 2, Folder "Delhi Zoological Park 1961–63."

108. "New Delhi Notebook," *The Statesman* (March 12, 1962), in NPD, Box 2, Folder "Delhi Zoological Park 1961–63."

109. Ibid.; "National Zoo to Embassy of India," (Rough Draft, ca. 1950), NZP, Box 159, Folder 9.

110. "Graham to Bean," (April 24, 1948), in *Rhino! Rhino!*

111. "Marlin Perkins to Blaine Hoover," (December 2, 1942), MP, Series 4, Box 3, Folder 49.

112. M. W. Childs, "The Snake Man's Fight for Life," *St. Louis Post-Dispatch* (Sunday magazine, February 17, 1929), MP, Series 3, Box 2, Folder 27.

113. Ibid.

114. "Forced Feeding for a Zoo Python," *St. Louis Post-Dispatch* (n.d.), Marlin Perkins Papers (sl 516), MP, Series 3, Box 2, Folder 28; Marlin Perkins, *My Wild Kingdom* (New York: E.P. Dutton, 1982), 35–40; "2,000 People See Python Get Fed through Hose," *Life Magazine* (October 4, 1937): 90.

115. Perkins, *My Wild Kingdom,* 45–47.

116. "Herpetologist Perkins Takes Over Park Zoo," *Chicago Daily Tribune* (December 11, 1944): 14; "Radio-Television: 'Zoo Parade's' New Safari," *Variety* (Jun 27, 1956): 23.

117. Perkins, *My Wild Kingdom,* 113.

118. Ibid., 109–115.

119. Let's Visit the Zoo Radio Program," (1969), PZ, Box Fin AA1.

120. "AAZPA Educational Committee," (ca. 1952), AZA, Roll 100.

121. "Zoorama," *Zoonooz* (March 1955): 11.

122. Perkins, *My Wild Kingdom,* 116–117. "Special Things a Zoo Man Must Know," (September 11, 195); "Animal Health," (December 5, 1949); "Monkey House," (October 18, 1948), all *Zoo Parade.*

123. "Evaluation of Educational Work in Zoological Parks and Aquariums as Revealed by Questionnaire Survey," (1957), AZA, Roll 103.

124. Perkins, *My Wild Kingdom,* 118–119.

125. "Perkins & Pals at Zoo Outrate TV Glamor Gals," *Chicago Sunday Tribune* (January 28, 1951), MP, Series 3, Box 2, Folder 30.

126. *Marlin Perkins' Zoo Parade* (New York: Dell, 1955), MP, Series 10, Box 10, Folder 257.

127. Gregg Mitman, *Reel Nature: America's Romance with Wildlife on Film* (Cambridge, MA: Harvard University Press, 1999), 134–156.

128. Perkins, *My Wild Kingdom,* 257.

129. Ibid., 161–162

130. Ken Stott Jr., "Bushwhackers," *Zoonooz* (October 1961): 10–15.

131. "Minutes of the Meetings of the American Association of Zoological Parks and Aquariums," (September, 1953), AZA, Roll 100; Lee S. Crandall, *A Zoo Man's Notebook* (Chicago: University of Chicago Press, 1966).

8. My Animal Babies

1. Belle J. Benchley, *My Animal Babies* (London: Faber and Faber, 1946), 14.

2. Ibid., 17–19. On histories of caring work, including in homes, see Eileen Boris and Jennifer Klein, *Caring for America: Home Health Care Workers in the Shadow of the Welfare State* (New York: Oxford University Press, 2012).

3. Benchley, *My Animal Babies,* 19–20.

4. Monique Berlier, "*The Family of Man:* Readings of an Exhibition," in Bonnie Brennan and Hanno Hardt, eds. *Picturing the Past: Media, History, and Photography* (Urbana: University of Illinois Press, 1999), 206–241; Eric J. Sandeen, *Picturing an Exhibition: "The Family of Man" and 1950s America* (Albuquerque: University of New Mexico Press, 1995); *The Race Concept: Results of an Inquiry* (Paris: UNESCO, 1952), 5;

Michelle Brattain, "Race, Racism, and Antiracism: UNESCO and the Politics of Presenting Science to the Postwar Public," *American Historical Review* 112 (December 2007): 1386–1413.

5. Marcia Winn, "2 Baby Orangs Draw Crowds to Zoo," *Chicago Daily Tribune* (April 29, 1940): 1.

6. "Shipload of Monkeys, Snakes, and Birds Anything but Fun to Girl Custodian," *New York Times* (August 3, 1949), 25.

7. "Arthur Foehl, Wild Animal Dealer, Dies in Singapore," *Billboard* (February 7, 1948), 54.

8. "Trefflich's Bird and Animal Co. Inc.," (August 17, 1948); "Henry Trefflich to William Mann," (September 1, 1948), both in NZP, Box 164, Folder 12.

9. "Cupid Hides Out in Office of Doctor," *Los Angeles Times* (August 24, 1952): A1; "Animal Loads Big Business for Airlines," *Los Angeles Times* (October 2, 1955): C14; Henry Trefflich (as told to Baynard Kendrick), *They Never Talk Back* (New York: Appleton-Century-Crofts, 1954), 113–136.

10. Marté Latham, *My Animal Queendom: Adventures in Collecting and Selling South American Animals* (Philadelphia: Chilton, 1963), 108–116.

11. Peter Ryhiner as told to Daniel P. Mannix, *The Wildest Game* (Philadelphia: J.B. Lippincott Company, 1958), 290–291, 319–320.

12. "55 Births at Zoo Here This Year," *St. Louis Globe-Democrat* (November 19, 1940), and "30,000 See Zoo's Baby Hippopotamus," *St. Louis Post-Dispatch* (July 21, 1941), both in STZ, Series II, Scrapbook, Vol. 11.

13. "Second Chimp Ever Born at Zoo to Have Role in Summer Shows," *St. Louis Post-Dispatch* (April 7, 1944), STZ, Series III, Newsclippings, Vol. 3.

14. Proud Zoo Mother's Joy Turns to Wrath when Photographer's Flash Frightens Baby," *St. Louis Post-Dispatch* (March 2, 1941), STZ, Series II, Scrapbook, Vol. 11; "Zoo Snapshots Baby Orang—It's a Girl," *St. Louis Post-Dispatch* (May 17, 1939), STZ, Series II, Scrapbook, Vol. 10; Benchley, *My Animal Babies*, 49–54.

15. "Zoo Babies," (Coronet Films, 1955); "Father of Hippo Is Big-Mouthed Braggart," *Evening Bulletin* (March 16, 1962), PZ, Scrapbooks, PR R17.

16. "Zoo Snapshots Baby Orang."

17. Benchley, *My Animal Babies*, 86–87.

18. "And the Stork Flew Low," *Fauna* 7 (June 1945): 61; Rex Polier, "Many Animals at Zoo Are Parental Delinquents; Keepers Perform Mother Roles for 'Babies'," *Sunday Bulletin* (June 19, 1955), PZ, Scrapbooks, PR R14.

19. "Three New Tiger Cubs Born at Zoo, Deserted by Mother," *St. Louis Post-Dispatch* (October 29, 1954), STZ, Series III, Newclippings, Vol. 8.

20. Benchley, *My Animal Babies*, 53.

21. Herbert L. Ratcliffe, "Baby Black Leopards," *Fauna* 2 (December 1940): 85–87; "'Mother Love' Threatens Life of Baby Orang-Utan at Zoo," *Philadelphia Record* (November 20, 1944); Barbara Barnes, "Zoo Moms Not Always Young's Best Pal," *Evening Bulletin* (April 20, 1946), both in PZ, Scrapbooks, PR R9.

22. Ratcliffe, "Baby Black Leopards"; "Adult Dik-diks Shun Baby, Zoo Show It Love," *Philadelphia Inquirer* (August 2, 1963), PZ, Scrapbooks, PR R17.

23. "Lion Cub, Disowned by Mother, Cared for at Attendant's Home," *St. Louis Star-Times* (February 9, 1944), STZ, Series III, Newsclippings, Vol. 3. On the connection between

race and adoption in the postwar era, see Laura Briggs, "Mother, Child, Race, Nation: The Visual Iconography of Rescue and the Politics of Transnational and Transracial Adoption," *Gender & History* 15, no. 2 (2003): 179–200.

24. Elaine Tyler May, *Homeward Bound: American Families in the Cold War Era* (Basic Books, 2008); Fred Turner, "The Family of Man and the Politics of Attention in Cold War America," *Public Culture* 24, no. 1 (2012): 55–84.

25. "Baby Ape Thrives on Tender Care of Zoo Employe's [*sic*] Wife," *St. Louis Globe-Democrat* (February 20, 1947), STZ, Series III, Newclippings, Vol. 5.

26. Sarah Potter, *Everybody Else: Adoption and the Politics of Domestic Diversity in Postwar America* (Athens: University of Georgia Press, 2014); Laura Briggs, *Somebody's Children: The Politics of Transracial and Transnational Adoption* (Durham: Duke University Press, 2012).

27. "Baby Chimp from Zoo Is Spoiled Child in St. Louis Woman's Home," *St. Louis Post-Dispatch* (September 12, 1945); "Just Like a Baby," *St. Louis Globe-Democrat* (September 13, 1945); "Chimpanzee in the Home," *St. Louis Post-Dispatch* (September 23, 1945), STZ, Series III, Newsclippings, Vol. 4.

28. "Baby Chimp from Zoo Is Spoiled Child in St. Louis Woman's Home"; "Chimpanzee in the Home," *St. Louis Post-Dispatch* (September 23, 1945), STZ, Series III, Newsclippings, Vol. 4.

29. Anna Mae Noell, *The History of Noell's Ark Gorilla Show* (Tarpon Springs, FL: Noell's Ark, 1979); Trefflich, *They Never Talk Back*, 76–77.

30. Lucile Q. Mann, *From Jungle to Zoo: Adventures of a Naturalist's Wife* (New York: Dodd, Mead, 1934), 10–22.

31. Ibid., 23–27.

32. "Memorandum, National Zoological Park—Smithsonian Institution Certificates of Appreciation for the Foster Rearing of Infant Animals," (March 25, 1965); "Theodore Reed to Remington Kellogg," (March 6, 1959); "Theodore Reed to Remington Kellogg," (February 27, 1959), all in NZOP, Box 11, Folder 12.

33. William Bridges, *Zoo Babies* (New York: William Morrow, 1953), 14–21.

34. Ibid.

35. "Jacqueline Sabin to William Mann," (April 9, 1943), and "Mann to Sabin," (April 26, 1943), NZP, Box 243, Folder 1.

36. "William Mann to Beatrix T. Moore," (June 30, 1944), NZP, Box 243, Folder 2.

37. "Beatrix T. Moore to William Mann," (June 26, 1944), NZP, Box 243, Folder 2.

38. Harry M. Wegeforth and Neil Morgan, *It Began with a Roar: The Beginning of the World-Famous San Diego Zoo* (San Diego: Zoological Society of San Diego, 1990), 38–39; "Frank Buck to Emerson Brown," (January 30, 1924), PZ, Box PER AA2. See also: Vernon N. Kisling Jr., "Zoological Gardens of the United States," in Kisling, ed., *Zoo and Aquarium History: Ancient Animal Collections to Zoological Gardens* (Boca Raton, FL: CRC Press, 2001), 170–171

39. Belle J. Benchley, *My Life in a Man-Made Jungle* (Boston: Little, Brown, 1940), 6, 90–92.

40. Ibid., 119–120.

41. Ibid., 119–166.

42. "Female Touch: Gorillas' Manners Impeccable," *Los Angeles Times* (March 1, 1953): 36; Joan Morton Kelly, "Spare the Rod," *Zoonooz* (August 1953): 3–5.

43. Belle J. Benchley, *Shirley Visits the Zoo* (Boston: Little, Brown, 1946).

44. "This Is Good Time to See Zoo Animals," *Chicago Daily Tribune* (January 9, 1958): W1.

45. AAZPA, "A Zoological Park: Why, Where, How," (Lansing: Michigan State University, Cultural Experiment Station, ca. 1957), AZA, Roll 101.

46. "Casey to Washington, D.C. Lion Department," (May 19, 1958), NZP, Box 212, Folder 1. A typical adult letter—one of many—calling for the lion's death is "Josephine Tuttle to Theodore Reed," (May 17, 1958), NZP, Box 212, Folder 1; Lisa Uddin, *Zoo Renewal: White Flight and the Animal Ghetto* (Minneapolis: University of Minnesota Press, 2015), chapter 2.

47. "Zoo Takes More Steps for the Safety of Visitors," *Washington Post Times Herald* (May 22, 1958), NZP, Box 212, Folder 2.

48. "Margery Johnson to Washington Zoo," (May 25, 1958), NZP, Box 212, Folder 1.

49. "Peter Gawura to Washington Zoo," (May 17, 1958), NZP, Box 212, Folder 1. Emphasis is original.

50. Loring Lovett, "Daddy, Animals Are Wonderful!" *Zoonooz* (April 1956): 2–4.

51. Charles R. Schroeder, "Children's Zoos World Wide," *Zoonooz* (April 1956): 12–13; Raka Shome, "'Global Motherhood': The Transnational Intimacies of White Femininity," *Critical Studies in Media Communication* 28, no. 5 (2011): 388–406.

52. Cynthia Ketchum, "Children Learn from Animals," *Zoonooz* (April 1956): 14.

53. Clay Cowran, "You Can Pet a Lion at the Brookfield Zoo," *Chicago Daily Tribune* (May 27, 1957): 1.

54. Louise Hutchinson, "This Mrs. Noah Keeps Busy as Foster Mother to 300 Baby Animals at the Zoo," *Chicago Daily Tribune* (June 21, 1959): 22; "Superintendent of Children's Zoo Loves 'Mothering' Her Animals," *Sunday Bulletin* (April 21, 1957), PZ, Scrapbooks, PR R15.

55. Sylvia Shepherd, "Volunteers Keep Zoo's Young Clean," *Chicago Tribune* (May 17, 1964): A1; Marlin Perkins, "Zoo Parade: Children's Zoo," (June 4, 1951).

56. Gay Keuster, "Women in Cages," *The Keeper* (March/April 1973): 37–39, in AAZPA, Box 9, Folder "AAZK, The Keeper."

57. Ibid.

58. *The Keeper* (January/February, 1973), MJ.

59. "Report of the Conservation of Wildlife Committee, Theodore Reed, Chair," (1967), AZA, Series I, Roll 108.

60. Leonard Carmichael, Mozelle Bigelow Kraus, and Theodore Reed, "The Washington National Zoological Park Gorilla Infant, Tomoko," *International Zoo Yearbook* 3 (January 1962): 88–93.

61. Ibid. See also "Foster Mother Loses Tiny Tots," *Universal Newsreel*, Volume 35, Release 37 (May 3, 1962), National Archives and Records Administration, Series: Motion Picture Releases of the Universal Library, Collection: MCA/Universal Pictures Collection.

62. "Clifford C. Gregg to Robert Bean," (September 20, 1957), AZA, Series I, Roll 101; "First Baby Gorilla," *Universal Newsreel*, Volume 30, Release 36 (April 29, 1957), National Archives and Records Administration, Series: Motion Picture Releases of the Universal Library, Collection: MCA/Universal Pictures Collection.

63. Don G. Davis, "Breeding Animals in Zoos," *International Zoo Yearbook* 4 (1963): 72–75.

64. Fairfield Osborn, "A New Opportunity—Zoos Help Wildlife," *International Zoo*

Yearbook 4 (1963): 65–66; Philip Armstrong, "The Postcolonial Animal," *Society and Animals* 10, no. 4 (2002): 413–420.

65. "American Association of Zoological Parks and Aquariums to All Members," (June 3, 1958), and "United States Senate, Subcommittee on Agricultural Research and General Legislation of the Committee on Agriculture and Forestry," "Statement of Leonard J. Goss, President, AAZPA," (July 23, 1958), AZA, Series I, Roll 101.

66. Martin W. Holdgate, *The Green Web: A Union for World Conservation* (London: Earthscan, 1999); Robert Boardman, *International Organization and the Conservation of Nature* (Bloomington: Indiana University Press, 1981).

67. E. M. Lang, "What Are Endangered Species?" *International Zoo Yearbook* 17 (1977): 2–5; "International Union for Conservation of Nature and Natural Resources," *International Zoo Yearbook* 5 (1965): 325–328. I am indebted as well to my colleague Kenneth Iain MacDonald for sharing the text of "IUCN—The World Conservation Union: A History of Constraint," Text of an Address given to the Permanent Workshop of the Centre for Philosophy of Law, Higher Institute for Philosophy of the Catholic University of Louvain (February 6, 2003).

68. United States Department of the Interior, "Wildlife Imported into the United States in 1967," MJ.

69. "AAZPA Resolution on Certain Endangered Species of Animals," (October 23, 1967), AAZPA, Box 35; W. K. Van den Bergh, "The Zoological Garden—Noah's Ark of Future Generations," *International Zoo Yearbook* 4 (1963): 61–62; Geoffrey Schomberg, "The Responsibilities of Zoos, Shippers, and Dealers," *International Zoo Yearbook* 14 (1974): 1–2; F. Wayne King, "International Trade and Endangered Species," *International Zoo Yearbook* 14 (1974): 2–12.

70. "Theodore Reed to Marvin L. Jones," (July 10, 1973), and "John Perry to Marvin L. Jone," (April 4, 1973), MJ.

71. "Marvin Jones to Ronald Strahan," (July 7, 1970), Marvin L. Jones, "Studbooks and Their Importance in the Modern Zoo World, a Presentation at the First Annual Conference of the American Association of Zoo Keepers," (May 5, 1970), MJ; Marvin Jones, "The Pigmy Hippopotamus, *Choeropsis liberiensis, Morton*," (For Presentation at the Annual Conference, American Association of Zoological Parks and Aquariums, December 3–7, 1967), AZA, Series I, Roll 108. For discussion of postwar ideas of breeding, see Alexandra Minna Stern, *Eugenic Nation: Faults and Frontiers of Better Breeding in Modern America* (Berkeley: University of California Press, 2005).

72. "Marvin Jones to John Perry," (June 21, 1973), MJ.

73. Jones, "Studbooks and Their Importance in the Modern Zoo World."

74. Ibid.

75. "Wayne King to Earl Baysinger," (July 21, 1972); "Clyde Hill to Wayne King," (July 3, 1972); "Endangered Species Permit No. ES-241," (August 11, 1972); "William P. Braker to Marvin Jones," (October 24, 1973); "Report of the Wildlife Conservation Committee to the AAZPA Board of Directors," (October 1973), all in MJ. The Marvin Jones Papers contain numerous other examples of such endangered species permits and the debate with the Conservation Committee around each transaction. In most case, there was concern over missing documentation.

76. "Marvin Jones to John Perry," (June 21, 1973).

77. John Perry, Donald D. Bridgewater, and Dana Horsemen, "Captive Propagation: A Progress Report," *Zoologica* 576 (Fall 1972): 109–118, in MJ.

78. Jones, "Studbooks and Their Importance in the Modern Zoo World."

79. "Studbooks for Rare Species of Wild Animals in Captivity," *International Zoo Yearbook* 12 (1972): 408–410; "William Conway to John Perry," (January 31, 1969), MJ.

80. "First Symposium on the Przewalksi Horse," *American Association of Zoological Parks and Aquariums Newsletter* 1 (February 1960): 1, AZA, Series I, Roll 102.

81. John Perry, "The Role of Zoos in Saving Threatened Species," AAZPA Mexico City Conference, (March 1967), AZA, Series I, Roll 108.

82. "Marvin Jones to Peter Kibbee," (August 22, 1973), MJ; "The Animal Exchange List from the American Association of Zoological Parks and Aquariums," (May 1958), AZA, Series I, Roll 101.

83. John Perry and Peter B. Kibbee, "The Capacity of American Zoos," *International Zoo Yearbook* 14 (1974): 240–247.

84. Ibid.

85. Ed Roberts, "The Zoo's Keeper," 19–21, and Ken Kawata, "Beyond the Shovel and Bucket," 23–26, both in *AAZK Conference Proceedings* (Chicago, 1974); R. D. Martin, ed., *Breeding Endangered Species of Captivity* (New York: Academic Press, 1975).

86. Bob Truett, "Zoo Keeping Made Easy," *Animal Keepers' Forum* 2 (May 1975): 3–5.

87. James E. Fouts, "Studbooks," *Animal Keepers' Forum* 2 (May 1975): 5–6.

88. "Happier at the Zoo," *Toledo Herald-Journal* (November 21, 1971): 14; Glenous M. Liebherr, "Artificial Insemination in Chimpanzees," *AAZK Conference Proceedings* (Chicago, 1974), 10–13; C. J. Hardin, Liebherr, and Olen Fairchild, "Artificial Insemination in Chimpanzees," *International Zoo Yearbook* 16 (1975): 132–134.

89. Robert A. Brown, "Why Children's Zoos?" *International Zoo Yearbook* 13 (1973): 258–261.

90. http://www.toledozoo.org/site/page/zoo_pal_animal_adoption_program; http://www .clevelandzoosociety.org/files/adopt-brochure-current14.pdf; https://www.applyweb.com /public/contribute?s=STLZOOAD; http://www.detroitzoo.org/support/adopts. All accessed December 17, 2014.

9. Dangerous Safari

1. Lewis Cotlow, *Zanzabuku [Dangerous Safari]* (New York: Rinehart & Company, 1956), 257; Cotlow, *Zanzabuku* (Republic Pictures, 1956).

2. Robert Ruark, "Jungle Pictures Can Be Dangerous," *Sarasota Herald-Tribune* (March 10, 1953): 4.

3. "Safari in Africa," *New York Times* (August 16, 1956): 30.

4. Lisa Uddin, *Zoo Renewal: White Flight and the Animal Ghetto* (Minneapolis: University of Minnesota Press, 2015).

5. Roderick P. Neumann, *Imposing Wilderness: Struggles over Livelihood and Nature Preservation in Africa* (Berkeley: University of California Press, 2002); Gregory Maddox, James Giblin, and Isaria N. Kimambo, eds., *Custodians of the Land: Ecology and Culture*

in the History of Tanzania (Athens: Ohio University Press, 1996); Jonathan S. Adams and Thomas O. McShane, *The Myth of Wild Africa: Conservation without Illusion* (New York: W.W. Norton, 1992).

6. "Wild Spots Have Good Points, Veteran Explorer Declares," *Reading Eagle* (August 17, 1956): 8; "The Last of the Great Explorers: The Lewis Cotlow Story," *Variety* (February 5, 1986): 110; Amy J. Staples, "The Last of the Great (Foot-Slogging) Explorers: Lewis Cotlow and the Ethnographic Imaginary in Popular Travel Film" in Jeffrey Ruoff, ed., *Virtual Voyages: Cinema and Travel* (Durham: Duke University Press, 2006), 195–216.

7. "Menu for a Black-Tie Dinner: Roast Lion and Marinated Boar," *New York Times* (April 16, 1977): 19.

8. "Dine on Jungle Dishes Here," *New York Times* (October 20, 1938): 25. Cotlow, Lewis. *In Search of the Primitive* (Boston: Little, Brown, 1966); Cotlow, *The Twilight of the Primitive* (New York, Macmillan, 1971); Cotlow, *Passport to Adventure* (New York: Bobbs-Merrill, 1942); John Platero, "Primitive No More," *Washington Post* (July 1, 1977): 113.

9. Cotlow, *Zanzabuku*, 322–323.

10. Roland Lindeman, "Africa—The Problem Land," *Parks and Recreation* 31 (July 1960): 343–345, 349–350, 356.

11. Cotlow, *Zanzabuku*, 170–178; Cotlow, *The Twilight of the Primitive* (New York: Macmillan, 1971), xi.

12. "James Dolan to Bernhard Grzimek," (May 5, 1970), MJ.

13. Brian Morris, "Wildlife Conservation in Malawi," *Environment and History* 7 (August 2001): 357–372; Abdallah Mkumbukwa, "The Evolution of Wildlife Conservation Policies in Tanzania During the Colonial and Post-Independence Periods," *Development Southern Africa* 25 (December 2008): 589–600; JoAnn McGregor, "Crocodile Crimes: People versus Wildlife and the Politics of Postcolonial Conservation on Lake Kariba, Zimbabwe," *Geoforum* 36, no. 3 (2005): 353–369.

14. Cotlow, *Zanzabuku*, 258–261.

15. Brian Herne, *White Hunters: The Golden Age of African Safaris* (New York: Holt, 2014), 310–311; Marie Smith, "Washington Society Tames Tamer," *Washington Post* (June 29, 1956): 37; "Hunter Says Herds Dwindle in Africa," *Los Angeles Times* (July 9, 1956): 23.

16. "John Seago to William Mann," (December 26, 1953), NZP, Box 164, Folder 1.

17. Cotlow, *Zanzabuku*, 183–185; "Tells Parliament 686 Africans Are Executed," *Atlanta Daily World* (October 26, 1954): 4.

18. Cotlow, *Zanzabuku*, 189.

19. Ibid., 46.

20. Jean Hartley, *Africa's Big Five and Other Wildlife Filmmakers: A Centenary of Wildlife Filming in Kenya* (Nairobi: Twaweza, 2010), 62; Dennis Hickey and Kenneth C. Wylie, *An Enchanting Darkness: The American Vision of Africa in the Twentieth Century* (East Lansing: Michigan State University Press, 1993).

21. "George Speidel to Board Members," (April 16, 1962), and "John Seago to Speidel," (February 21, 1961), ZSM, Box 14, Folder 4.

22. "AAZPA Resolution on Certain Endangered Species of Animals," (October 23, 1967), AAZPA, Box 35.

23. "George Speidel to Board Members," (April 16, 1962), and "Speidel to Board Members," (April 24, 1962), ZSM, Box 14, Folder 4.

24. "John Seago to Theodore Reed," (February 27, 1959), NZP, Box 163, Folder 11.

25. "John Seago to Theodore Reed," (June 17, 1958), NZP, Box 163, Folder 11.

26. "John Seago to Theodore Reed," (January 21, 1961), NPD, Box 3, Folder: John Seago.

27. "John Seago to Theodore Reed," (February 27, 1962), NPD, Box 3, Folder: John Seago.

28. "John Seago to Theodore Reed," (September 2, 1965), NPD, Box 3, Folder: John Seago.

29. "Seago to Reed," (September 2, 1965).

30. "Seago to Reed," (February 27, 1959).

31. "John Seago to Theodore Reed," (September 8, 1958), NZP, Box 163, Folder 11.

32. "John Seago to George Speidel," (February 6, 1961), ZSM, Box 14, Folder 4.

33. Theodore Reed, "The Bongo Story," (ca. 1968), NZA, Box 1, Folder: Bongo.

34. Theodore Reed, "Project: Bongo," (March 8, 1968), and "Theodore Reed to John Seago," (February 15, 1868), both in NZA, Box 1, Folder: Correspondence re: Bongo (1968–1969).

35. "John Seago to Theodore Reed," (December 3, 1968), and "List of Animals Caught in Trap and Released," (February 6, 1969), both in NZA, Box 1, Folder: Correspondence re: Bongo (1968–1969).

36. "John Seago to Theodore Reed," (June 27, 1968); "Cable Received," (June 11, 1969); "John Seago to Theodore Reed," (October 21, 1969), all in NZA, Box 1, Folder: Correspondence re: Bongo (1968–1969); Reed, "The Bongo Story."

37. "Theodore Reed to John Seago," (November 26, 1969), NZA, Box 1, Folder: Correspondence re: Bongo (1968–1969).

38. "Theodore Reed to John Seago," (October 26, 1970), NZA, Box 1, Folder: Correspondence re: Bongo (1970–1971).

39. "John Seago to Theodore Reed," (January 2, 1962), and "Seago Account—Partial Payment," (March 9, 1962), both in NPD, Box 3, Folder: John Seago.

40. William G. Conway, "How to Exhibit a Bullfrog: A Bed-Time Story for Zoo Men," *International Zoo Yearbook* 13 (January 1973): 221–226.

41. "Rockbound Tasha Shuns Zoo's Island Retreat," *Philadelphia Daily News* (April 9, 1968), PZ, Box AC A1.

42. "Sally and Susan Carr to Philadelphia Zoo," (April 14, 1968); "V. M. McCallum to Philadelphia Zoo," (April 16, 1968); and "Hemie Reimer to Philadelphia Zoo," (n.d.), PZ, Box AC A1.

43. "E. B. J. to Philadelphia Zoo," (n.d.), PZ, Box AC A1.

44. "Confidential Memorandum: Condition and Status Report Regarding Tasha," (June 5, 1968), PZ, Box AC A1.

45. "The Financial Headaches Running the Phila. Zoo," *Delaware Valley Business Fortnight* (June 14, 1971), and "Zoo to Close 2 Weekdays to Save Money," *Philadelphia Inquirer* (October 27, 1971), in PZ, Scrapbooks, Box PR L3.

46. Paul B. Wiener, "The Zoo—Neurotic Animals Going through Changes in Their Cages," *Thursday's Drummer* (October 22, 1970), in PZ, Scrapbooks, PR L3.

47. "Tasha Tigress Down for the Count" and "Haul That Tiger!" (April 26, 1968); "Public Relations Department: Tasha," (N.D.), PZ, Box AC A1; Bruce Bushel, "Atmospheric Contamination Probably Killed Tiger," in PZ, Scrapbooks, Box PR L3.

48. Wiener, "The Zoo—Neurotic Animals."

49. Desmond Morris, "Shame on the Naked Cage," *Life* 65 (November 8, 1968): 70–86;

D. J. Osborn, "Dressing the Naked Cage," *Curator: The Museum Journal* 14 (September 1971): 194–199; Heini Hediger, *Man and Animals in the Zoo* (New York: Delacorte Press, 1969); George Leposky, "Getting Back to Nature at Brookfield Zoo," *Inland Architect* 16 (1972): 22–26; John Perry, *The World's a Zoo* (New York: Dodd, Mead, 1969).

50. Bernard Fensterwald, "Time to Phase Out Zoos," *Washington Post* (February 7, 1974): A18.

51. Jack Anderson, "Some Zoos Called Dens of Horror;" "Most Large Zoos Obey Law, Phila. Spokesman Contends," *Evening Bulletin* (July 26, 1971), in PZ, Scrapbooks, Box PR L3.

52. Conway, "How to Exhibit a Bullfrog," 221.

53. "Precautions Make Phila. Zoo One of Safest," *Philadelphia Inquirer* (May 17, 1959), PZ, Scrapbooks, PR R16.

54. Charles MacNamara, "The Territorial Imperative," *Philadelphia Magazine* (October 1970): 94–95, 108–111.

55. "Zoo's PR Man Quits, Cites Laxity of Board," *Philadelphia Daily News* (August 8, 1970), "Aide Assails Zoo as Cruel to Animals; Chief Denies Charge," *Philadelphia Inquirer* (August 9, 1970), all in PZ, Scrapbooks, Box PR L3.

56. Roger Conant, *A Field Guide to the Life and Times of Roger Conant* (Toledo: Toledo Zoological Society, 1997), 242.

57. For a classic interpretation of urban segregation and poverty, see Thomas Sugrue, *The Origins of the Urban Crisis: Race and Inequality in Postwar Detroit* (Princeton: Princeton University Press, 1996). See Clark, quoted in Douglas S. Massey and Nancy A. Denton, *American Apartheid: Segregation and the Making of the Underclass* (Cambridge, MA: Harvard University Press, 2003), 3.

58. Morris, "Shame of the Naked Cage," 86; Uddin, *Zoo Renewal.*

59. "Lung Cancer Deaths Increase at Zoo; Air Pollution Suspected as Cause," *Evening Bulletin,* PZ, Scrapbooks, Box PR R17.

60. Desmond Morris, *The Human Zoo* (New York: McGraw Hill, 1969), 8–10.

61. Conant, *The Life and Times,* 248; James Stewart-Gordon, "There's No Business Like Zoo Business," *Reader's Digest* (May 1968): 209–216; "Mice Reflect City Problems," *Philadelphia Inquirer* (July 7, 1968), all in PZ, Scrapbooks, Box PR L2.

62. Conant, *The Life and Times,* 247.

63. Ibid., 251.

64. "Urban Life, Sex Frustration, Causing Heart Disease," *Philadelphia Evening Bulletin* (November 10, 1967), PZ, Scrapbooks, Box PR L2.

65. Conant, *The Life and Times,* 249, 250.

66. Daniel Moynihan, *The Negro Family: The Case for National Action* (Washington, DC: Office of Policy Planning and Research, 1965).

67. Morris, *The Human Zoo,* 136–148.

68. "White Citizen," "How Far for the Negro?" *Evening Bulletin* (April 8, 1965), in PZ, Scrapbooks, Box PR L1.

69. "Zoo's PR Man Quits"; "Territorial Imperative"; "Young Robber Knocks Out Guard at Zoo with Turtle," *Evening Bulletin* (March 27, 1962), in PZ, Scrapbooks, Box PR R17.

70. "Gangs at Zoo Rob, Attack Children," *Courier-Post* (May 22, 1969), in Philadelphia Zoo Records, Scrapbooks, Box PR L2; "Zoo 'Slum Clearance' Loses Out—to Humans,"

Philadelphia Daily News (October 14, 1961), in PZ, Scrapbooks, Box PR R16; Hugh Scott, "Beasts Suffer from Urban Pressures," *Philadelphia Inquirer* (July 11, 1965), in PZ, Scrapbooks, Box PR L1.

71. Conant, *The Life and Times,* 253.

72. "America's Zoos—Living Institutions Protecting Living Things through Education, Information, and Public Awareness," (ca. 1976), AAZPA, Box 36.

73. *Official Guide Book Milwaukee County Zoo* (1967), ZSM, Box 22, Folder 3. Emphasis is original.

74. Morris, "Shame of the Naked Cage," 78.

75. "Signs of Headway at the San Diego Wild Animal Park," *Zoonooz* (November 1971): 8–13.

76. Ibid., 13.

77. "On the Grand Tour," *Zoonooz* (May 1972): 9–19.

78. "Southern White Rhinoceros," *Zoonooz* (May 1971): 12–16.

79. "H. S. Sandruna, Chief Game Warden, to Theodore Reed," (May 20, 1959), NZP, Box 163, Folder 11.

80. "Africans Complain We Have False Idea of Country," *Philadelphia Inquirer* (September 7, 1968), PZ, Scrapbooks, Box PR L2.

81. "On the Grand Tour," 9–11.

82. Ibid., 9–19.

83. Zoological Society of San Diego, *Wild World of Animals* (San Diego: Zoological Society of San Diego, 1973).

84. "Seven Keys to Singapore," (1982), Episode 2: *Bring 'Em Back Alive* (CBS Television).

Conclusion

1. "Since 1870 . . ." (written notes on yeti expedition, ca. 1962), MP, Series 9, Box 9, Folder 235; Sir Edmund Hillary and Desmond Doig, *High in the Thin Cold Air* (Garden City, NY: Doubleday, 1962), iii–50.

2. See, for example, Ivan T. Sanderson, "The Strange Story of America's Abominable Snowman," *True: The Man's Magazine* (December 1959): 40–43, 123–126, in MP, Series 9, Box 9, Folder 231; and Norman G. Dyhrenfurth, "The Search for the Snowman," (Screen treatment, n.d.), MP, Series 11, Box 11, Folder 271.

3. "World Book Encyclopedia Scientific Expedition to the Himalayas Led by Sir Edmund Hillary," (n.d., ca. 1962), MP, Series 9, Box 9, Folder 234.

4. "Capture Gun (for yeti search)," (Photo, n.d., ca. 1960–61), MP, Series 26, Box 38, Notebook 2.

5. Marlin Perkins, *My Wild Kingdom, An Autobiography* (New York: E.P. Dutton, 1982), 139–141.

6. Ibid., 143–147.

7. "Since 1870 . . ."

8. "Publicity—R. Marlin Perkins," (ca. 1962), MP, Series 9, Box 9, Folder, 232; "Since 1870 . . ."

9. "Don Brewer to Marlin Perkins," (June 8, 1960); "George A. Agogino to Bailey Howard," (January 3, 1960); "Agogino to John W. Dienhart," (February 8, 1960); and "Grace Decker to Marlin Perkins," (August 27, 1960), MP, Series 4, Box 3, Folder 53.

10. "'Bob' [Robert Cooper] to Marvin Jones," (September 28, 1968), MJ.

11. "Marvin Jones to 'Bob' [Robert Cooper]," (October 4, 1968), MJ

12. "Marvin Jones to 'Bob' [Robert Cooper]," (November 26, 1968), MJ.

13. "Henry Trefflich to Zoo Directors," (October 24, 1977), NPD, Box 4, Folder: Trefflich.

14. "Henry Trefflich to William A. Xanten Jr.," (March 4, 1978), NPD, Box 4, Folder: Trefflich.

15. "Henry Trefflich to Zoo Directors," (January 17, 1978), NPD, Box 4, Folder: Trefflich.

16. "Ildiko P. DeAngelis to Judith A. Block," (June 17, 1985), NPD, Box 4, Folder: Trefflich.

17. Fred J. Zeehandelaar, "If Animals Could Talk," *The Keeper* 6 (January–February 1973): 7–15.

18. "Marvin Jones to Robert Cooper," (March 14, 1973), MJ.

19. Ibid.; Jennie Erin Smith, *Stolen World: A Tale of Reptiles, Smugglers, and Skullduggery* (New York: Crown Publishing, 2011), 57–58; "United States vs. Fred J. Zeehandelaar, 498 F.2d 352" (2d. Circuit Court, 1974).

20. "A Walk on the Wild Side," *The Rotarian* (January 1977): 20–21.

21. Ed Roberts, "The Zoo's Keeper," *AAZK Conference Proceedings* (Chicago, 1974): 19–20.

22. Ibid., 20.

23. Ibid., 20–12; Gay Kuester, "Keepers as Educators?—Conservationists?" *Animal Keepers' Forum* 2 (July 1976): 4.

24. Kuester, "Keepers as Educators?—Conservationists?" 4.

25. Oliver M. Claffey, "Talking Zookeeper Blues," (1983), in Rachel Watkins Rogers, *Zoo and Aquarium Professionals: The History of AAZK* (Topeka, KS: American Association of Zoo Keepers, 1992), 206; Vicki Croke, *The Modern Ark: The Story of Zoos: Past, Present, and Future* (New York: Scribner, 1997).

26. Jo Nugent, "What's New at the Zoo?" *The Rotarian* (January 1977): 14.

27. "Symposium: What's New at the Zoo?" *The Rotarian* (January 1977): 15–19.

28. David Simon, "What Is a Zoo? Where Are Zoos Going?" *Animal Keepers' Forum* 11(June 1975): 6–7.

29. Leslie Kaufman, "Date Night at the Zoo, If Rare Species Play Along," *New York Times* (July 5, 2012): A1.

30. Leslie Kaufman, "To Save Some Species, Zoos Must Let Others Die," *New York Times* (May 27, 2012): A1.

31. "Symposium: What's New at the Zoo?"

32. Robert L. Wolf and Barbara L. Tymitz, "'Do Giraffes Ever Sit?': A Study of Visitor Perceptions at the National Zoological Park, Smithsonian Institution," (March 1979), 17; Clyde A. Hill, "An Analysis of the Zoo Visitor," *International Zoo Yearbook* 11 (January 1971): 158–165.

33. Wolf and Tymitz, "'Do Giraffes Ever Sit?'" 14. For examples about the ongoing debate among zoo officials about zoos' future, see Bryan G. Norton, Michael Hutchins, Elizabeth F. Stevens, and Terry L. Maple, eds., *Ethics on the Ark: Zoos, Animal Welfare, and Wildlife Conservation* (Washington, DC: Smithsonian Institution Press, 1995); G. B. Rabb and C. D. Saunders, "The Future of Zoos and Aquariums: Conservation and Caring," *International Zoo Yearbook* 39 (January 2005): 1–26; M. Hutchins, "Zoo and Aquarium Animal Management and Conservation: Current Trends and Future Challenges," *International Zoo Yearbook* 38 (January 2003): 14–28; William G. Conway, "Where We Go

from Here?" *International Zoo Yearbook* 20 (January 1980): 184–189; J. M. Knowles, "Zoos and a Century of Change," *International Zoo Yearbook* 38 (January 2003): 28–34; http://www.canisiusishar.org/symposia/future_of_zoos.htm. Accessed December 30, 2014.

34. Wolf and Tymitz, "'Do Giraffes Ever Sit?'" 17–19.

35. Ibid., 17, 25.

36. Ibid., 22, 24.

37. Ibid., 21.

38. Ibid., 22, 30.

39. Jordan Carlton Schaul, "A Critical Look at the Future of Zoos—An Interview with David Hancocks," (March 13, 2012), http://voices.nationalgeographic.com/2012/03/13/39842/. Accessed December 30, 2014.

40. David Hancocks, *A Different Nature: The Paradoxical World of Zoos and their Uncertain Future* (Berkeley: University of California Press, 2001), 111–148.

41. Mark D. Irwin, John B. Stoner, and Aaron M. Cobaugh, eds., *Zookeeping: An Introduction to the Science and Technology* (Chicago: University of Chicago Press, 2013); Geoff Hosey, Vicky Melfi, and Sheila Pankhurst, *Zoo Animals: Behaviour, Management, and Welfare*, 2nd ed. (Oxford: Oxford University Press, 2013).

42. Irus Braverman, *Zooland: The Institution of Captivity* (Stanford: Stanford University Press, 2013).

43. William Cronon, *Nature's Metropolis: Chicago and the Great West* (New York: W.W. Norton, 1991); N. Katherine Hayles, "Simulated Nature and Natural Simulations: Rethinking the Relation between the World," in William Cronon, ed., *Uncommon Ground: Rethinking the Human Place in Nature* (New York: W.W. Norton, 1996), 409–425; Hancocks, *A Different Nature;* Ken Kawata, "Of Circus Wagons and Imagined Nature: A Review of American Zoo Exhibits, Part II," *Der Zoologische Garten* 80 (January 2011): 352–365.

44. Wolf and Tymitz, "'Do Giraffes Ever Sit?'" 21.

45. Kendall Hamilton, "Hero of the Year," *Newsweek* (November 1996): 40; Kathy Sawyer, "When a Gorilla Reaches Out to Help, We Humans Go Ape," *The Sacramento Bee* (August 25, 1996): D6.

46. Stacey Singer, "Zoo's New Top Banana Binti-Jua's Rescue of Boy Thrills Millions," *Chicago Tribune* (August 18, 1996): 1; Mark Weiner, "Gorilla's Gentleness Came as No Surprise; Burnet Zoo Director Has Known Binti Jua for Years," *Syracuse Herald American* (August 25, 1996): A1; Kathy Boccella, "Animal Behavior: Binti Jua's Instincts Could Have Taken Over," *Dayton Daily News* (August 30, 1996): 1C.

47. Frans de Waal, "Is a Gorilla Capable of Humanity?" *New York Times* (August 25, 1996): F9; Robert Allison and Charles Hirshberg, "Primal Compassion," *Life* 19 (November 1996): 78; Frans de Waal, *The Ape and the Sushi Master: Cultural Reflections of a Primatologist* (New York: Basic Books, 2001).

ON ZOO SOURCES

I went looking for zoos and their history in the usual places historians go to understand the past. In search of archives, the research process, however frustrating, became central to the analysis of the book. What makes the history of zoos so volatile in the present that many zoos close their archives and try to curate their own histories?

At first the search for zoos with open archives reached only dead ends. I began to wonder if I would even be able to write a book with the archival depth I expected. I realized how much history, in fact, matters to zoos. Zoos might serve the public, but their administrators and managers think of them as private institutions. Early on, I spoke to a librarian at one of the nation's largest zoos, and she told me flatly: "You will never see a single page in our archive." That is why several major zoos, including San Diego and Brookfield, have very little that is scholarly written about them.

Paradoxically, this strange lack says something about how conscious (and wary) zoos are of their own past. They would rather write their own histories than open archives to outside researchers. And such histories, often meant to appeal to donors, follow an almost identical narrative: begin with anecdotes of the founders, then describe the zoo's early struggles and the rebuilding of enclosures during the Great Depression. Acknowledge errors from the past, but highlight the singular importance of animal display for the conservation of endangered species. Above all, describe conservation as the motivation for zoos from beginning to present.

There is, of course, reason for zoos to worry about what lives (or has died) in their archives. Zoos are, and probably always will be, intensely contested places, with critics calling, at the very least, for the closure of exhibits, notably of elephants. There is legitimate concern that older anecdotes can become fodder for today's struggles. True enough, some critics would seek the animal skeletons in the zoo's closet. It is simply easier to keep the past behind locked doors.

If the vast majority of Americans visited zoos, then, I'd argue, zoo history is *their* history, not a proprietary belonging. And if American zoos engaged the lives and deaths of animals and people on a global scale, then their past is the world's history. It is easy to see the contrast between the zoos described in this book and those we can visit today. Most notably, the zoos of yesteryear were defined heavily by the animal trade and the wild adventure it encouraged. Today's zoos acknowledge the grim reality of endangered species. Yet zoos have changed over time, not abruptly switched course and disengaged from past practices. Older ways of thinking about animals and peoples, as well as exotic lands, creep into the ways in which we seek to save some animals and let others disappear. Ideas of the exotic and colonized jungle or veldt still help determine the ways we imagine the natural lives of animals we see on display. History, like biology, helps determine our future environments.

Private institutions or not, the story of zoos is also the story of the nation and how its peoples related to overseas empires. Recognizing that reality through a scholarly engagement with the past can help transform animal conservation. Can we think about preserving biodiversity as part of a campaign for global equity rather than a vilification of indigenous populations as poachers-in-waiting? Those zoos—and I would highlight and celebrate the National, St. Louis, Milwaukee, Bronx, and Philadelphia Zoos—that keep their archives open to the public help prove that past practices can inform better futures.

No archive is ever fully closed even if the doors are locked. There is much we can learn about the histories of private, secretive zoos in the histories and papers of other institutions. Zoo directors wrote to each other, and thus while researchers are effectively barred from the San Diego or Brookfield Zoo archives (for example), there is a great deal we can learn about those zoos, their workers, their animals, and their visionaries. Moreover, the irony of closed archives is the reality of public fascination with the zoos and the animals that populated them. So animal collectors, zoo directors, and even zookeepers wrote. Raymond Ditmars and Belle Benchley penned fiction. Frank Buck, Charles Mayer, William Mann, Peter Ryhiner, Wynant Hubbard, Marté Lapham, William "Slim" Lewis, Henry Trefflich, and Marlin Perkins wrote memoirs. Buck published comic books. Many others wrote children's books and, naturally, zoo and animal guides. William Beebe, Mann, and William T. Hornaday published animal studies. Zoos and their directors talked to each other in journals, including *Parks and Recreation* and, more recently, *International Zoo Yearbook*. A few individuals, like Hornaday, Perkins, and Marvin Jones, left behind personal papers.

Visitors, as well, left historical traces. On eBay and other sites, I collected a large if idiosyncratic archive of the material culture of zoos that ordinary visitors saved.

Sometimes visitors wrote revealing notes on the back of postcards. Other times they bought Frank Buck paraphernalia, including his pith helmet and signed sundial. They played with a model zoo train car. A giraffe head and neck pop out at the push of a button. Visitors saved movie posters, collected playing cards, and read comic books.

This collection, at least, is open for viewing. I have donated it all to the library at the University of Toronto Scarborough, and it's available digitally. Experience life in the animal game as an animal collector while playing (the online version of) Marlin Perkins's Zoo Parade board game. Along the way, step off the game trails and join expeditions back into the past of American zoos. What was it like to visit the first American zoos? Join Frank Buck on an animal-buying expedition to Singapore's animal shops. Visit the Great Depression–era monkey islands, the postwar children's zoos, or the more recent zoo safari parks. Follow the changing nature of the animal trade through the careers of Wynant Hubbard, Frank Buck, and Fred J. Zeehandelaar. Walk the picket lines with striking zookeepers, or search for the yeti with Perkins and Sir Edmund Hillary. Imagine the life of zoo animals like Gunda, or experience the excitement of wild animal movies.

To play the board game:

    https://digitalscholarship.utsc.utoronto.ca/ZooParade/

To browse the digital collection of zoo history:

    http://digitalscholarship.utsc.utoronto.ca/projects/islandora/object/animal empire%3Aroot

ILLUSTRATION CREDITS

ACKNOWLEDGMENTS

This book is ultimately about the intimate relations we (as humans) forge with animals and with each other through animals. Watching keepers, animals, children, parents, and other visitors formed many of the questions I've tried to ask and answer. Animal and child, babe and adult showed disappointment, boredom, anger, love, frustration, and fatigue, and in the process, they shaped this book. I owe special thanks to zookeepers and their animals, especially at the St. Louis, Bronx, Potter Park (Lansing, MI), Brookfield, National (Washington, DC), Lincoln Park, Philadelphia, Pittsburgh, and Milwaukee County Zoos. Zoos may be particularly reluctant to reveal their pasts; those zoos that have either opened their archives or donated their papers deserve special thanks. So, too, do their especially generous archivists and librarians. I am grateful for the help of the staff at the Wildlife Conservation Society, Milwaukee County Zoo, and Philadelphia Zoos. My thanks as well to librarians and archivists at the Smithsonian Institution, Western Historical Manuscripts Collection (University of Missouri–St. Louis), University of Wisconsin–Milwaukee, National Archives (Washington, DC), National Archives of Singapore, British Library, Tamiment-Wagner Labor Archives, New York Public Library, Library of Congress, and University of Arizona.

I have been extraordinarily privileged to work with a brilliant cohort of graduate students. Some of them worked on this project as research assistants. Others constantly expanded my range as a historian and encouraged me, especially, to think more about empire and the environment. My thanks to Rebecca O'Neill, Camille Bégin, Will Riddell, Paul Lawrie, Nathan Cardon, Ian Rocksborough-Smith, Holly Karibo, and Jodi Giesbrecht. Amanda Wedge, now a PhD student who began working with me as a gifted undergraduate in my class on the history of animals and people, deserves a special, heartfelt thanks. Amanda is a collaborator on the digital element of this book. She's also a gifted scholar of animals and people and is writing a fascinating dissertation. I have learned so much from you all (and from a new generation of graduate scholars in the program today).

I am just as lucky to have had the opportunity to collaborate with an incredibly dedicated staff of librarians at the University of Toronto and the University of Toronto Scarborough. Thanks especially to Whitney Kemble, the liaison librarian for history. I am grateful to the staff at the Digital Scholarship Unit at the University of Toronto Scarborough Library, particularly Kirsta Stapelfeldt and Sarah Forbes, who encouraged me to think about the potential of digital work to reach new audiences. I am indebted to them for their hard work in digitizing my eclectic collection of ephemera and in programming the Zoo Parade game.

I have worked for more than a dozen years at the University of Toronto in the tricampus graduate Department of History and the undergraduate Department of Historical and Cultural Studies. I have been especially lucky to teach a class on the history of animals and people. Students' thoughts, questions, insights, and interests have profoundly influenced this book. My colleagues, as well, have been extraordinarily generous. I am lucky to have benefited especially from the brilliance of Kevin Coleman, Donna Gabaccia, Rick Halpern, Ken MacDonald, Sean Mills, Steve Penfold, Jeffrey Pilcher, Natalie Rothman, and Jo Sharma. Beyond the U of T, my thanks for the insights and advice of Timothy Barnard, John Beck, Eileen Boris, Jason Colby, Jon Coleman, Donna Haverty-Stacke, Ken Kawata, Jennifer Klein, Geoffrey Leonardelli, Jana Lipman, David Offenhall, Bill Rapley, Krishnendu Ray, Fiona Tan, and two anonymous readers from Harvard University Press. Kathleen McDermott at Harvard University Press has been a tremendous supporter since we first met at a conference in Hartford. Audiences at the Ambedkar University, Delhi University, Hunter College, Indian Institute of Technology–Guwahati, iSchool (University of Toronto), Michigan State University, National University of Singapore, Tulane University, and University of Toronto Mississauga all offered useful leads and encouragement. Research for this book has been funded by the Canada Research Chair Program/Programme des Chaires de recherche du Canada.

I was raised in an environment where books—and zoos—mattered. More than I can ever express in words, I owe so much to my parents, Carl and Jessica Bender (and to my mother, Jessica, for reading and editing the entire manuscript). So many childhood visits to the St. Louis Zoo were always on my mind as I was writing. My thanks and love as well to my brother Michael Bender and my sister-in-law Tamara Koss and their kids, Stephen and Alexandra. This book would be just words on paper without Jo Sharma and our darling daughter, Piya Rose Sharma Bender. Instead, it's a product of my true love for them and of them for me. Jo's a colleague, my best advisor, and the historian I admire most. She's also my partner in raising Piya, over the course of writing this book, from wee infant to big girl.

Piya is the inspiration for this book. She was my traveling companion during many zoo visits. I've watched her blossom into a beautiful person, and her love for Ditto, Spotsy, Franklin Jacob, the tigers, and more helped guide my thinking. I'm writing with her and her future in my mind and in my heart.

INDEX

AAZK. *See* American Association of Zoo Keepers

AAZPA. *See* American Zoo and Aquarium Association

Abreu, Rosalià: ape collection of, 153; Yerkes visiting, 153

Adam Forepaugh circus, 43

Adoption, animal: new form of, 271; of Oofy, 248–249; in postwar years, 248; zoo wives and, 251–252. *See also* Surrogate mother

Adventurers' Club handbook, 107, 109

African plains, exhibit: at Bronx Zoo, 212–214; segregation at, 214; visitors at, 213; World War II and, 214

Akeley, Carl: gorillas and, 146; Gunda killed by, 50

Akeley, Mary Jobe, 168; Ali and, 99–100, 109, 111

American Association of Zoo Keepers (AAZK): care work of, 269; conservation and, 199, 309; founding of, 198; Roberts in, 309; Sweeney and, 198; Zeehandelaar keynote for, 308

American zoo: children at, 3–4; crime at, 295; European zoos inspiring, 11; growth of, 22–23; history of, 12, 16; 1960s and 1970s changing, 274; Philadelphia as first, 12, 16; during postwar years, 6; societal problems reflected in, 292–293; television and radio programs of, 230

American Zoo and Aquarium Association (AAZPA), 196, 264; conservation resolution of, 265

Animal diplomacy, 226–227

Animals: birds, 41, 56–57; bison, 36–37, 38–40; death rate of zoo, 45; electric eel, 139–140; escapes by, 45; family structure of, 244–245; giant panda, 139–140; giraffe, 68–69, 71, 74–75; grizzly bear, 53; hippopotamus, 74; jaguar, 46; leopard, 4, 246; motherhood of, 244–245; orangutan, 61, 62, 240, 248; snakes, 45, 56, 125–126, 210, 215–216, 229, 232; urban neurosis of, 288–289; warthogs, 237–238; Yeti, 302–305. *See also* Apes; Breeding; Chimpanzee; Elephant; Gorilla; Monkeys; Rhinoceros; Tiger

Animal trade, 66; Camp in, 70; capture techniques in, 77; Chinese traders and, 61–62; CITES impacting, 265; declining, 264; decolonization in, 283–284; domestication and, 77–78; empire in, 53–55, 57, 64, 68, 70–72, 74–75, 167, 273; fictitious stories of, 80–81; films and, 95, 102–107, 136, 164–166, 208, 233–234, 272–273, 277, 279, 281–282, 289; Germans in, 56–58; Hagenbeck family in, 57–58, 69; Mann family in, 51–53, 56, 58–59, 67–70; native workers in, 71–74, 86, 88, 98; orangutan price in, 61; permits for, 64, 70–75, 167; popular culture and, 95; postwar, 223; Rosefelt in, 223; Seago in, 283; Singapore and, 52, 58–59, 60; U.S. empire and,